DRUG–DRUG INTERACTIONS FOR THERAPEUTIC BIOLOGICS

DRUG–DRUG INTERACTIONS FOR THERAPEUTIC BIOLOGICS

Edited by

HONGHUI ZHOU
Senior Scientific Director and Janssen Fellow
Head of Pharmacokinetics and Pharmacodynamics
Biologics Clinical Pharmacology
Janssen Research and Development, LLC

BERND MEIBOHM
Professor of Pharmaceutical Sciences
Associate Dean for Graduate Programs and Research
College of Pharmacy
The University of Tennessee Health Science Center

WILEY

Published by John Wiley & Sons, Inc., Hoboken, New Jersey
Published simultaneously in Canada

For general information on our other products and services or for technical support, please contact our Customer Care Department within the United States at (800) 762-2974, outside the United States at (317) 572-3993 or fax (317) 572-4002.

Wiley also publishes its books in a variety of electronic formats. Some content that appears in print may not be available in electronic formats. For more information about Wiley products, visit our web site at www.wiley.com.

Library of Congress Cataloging-in-Publication Data:

Drug-drug interactions for therapeutic biologics / edited by Honghui Zhou, Bernd Meibohm.
p. ; cm.
Includes bibliographical references and index.
ISBN 978-1-118-03216-9 (cloth)
I. Zhou, Honghui. II. Meibohm, Bernd.
[DNLM: 1. Drug Interactions. 2. Drug Discovery. 3. Proteins–therapeutic use. QV 37.5]
615.7′045–dc23

2013000334

10 9 8 7 6 5 4 3 2 1

CONTENTS

PREFACE

In the past two decades we have seen tremendous progress in the area of therapeutic biologics. With more and more therapeutic proteins being used in poly-pharmacy settings and the potential toxicity risk of drug-drug interactions, there is during drug development a need for a thorough review of potential drug-drug interactions involving therapeutic biologics. However, literature references on this topic have so far been scarce. Thus, we feel the scientific community would benefit from a systemic presentation of the current status of knowledge on this topic. The proposed book project is intended to fill this void.

The book is expected to greatly benefit scientists and researchers in the pharmaceutical and biotech industry as well as academia who are involved in drug development for both therapeutic biologics and traditional small molecule drugs. The expected audience will be pharmaceutical and biotech scientists, clinical pharmacologists, medicinal chemists, and toxicologists. Scientists and clinicians in pharmaceutical and biotech industry can utilize the book as a resource to strategize, plan and implement drug-drug interaction assessments involving therapeutic biologics. Academic pharmacokinetics, pharmacology, and biochemistry scientists working on mechanisms for biologic drug-drug interactions will also find this book very useful as a compilation of the current state-of-the-art.

The current book focuses on both theoretical and practical aspects of drug-drug interaction assessments for therapeutic biologics in drug development. We are fortunate that many of the experts and opinion leaders from various areas of therapeutic biologics drug development and drug-drug interactions have participated in the writing of this book, and we are indebted to them for their time and dedication to participate in this project. The content includes topics such as drug-drug interaction risks (both theoretical and observed) in combination with small molecules and with other biologics, pharmacokinetic drug-drug interactions, pharmacodynamic drug-drug interactions, utility of *in vitro* methods in drug-drug interaction assessment and prediction, modeling-independent and modeling-based methods to assess potential drug-drug interactions, risk-based strategies for evaluating biologic drug-drug interactions, strategies to minimize drug-drug interaction risk and mitigate toxic interactions, and regulatory perspectives on biologic drug-drug interaction assessments.

Though there are several books covering drug-drug interactions for conventional small molecules, a book that is comprehensive with all the above topics for biotherapeutics is not currently available. Thus, we are convinced that that textbook addresses a currently unmet need in drug development sciences and we are confident that the scientific community will benefit from the experience and expertise of the contributors to this book project.

HONGHUI ZHOU
BERND MEIBOHM

Spring House, PA, and
Memphis, TN
August 2012

ABOUT THE EDITORS

Honghui Zhou, PhD, FCP

Honghui Zhou is currently a Senior Scientific Director at Janssen Research and Development, LLC, Johnson & Johnson and is heading the Pharmacokinetics and Pharmacodynamics Department within Biologics Clinical Pharmacology.

Prior to joining Centocor, Dr. Zhou was a Director of Clinical Pharmacology at Wyeth Research (now Pfizer). He also worked for Novartis Pharmaceuticals Corp. and Johnson & Johnson Pharmaceutical Research and Development in the area of clinical pharmacology and pharmacokinetics/pharmacodynamics (PK/PD) in both small molecular drugs and therapeutic proteins. In 2012, Honghui was elected as a Janssen Fellow.

Dr. Zhou has authored more than 150 original peer-reviewed scientific papers, book chapters, and conference abstracts in PK/PD and drug–drug interactions. He has also been an invited speaker in many national and international conferences. He is board certified by American Board of Clinical Pharmacology (ABCP) and is Fellow of Clinical Pharmacology (FCP) in ACCP. He currently serves as a section editor for Biologics for the *Journal of Clinical Pharmacology*. He also serves as Board of Reagents of ACCP (2009–2014). He co-chairs the IQ Therapeutic Protein–Drug Interaction Working Group (previously Pharma/FDA/Academia Therapeutic Protein–Drug Interaction Steering Committee). Honghui is a graduate of the China Pharmaceutical University, BS in Pharmacology, and the University of Iowa, PhD in Pharmaceutics.

Bernd Meibohm, PhD, FCP

Bernd Meibohm is a Professor of Pharmaceutical Sciences and Associate Dean for Research and Graduate Programs at the College of Pharmacy, the University of Tennessee Health Science Center, Memphis.

Prior to joining the University of Tennessee, Dr. Meibohm conducted research at the University of South Carolina and the University of Florida. Dr. Meibohm's scientific interests include chronic inflammatory pulmonary diseases, pediatric pharmacotherapy, and the application of quantitative modeling and simulation techniques in preclinical and clinical drug development, with a specific focus on biotech drugs. His research has resulted in two textbooks, over 200 peer-reviewed scientific papers, book chapters, and conference abstracts, and over 100 invited scientific presentations to national and international audiences.

Dr. Meibohm is a Fellow of the American Association of Pharmaceutical Scientists (AAPS) and American College of Clinical Pharmacology (ACCP). He was the 2010 Chair for the Pharmacokinetics, Pharmacodynamics and Drug Metabolism (PPDM) section of AAPS and currently serves as the President-Elect for ACCP. Dr. Meibohm is also serving as associate editor for *The AAPS Journal* and as section editor for Pharmacokinetics and Pharmacodynamic for the *Journal of Clinical Pharmacology*; he is a member of the editorial boards of the *Journal of Pediatric Pharmacology and Therapeutics*, the *Journal of Pharmacokinetics and Pharmacodynamics*, *Les Annales Pharmaceutiques Françaises*, and *Die Pharmazie*.

CONTRIBUTORS

JEFFREY S. BARRETT, Laboratory for Applied PK/PD, Division of Clinical Pharmacology and Therapeutics, The Children's Hospital of Philadelphia, Philadelphia, PA, USA; School of Medicine, University of Pennsylvania, Philadelphia, PA, USA

BARBARA J. BRENNAN, Hoffmann-La Roche Inc., Nutley, NJ, USA

SOUVIK CHATTOPADHYAY, Drug Metabolism and Pharmacokinetics, Drug Safety Sciences, Janssen Research and Development, LLC, Spring House, PA, USA

SHANNON DALLAS, Drug Metabolism and Pharmacokinetics, Drug Safety Sciences, Janssen Research and Development, LLC, Spring House, PA, USA

LESLIE J. DICKMANN, Pharmacokinetics and Drug Metabolism, Amgen Inc, Seattle, WA, USA

MARTIN E. DOWTY, Pfizer Inc., Andover, MA, USA

RAYMOND EVERS, Department of Pharmacokinetics, Pharmacodynamics, and Drug Metabolism, Transporters and In Vitro Technologies, Merck & Co., Inc., Rahway, NJ, USA

UWE FUHR, Department of Pharmacology, Clinical Pharmacology, University Hospital of Cologne, Köln, Germany

SANDHYA GIRISH, Department of Clinical Pharmacology, Genentech, Inc., South San Francisco, CA, USA

CHUANPU HU, Pharmacokinetics and Pharmacodynamics, Biologics Clinical Pharmacology, Janssen Research & Development, LLC, Spring House, PA, USA

Shiew-Mei Huang, Office of Clinical Pharmacology, Office of Translational Sciences, Center for Drug Evaluation and Research, Food and Drug Administration, Silver Spring, MD, USA

Alexander Jetter, Department of Clinical Pharmacology and Toxicology, University Hospital Zürich, Zürich, Switzerland

Amita Joshi, Department of Clinical Pharmacology, Genentech, Inc., South San Francisco, CA, USA

Tarundeep Kakkar, Genomics Institute of the Novartis Research Foundation, BDU Translational Sciences, San Diego, CA, USA

Simone Kasek, College of Pharmacy, The University of Tennessee Health Science Center, Memphis, TN, USA

Narendra Kishnani, Department of Biotransformation, Pharmaceutical Candidate Optimization, Bristol-Myers Squibb Co., Princeton, NJ, USA

Joan Korth-Bradley, Clinical Pharmacology, Pfizer Inc., Collegeville, PA, USA

Eugenia Kraynov, Pfizer Inc., San Diego, CA, USA

Jocelyn Leu, Janssen Research & Development, LLC, Spring House, PA, USA

Christine Li, College of Pharmacy, The University of Tennessee Health Science Center, Memphis, TN, USA

Wararat Limothai, College of Pharmacy, The University of Tennessee Health Science Center, Memphis, TN, USA

Dan Lu, Department of Clinical Pharmacology, Genentech, Inc., South San Francisco, CA, USA

Dora Babu Madhura, University of Tennessee Health Science Center, Memphis, TN, USA

Bernd Meibohm, College of Pharmacy, The University of Tennessee Health Science Center, TN, USA

Theresa Nguyen, Department of Pharmacokinetics, Pharmacodynamics, and Drug Metabolism, Transporters and In Vitro Technologies, Merck & Co., Inc., Rahway, NJ, USA

Chetan Rathi, College of Pharmacy, The University of Tennessee Health Science Center, Memphis, TN, USA

Sumit Rawal, College of Pharmacy, The University of Tennessee Health Science Center, Memphis, TN, USA

Kellie Reynolds, Office of Clinical Pharmacology, Office of Translational Sciences, Center for Drug Evaluation and Research, Food and Drug Administration, Silver Spring, MD, USA

Josiah Ryman, College of Pharmacy, The University of Tennessee Health Science Center, Memphis, TN, USA

Carlo Sensenhauser, Drug Metabolism and Pharmacokinetics, Drug Safety Sciences, Janssen Research and Development, LLC, Spring House, PA, USA

Jose Silva, Drug Metabolism and Pharmacokinetics, Drug Safety Sciences, Janssen Research and Development, LLC, Spring House, PA, USA

J. Greg Slatter, Pharmacokinetics and Drug Metabolism, Amgen Inc., Seattle, WA, USA

Yu-Nien (Tom) Sun, Quantitative Pharmacology, Department of Pharmacokinetics and Drug Metabolism, Amgen Inc., Thousand Oaks, CA, USA

Frank-Peter Theil, Department of Clinical Pharmacology, Genentech, Inc., South San Francisco, CA, USA

Margaret Thomson, College of Pharmacy, The University of Tennessee Health Science Center, Memphis, TN, USA

Ashit Trivedi, College of Pharmacy, The University of Tennessee Health Science Center, Memphis, TN, USA

Jian Wang, Office of Translational Sciences, Office of Clinical Pharmacology, Center for Drug Evaluation and Research, Food and Drug Administration (FDA), Silver Spring, MD, USA

Yow-Ming C. Wang, Office of Clinical Pharmacology, Office of Translational Sciences, Center for Drug Evaluation and Research, Food and Drug Administration, Silver Spring, MD, USA

Larry C. Wienkers, Pharmacokinetics and Drug Metabolism, Amgen Inc, Seattle, WA, USA

Di Wu, Laboratory for Applied PK/PD, Division of Clinical Pharmacology and Therapeutics, The Children's Hospital of Philadelphia, Philadelphia, PA, USA

Lei Zhang, Office of Clinical Pharmacology, Office of Translational Sciences, Center for Drug Evaluation and Research, Food and Drug Administration, Silver Spring, MD, USA

Xiaoping Zhang, Hoffmann-La Roche Inc., Nutley, NJ, USA

Hong Zhao, Office of Clinical Pharmacology, Office of Translational Sciences, Center for Drug Evaluation and Research, Food and Drug Administration, Silver Spring, MD, USA

Honghui Zhou, Pharmacokinetics and Pharmacodynamics, Biologics Clinical Pharmacology, Janssen Research and Development, LLC, Spring House, PA, USA

Min Zhu, Quantitative Pharmacology, Department of Pharmacokinetics and Drug Metabolism, Amgen Inc., Thousand Oaks, CA, USA

CHAPTER 1

DRUG INTERACTIONS FOR THERAPEUTIC PROTEINS: A JOURNEY JUST BEGINNING

HONGHUI ZHOU and BERND MEIBOHM

1.1 INTRODUCTION

Over the last three decades, therapeutic proteins, in particular, antibody-based biotherapeutics, have played an increasingly important role in pharmacotherapy, and in some therapeutic areas, such as immune-mediated inflammatory diseases (IMIDs) and oncology, therapeutic proteins have fundamentally changed the therapeutic paradigm. Therapeutic proteins have also presented enormous commercial potential. For example, the top 10 antibody-based biotherapeutics accounted for around \$50 billion of worldwide sales in 2011.[1] The majority of these are either in IMID (adalimumab, etanercept, infliximab, rituximab, natalizumab, omalizumab) or in oncology (rituximab, bevacizumab, trastuzumab, cetuximab) therapeutic areas. Hundreds of investigational antibody-based and other protein therapeutics are currently under development at different stages, spanning discovery to phase III clinical investigations.

Owing to an expected increase in the coadministration of biotherapeutic agents with established pharmacotherapy regimens, there is an increasing likelihood for the occurrence of clinically relevant drug interactions. Therapeutic proteins, however, have long been perceived to have a very low propensity for drug–drug interactions because they are eliminated via catabolic routes, either nonspecific pathways or target-mediated pathways, that are independent from the elimination pathways of small molecules, which are usually eliminated by noncatabolic pathways such as hepatic metabolism via cytochrome P450 (CYP), renal excretion, and biliary excretion. Though it has been known for decades that some cytokines such as interferons, tumor necrosis factor α (TNF-α), and

Drug–Drug Interactions for Therapeutic Biologics, First Edition.
Edited by Honghui Zhou and Bernd Meibohm.

interleukin 6 (IL-6) can down-regulate CYPs,[2] very few drug–drug interactions had been reported for biotherapeutics until 2007, when two review articles containing examples of drug interactions involving therapeutic proteins were published.[3,4] The majority of reported drug interactions associated with therapeutic proteins seem to be indirect; however, a mechanistic understanding for many of the observed interactions is still lacking.[5–7]

1.2 SCIENTIFIC/REGULATORY LANDSCAPE OF THERAPEUTIC PROTEIN–DRUG INTERACTIONS

To help assess the common practice of evaluating therapeutic protein–drug interactions across the biotech/pharma industry and to shed some light on how and when a sensible therapeutic protein–drug interaction assessment strategy should be incorporated into therapeutic protein drug development, a survey was conducted within the Biotechnology Industry Organization (BIO) member companies in 2010. It is not surprising that a majority of the responder companies did not have internal strategies for evaluating therapeutic protein–drug interactions at the time of the survey. Nevertheless, the most favored approach employed to address potential drug–drug interactions of therapeutic proteins at that time was a tailored and integrated (i.e., case-by-case) strategy that addressed the possibility of the therapeutic protein acting as either an initiator (perpetrator) or target (victim) of the interaction. Despite the fact that many of the companies responding to the survey reported drug–drug interactions involving therapeutic proteins, the majority of the clinical therapeutic protein–drug interactions studied did not warrant dose adjustment. In other words, most of the observed clinical therapeutic protein–drug interactions did not reach a clinically significant level. Routine *in vitro* screening and preclinical drug–drug interaction studies were not widely used for the evaluation of therapeutic proteins. For clinical development, dedicated clinical pharmacology drug–drug interaction studies were the most frequently used methodology, followed by population pharmacokinetics-based and clinical cocktail approaches.[8]

The BIO survey results indicated that there was a pressing need to have a science-driven and risk-based assessment strategy for therapeutic protein–drug interactions (TP-DIs). A closer collaboration among scientists from the biotech/pharma industry, regulatory agencies, and academia appeared to be essential in reaching that goal. As a result, a TP-DI steering committee from industry, the FDA, and academia was founded in 2009 to address this challenge. The initial scope of this committee was focused only on pharmacokinetics (PK) and metabolism-based drug–drug interactions for the major classes of therapeutic proteins, including monoclonal antibodies, fusion proteins, cytokines (excluding antibody–drug conjugates). The committee intended to investigate the potential for therapeutic proteins to interact, either as initiators or targets, with drugs that are metabolized via CYP enzyme pathways. Two major focus areas the committee concentrated on were (1) to critically assess standard *in vitro* screening techniques and methodologies (e.g., for cytokine-related drug–drug and drug–disease interactions) and (2) to provide guidance for study designs with consideration of specific disease area (e.g., oncology) issues and timings.

Several scientific knowledge gaps were identified from a 2010 American Association of Pharmaceutical Scientists (AAPS) workshop on Strategies to Address Therapeutic Protein-Drug Interactions during Clinical Development.[9] One gap was associated with the relevance of *in vitro* systems to assess potential therapeutic protein–drug interactions, and another gap was a lack of best practices for using population PK-based approaches to assess potential therapeutic protein–drug interactions. The steering committee also identified similar gaps and consequently formed two working groups to specifically tackle them.

During the same time period, scientists from the FDA published two important review articles on TP-DI, but these were mostly from a regulatory perspective.[10,11] In 2012, a draft of a new drug–drug interaction guidance document was made available by the FDA for public comments.[12] That draft included a dedicated section on therapeutic protein–drug interaction to address specifically the newly emerging area of drug–drug interactions with therapeutic proteins.

The Workshop on Recent Advances in the Investigation of Therapeutic Protein Drug-Drug Interactions: Preclinical and Clinical Approaches was held on June 4–5, 2012. The workshop, co-sponsored by the FDA Office of Clinical Pharmacology and the Drug Metabolism and Clinical Pharmacology Leadership Group of the IQ Consortium, was intended to facilitate a better understanding of the current science, investigative approaches, knowledge gaps, and regulatory requirements related to the evaluation of therapeutic protein–drug interactions. The workshop also provided an opportunity to discuss the current views from the two (*in vitro* and population PK approaches) therapeutic protein–drug interaction working groups. The proceedings from this workshop are being compiled with the intent of issuing white papers in these subject areas. It is anticipated that the recommendations from both white papers will soon provide pharmaceutical scientists with sensible and scientifically sound best practices and an assessment framework for using *in vitro* and population PK-based approaches for evaluating therapeutic protein–drug interactions.

Our current understanding of the mechanisms of many therapeutic protein–drug interactions is still in its infancy. Much basic research needs to be conducted to verify several existing hypotheses related to therapeutic protein–drug interactions. Continued close collaborations among fellow scientists in industry, academia, and regulatory agencies will be vital to generate more plausible mechanistic hypotheses and collectively address the many challenges in this area. Through these collaborative efforts, the knowledgebase on therapeutic protein–drug interactions will likely be largely expanded in the near future, and it is hoped and anticipated that over the next decade a similar level of mechanistic understanding and systemic assessment methodology will be achieved and developed for drug interactions with protein therapeutics as it has been established in the last two decades for small molecule drugs. The journey toward that goal has just begun.

REFERENCES

1. R&D Pipeline News, Top 30 Biologics 2011, April 25, 2012. Available at www.pipelinereview.com.
2. Morgan ET. Regulation of cytochrome P450 by inflammatory mediators: why and how? *Drug Metab Dispos* **29**, 207–12 (2001).
3. Seitz K, Zhou H. Pharmacokinetic drug-drug interaction potentials for therapeutic monoclonal antibodies: reality check. *J Clin Pharmacol* **47**, 1104–18 (2007).
4. Mahmood I, Green MD. Drug interaction studies of therapeutic proteins or monoclonal antibodies. *J Clin Pharmacol* **47**, 1540–54 (2007).
5. Zhou H, Mascelli MA. Mechanisms of monoclonal antibody-drug interactions. *Annu Rev Pharmacol Toxicol* **51**, 359–72 (2011).
6. Kraynov E, Martin SW, Hurst S, et al. How current understanding of clearance mechanisms and pharmacodynamics of therapeutic proteins can be applied for evaluation of their drug-drug interaction potential. *Drug Metab Dispos* **39**, 1779–83 (2011).
7. Meibohm B. Mechanistic basis for potential drug-drug interactions with therapeutic proteins. Paper presented at the Workshop on Recent Advances in the Investigation of Therapeutic Protein Drug-Drug Interactions: Preclinical and Clinical Approaches. Silver Spring, MD, June 4–5, 2012.
8. Lloyd P, Zhou H, Theil FP, et al. Highlights from a recent BIO survey on therapeutic protein-drug interactions. *J Clin Pharmacol* **52**, 1755–63 (2012).
9. Girish S, Martin SW, Peterson MC, et al. AAPS workshop report: strategies to address therapeutic protein-drug interactions during clinical development. *AAPS J* **13**, 405–16 (2011).
10. Huang SM, Zhao H, Lee JI, et al. Therapeutic protein-drug interactions and implications for drug development. *Clin Pharmacol Ther* **87**, 497–503 (2010).
11. Lee JI, Zhang L, Men AY, et al. CYP-mediated therapeutic protein-drug interactions: clinical findings, proposed mechanisms and regulatory implications. *Clin Pharmacokinet* **49**, 295–310 (2010).
12. U.S. Department of Health and Human Services, FDA, and Center for Drug Evaluation and Research. Guidance for industry: drug interaction studies–study design, data analysis, implications for dosing, and labeling recommendations. February 2012. Available at www.fda.gov/downloads/Drugs/GuidanceComplianceRegulatoryInformation/Guidances/UCM292362.pdf.

CHAPTER 2

PHARMACOKINETIC AND PHARMACODYNAMIC-BASED DRUG INTERACTIONS FOR THERAPEUTIC PROTEINS

DAN LU, SANDHYA GIRISH, FRANK-PETER THEIL, and AMITA JOSHI

2.1 INTRODUCTION

Therapeutic proteins (TPs) are protein products manufactured for pharmaceutical use. They include monoclonal antibodies (mAbs), antigen-binding fragments, antibody–drug conjugates (ADCs), cytokines, enzymes, growth factors, and miscellaneous proteins (e.g., fusion proteins and recombinant proteins). The development of therapeutic biologics, including TPs, is increasingly important in the pharmaceutical industry.[1] To achieve greater clinical benefits, TPs are often being combined with other TPs and small molecule drugs (SMDs). Whether drug interactions (DIs) in combination therapy result in an undesirable impact on efficacy and safety needs evaluation. To date, for the observed therapeutic protein–drug interactions (TP-DIs) that affect the exposure of TPs, only a modest change in exposure is observed and no impact on safety or efficacy has been documented, suggesting a limited clinical relevance.[2] This might be because most TPs have a relatively large therapeutic range compared to the majority of traditional SMDs. However, TP-DIs that affect the exposure of some drugs with a narrow therapeutic range (NTR), such as some SMDs and ADCs, may have an impact on efficacy and safety. The TP-DIs that result in enhanced toxicity due to undesirable pharmacodynamic (PD) interactions without a direct impact on exposures may also be clinically relevant. Thus the evaluation of TP-DIs is an important and evolving topic for the development of TPs in combination with other drugs.

This chapter reviews the major absorption, distribution, metabolism, and excretion (ADME) pathways of TPs, summarizes the potential mechanisms of

Drug–Drug Interactions for Therapeutic Biologics, First Edition.
Edited by Honghui Zhou and Bernd Meibohm.

pharmacokinetic (PK) and PD-based TP-DIs, and recommends a question-based TP-DI risk assessment strategy during clinical development. The DIs for some nonprotein biologics such as nucleic acid–based derivatives are reviewed in other chapters.

2.2 DISTRIBUTION, CATABOLISM/METABOLISM, AND EXCRETION MECHANISMS OF THERAPEUTIC PROTEINS

ADME processes determine the PK properties of SMDs and TPs. In drug combinations, one drug may impact the ADME processes of another drug, leading to a change in its exposure. For SMDs, absorption is mainly mediated by the solubility and permeability of a SMD and its interaction with transporters. Distribution of SMDs is mediated by several key processes, such as blood perfusion, permeability across membrane barriers, and nonspecific binding. Metabolism of SMDs is mainly mediated by cytochrome P450 (CYP) and non-CYP enzymes (such as *N*-acetyl and glucuronyl transferase). Excretion of SMDs mainly occurs via renal filtration or renal and biliary secretion mediated by transporters.[3] Figure 2-1*a* depicts the typical clearance pathways for SMDs.

For TPs, ADME processes are different from SMDs.[4–6] Owing to high gastrointestinal enzyme activity and low permeability through the gastrointestinal mucosa, most TPs are not therapeutically active on oral administration. Consequently other routes of administration, such as intravenous, subcutaneous, and intramuscular routes of injection are used for TPs.[6] For subcutaneous injections of TPs with large molecular weight, convective transport across local lymphatic vessels is the major mechanism of absorption from the injection site.[7] The processes of distribution, catabolism, and excretion of TPs are reviewed in detail in this chapter. As illustrated in Figure 2-1*b*, the catabolism of TPs are mainly mediated by nonspecific clearance pathways. Target-mediated drug disposition (TMDD) and immunogenicity-mediated pathways also play roles in the clearance of some TPs. ADCs belong to a more complex group of TPs, made up of both a mAb and a small molecule cytotoxic agent. Their PK properties are also reviewed here.

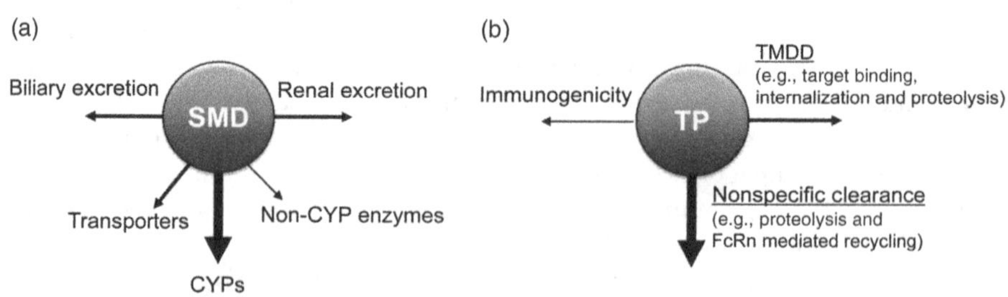

FIGURE 2-1 Comparison of clearance mechanisms of (a) a SMD and (b) a TP. CYP: cytochrome P450; FcRn: neonatal Fc receptor; SMD: small molecule drug; TMDD: target-mediated drug disposition; TP: therapeutic protein. (See the color version of this figure in Color Plates section.)

2.2.1 Distribution of Therapeutic Proteins

Distinct from most SMDs that widely distribute to various tissues and organs after administration, distribution of mAbs and large TPs is usually confined by their large size; consequently the molecules have limited mobility through membranes. This often results in a relatively small volume of distribution. The volume of distribution of mAbs and ADCs at steady state is often a low multiple (1 to 2) of physiologic plasma volume (approximately 50 mL/kg). This is similar to the distribution characteristics for an endogenous immunoglobulin G (IgG). The distribution of TPs outside the systemic circulation is mediated by limited interstitial penetration in various organs, convection-dominated lymphatic drainage, specific and nonspecific binding to peripheral tissues, and target-mediated cellular uptake. For TPs with relatively low molecular mass, preclinical study results have demonstrated better tissue penetration.[6] Unlike SMDs, transporters usually do not play a role in the distribution of large TPs.

2.2.2 Catabolism of Therapeutic Proteins

Most TPs are mainly catabolized by proteolytic degradation in cellular lysosomes through nonspecific pathways, resulting in peptides and amino acids that are reutilized for protein synthesis.[4,6] It is generally believed that nonspecific catabolism of TPs may take place predominantly in the lysosomes of endothelial cells and the mononuclear phagocyte system (MPS). TPs, such as mAbs and some fusion proteins containing a fragment crystallizable region (Fc region), interact with neonatal Fc receptors (FcRn) similar to endogenous IgGs. In adults, FcRn is primarily expressed in the vascular endothelial cells. FcRn is also detectable on monocytes, tissue macrophages, and dendritic cells. The FcRn-mediated recycling protects IgG type of proteins (e.g., endogenous IgGs, mAbs, and Fc fusion proteins) from proteolytic degradation in lysosomes, consequently delaying their catabolism and prolonging their half-lives compared to other types of proteins that are not rescued by FcRn-mediated recycling.[4,5] As a result, endogenous IgGs, mAbs, and Fc fusion proteins usually have relatively long half-lives, ranging from several days to weeks. The pathways of nonspecific clearance and FcRn-mediated recycling are typically low-affinity and high-capacity pathways, which are usually nonsaturable at therapeutically relevant doses. For mAbs, relatively constant values of nonspecific clearance are found in each species. In humans, this value is 3–6 mL/day/kg and is affected by multiple pathophysiological and demographical covariates.[8]

In addition to the nonspecific clearance pathways, TMDD may also play a role in the clearance of target-binding proteins (e.g., mAbs, Fc fusion proteins, recombinant proteins). By this mechanism, a TP is cleared from the systemic circulation by binding to its target antigen followed by proteolytic degradation. The target antigens can be cell-surface receptors or soluble antigens. For targets that are cell-surface receptors, a TP is cleared after the TP–antigen complex is internalized and degraded in the lysosomes of target cells or when the TP-opsonized cell engages in immune effector function, which triggers apoptosis of the target cells by complement-dependent cytotoxicity and antibody-dependent cellular cytotoxicity followed by degradation of the TP. For targets that are soluble antigens, a TP is cleared after the TP–antigen complex is eliminated via phagocytosis and proteolysis by endothelial cells and MPS. TMDD is typically a high affinity, low capacity and saturable process. When TMDD plays an important role in TP clearance, the PK parameters of the TP is concentration and dose dependent and may show a time-dependent decrease of clearance if receptor capacity is decreased after repeated treatment. For example, efalizumab[9] and panitumumab[10] show higher clearance at low concentrations and doses in clinical applications. The clearances of gemtuzumab and rituximab decrease after the second dose compared to the first dose, which may result from the decrease of target-mediated clearance after a reduction in target tumor cell number after the first dose of treatment.[11] For most TPs with TMDD involvement, the TMDD pathway is usually more dominant at low doses and low concentrations of the TPs when this pathway is not saturated. At therapeutic doses of these TPs, the therapeutic protein is often in great excess compared to the expression level of the respective target antigen available for binding under equilibrium conditions; consequently, the nonspecific clearance pathways play a dominant role. For these TPs at their prescribing doses (e.g., pertuzumab,[12] bevacizumab,[13] and trastuzumab[14]), changes of target antigen levels generally have a minimal impact on their clearance, and their PK parameters are concentration and dose independent.

The ability of TPs to elicit humoral responses, i.e., immunogenicity, can often modulate the clearance of TPs. The humoral response leads to the formation of antitherapeutic antibodies (ATAs), which may form immunocomplexes with TPs and consequently affect the clearance rates by affecting the binding of a TP to its target or affecting the nonspecific clearance pathways. For example, accelerated clearance of infliximab and adalimumab has been reported after development of ATA in rheumatoid arthritis (RA) patients.[15,16]

2.2.3 Excretion of Therapeutic Proteins

Excretion mechanisms for TPs also differ from those for SMDs. Renal clearance is generally negligible when the molecular size of a TP exceeds the cutoff size for renal filtration of approximately 45 kDa.[17] Tubular secretion does not occur to any significant extent for large TPs. The peptides resulting from TP catabolism may be partially reabsorbed in the proximal or distal tubule of the nephron or are further catabolized in kidney. Biliary excretion of TPs has been reported for only some fragment peptides and proteins such as immunoglobulin A and octreotide,[6,18] which are subsequently degraded in the gastrointestinal tract.

2.2.4 Pharmacokinetic Properties of Antibody–Drug Conjugates

ADCs, as an emerging class of TPs, have the PK properties of both mAbs and SMDs. ADCs are composed of a potent cytotoxic agent conjugated to a mAb via various types of linkers.[19,20] ADCs bind to their target receptors on the surface of tumor cells. The complexes are internalized and degraded and subsequently release the cytotoxic agents to kill tumor cells. Usually the PK properties of multiple analytes, such as the conjugate and the unconjugated cytotoxic agent, are assessed after administration of an antibody–drug conjugate.

To date all ADCs are administered intravenously.[20] The distribution of ADCs is similar to their unconjugated mAbs. For example, in a preclinical *in vivo* study, it was found that trastuzumab emtansine (T-DM1), an ADC for the treatment of human epidermal growth receptor 2 (HER2) positive solid tumors, had similar tissue distribution to that of trastuzumab, the mAb component of T-DM1, indicating that conjugation does not impact the distribution of trastuzumab.[21] ADCs are catabolized by similar pathways as mAbs, including nonspecific proteolytic degradation and TMDD pathways. Immunogenicity may also play a role in ADC clearance.

In addition, the processes of linker chemistry-determined deconjugation in plasma and tissue are also involved in the catabolism and clearance of ADCs. The formation rate of the small molecule cytotoxic component by catabolism of the ADC is usually much slower than the elimination clearance of the small molecule cytotoxic component itself, resulting in formation rate-limited pharmacokinetics. Upon formation, these unconjugated cytotoxic molecules undergo typical clearance pathways of SMDs, such as hepatic metabolism and renal and biliary excretion, as mediated by CYP, non-CYP enzymes, and transporters.[21,22] The low dose of the SMD component of an ADC and relatively slow formation rate combined with a relatively fast elimination rate of the unconjugated SMD molecules may explain the observed relatively low systemic exposure of the unconjugated cytotoxic agent. For example, the average maximal concentration of the derivative of maytansine (DM1) is ~5 ng/mL after the administration of 3.6 mg/kg of T-DM1 every 3 weeks.[22] The average maximal free monomethyl auristatin E (MMAE) concentrations are 5–7 ng/mL after the every-3-week administration of 1.8–2.7 mg/kg of brentuximab vedotin,[23] a MMAE-containing ADC.[24]

2.3 MAJOR MECHANISMS OF THERAPEUTIC PROTEIN–DRUG INTERACTIONS

We are categorizing DIs as either PK based or PD based. PK-based DIs are those resulting from direct competition, inhibition, or induction of drug ADME mechanisms without involvement of the therapeutic targets. PD-based DIs are those resulting from modulation of the systems or target biology via the PD effects of drugs in combination. Both PK- and PD-based DIs may result in relevant changes in exposure and lead to a potential impact on safety and efficacy outcomes, especially for drugs with a NTR. PD-based DIs may also cause undesirable toxicity without an impact on exposure. Unlike SMDs, which are often susceptible to PK-based DIs due to an alteration in CYP and transporter-mediated ADME processes by drug combinations,[3,25] TP-DIs are often mechanistically different.

2.3.1 Impact of Pharmacokinetic-Based Therapeutic Protein–Drug Interactions on the Exposures of Therapeutic Proteins and Small Molecule Drugs

PK-based TP-DIs are not common because TPs and SMDs have distinct PK properties. The nonspecific clearance pathways for TPs are usually unsaturable at therapeutic concentrations. Likewise, these pathways are unlikely to be saturated by the combination of two TPs. For example, the clearances of trastuzumab and bevacizumab are dominated by nonspecific pathways at their clinically efficacious doses. No alteration in PK properties is observed when they are given in combination.[26]

When a TP is combined with a SMD, there is usually no direct overlap and competition in the metabolism and clearance pathways, thus PK-based DIs are unlikely. For example, chemotherapeutic agents such as irinotecan, 5-fluorouracil, and platinum-based therapy (i.e., cisplatin, carboplatin) do not affect the PK properties of cetuximab in cancer patients.[27–29] Similarly, no PK-based DIs are observed between bevacizumab and any of the following agents: capecitabine, cisplatin, 5-fluorouracil, irinotecan, oxaliplatin, or paclitaxel.[30] A dedicated study was conducted to evaluate potential TP-DIs for the combination of bevacizumab and irinotecan (as part of the FOLFIRI regimen containing irinotecan, fluorouracil, and leucovorin). This study demonstrated that the 90% confidence interval of geometric mean ratios for exposure of irinotecan and SN-38 (the active metabolite of irinotecan) in the absence of versus in the presence of bevacizumab were both within the prespecified no effect boundaries, indicating no clinically relevant TP-DIs for this combination.[31] Additional examples of no TP-DIs for combinations of anticancer mAbs with chemotherapeutic and antineoplastic SMDs have been reviewed in recent publications.[2,30,32–35]

PK-based TP-DIs involving ADCs are theoretically possible because the cytotoxic component of the ADC, once deconjugated, may elicit PK-based DIs when the ADC is combined with other SMDs. The cytotoxic agent, which is often a CYP substrate, is likely a victim of DIs when combined with SMDs that are CYP inhibitors or inducers. However, the cytotoxic agent has a relatively low systemic exposure. Thus it is not expected to have any impact on CYP and transporter activities in clinical settings and is unlikely to be a perpetrator. Data for ADC-related DIs are limited to assessments for T-DM1 and brentuximab vedotin (Adcetris). When T-DM1 is given in combination with taxanes (paclitaxel or docetaxel), the PK properties of taxanes and DM1 remain unchanged because taxanes and DM1 are not potent CYP inhibitors or inducers at clinically relevant concentrations.[36,37] A dedicated study of brentuximab vedotin found that it does not affect the PK parameters of midazolam, a CYP3A4 substrate. In the same study, the unconjugated MMAE exposure increased ~34% when brentuximab vedotin was combined with ketoconazole (a potent CYP3A4 inhibitor) and decreased ~46% when brentuximab vedotin was combined with rifampin (a potent CYP3A4 inducer). Therefore it is recommended that patients who are receiving strong CYP3A4 inhibitors concomitantly with brentuximab vedotin should be closely monitored for MMAE-related adverse reactions. These results are expected because MMAE is a substrate of CYP3A4 but not a CYP inhibitor or inducer at clinically relevant concentrations.[23]

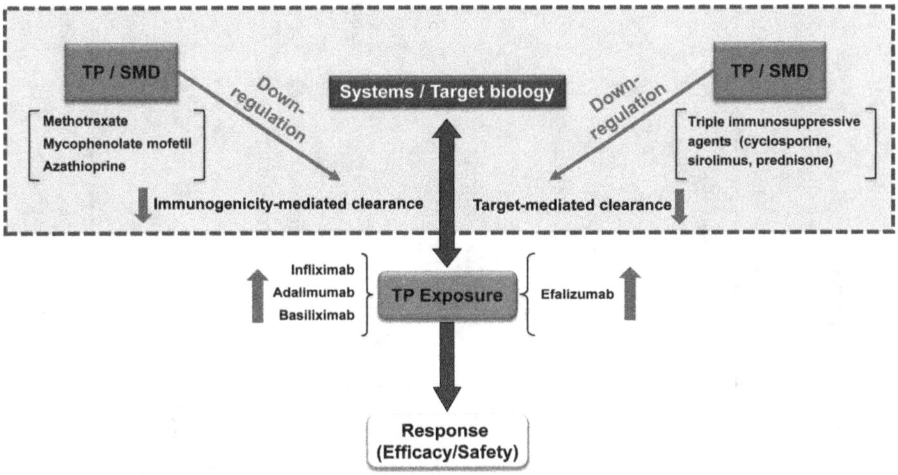

FIGURE 2-2 Theoretical mechanisms of PD-based TP-DIs that change exposure of the TP. SMD: small molecule drug; TP: therapeutic protein. (See the color version of this figure in Color Plates section.)

2.3.2 Impact of Pharmacodynamic-Based Therapeutic Protein–Drug Interactions on the Exposure of Therapeutic Proteins

Distinct from the less common cases of PK-based TP-DIs, there are several plausible mechanisms for PD-based TP-DIs that change the exposure of TPs. As shown in Figure 2-2, interaction between the biological systems or target biology with a TP may affect the TP's exposure through immunogenicity-mediated clearance or target-mediated clearance pathways. The TPs or SMDs given in combination may modulate these clearance pathways by their PD effect, leading to DIs. For example, immunosuppressants such as methotrexate (MTX), mycophenolate mofetil, and azathioprine increase the exposures of infliximab, adalimumab, and basiliximab, possibly due to the effect of the immunosuppressants on decreasing the immunogenicity rate of these mAbs when they are given in combination. In another case, triple immunosuppressive agents may decrease target (CD11a+ T-cells) level, subsequently decreasing target-mediated clearance of efalizumab and thus increasing its exposure.

2.3.2.1 Pharmacodynamic-Based Therapeutic Protein–Drug Interactions Owing to Changes in Immunogenicity-Mediated Clearance of Therapeutic Proteins Some immunosuppressive drugs may modulate the humoral immune response and decrease the immunogenicity of a TP, thus modulating its clearance. This is possible only for TPs that have a relatively high immunogenicity rate and when the clearances of TPs are impacted by immunogenicity. Examples include infliximab, adalimumab, and basiliximab, as listed in Table 2-1. Infliximab and adalimumab are both mAbs antagonizing tumor necrosis factor α (TNF-α)[44] and are often given in combination with immunosuppressive agents

TABLE 2-1 Examples and Clinical Relevance of PD-Based TP-DIs That Have an Impact on PK Properties of TPs

Drug Combination	Type of TP and Indication	Possible Mechanisms for TP-DIs	Change in PK Properties	Clinical Relevance	References
Infliximab + MTX	Infliximab: anti-TNFα mAb Indication: RA	MTX may inhibit humoral response; ↓ immunogenicity of infliximab.	MTX ↓ CL of infliximab.	No dose adjustment is recommended.	38,39
Adalimumab + MTX	Adalimumab: anti-TNFα mAb Indication: RA	MTX may inhibit humoral response; ↓ immunogenicity of adalimumab.	MTX ↓ CL of adalimumab by 29–44%.	No dose adjustment is recommended.	15,40
Basiliximab + CsA + corticosteroids + mycophenolate mofetil or azathioprine	Basiliximab: chimeric mAb against IL-2Rα on T-cells Indication: prevent transplant rejection	Mycophenolate mofetil or azathioprine may inhibit humoral response; ↓ immunogenicity of basiliximab.	Mycophenolate mofetil ↓ CL of basiliximab by 51%; azathioprine ↓ CL of basiliximab by 22%.	No dose adjustment is recommended.	41,42
Efalizumab + triple immunosuppressive agents (CsA, sirolimus, prednisone)	Efalizumab: humanized IgG1 anti-CD11a mAb Indication: psoriasis, prevent transplant rejection	Triple immunosuppressive agents may ↓ TMDD of efalizumab by suppression of CD11a+ T-cells.	Triple immunosuppressive agents ↓ CL of efalizumab by 50% in transplant patients compared to psoriasis patients.	No dose adjustment is recommended.	43

CL: clearance; CsA: cyclosporine; IL-2R: interleukin 2 receptor; mAb: monoclonal antibody; MTX: methotrexate; PD: pharmacodynamic; PK: pharmacokinetic; RA: rheumatoid arthritis; TMDD: target-mediated drug disposition; TNF: tumor necrosis factor; TPs: therapeutic proteins; TP-DIs: therapeutic protein–drug interactions.

such as MTX for the treatment of autoimmune diseases and prevention of rejection after organ transplantation. It was found that patients with positive ATA responses needed higher doses of infliximab to maintain the therapeutic concentration.[16] Similarly, serum trough adalimumab concentrations were dramatically lower in patients who developed antiadalimumab antibodies.[15] Concomitant dosing of MTX was associated with a decreased incidence of ATA formation against infliximab[38,45] and adalimumab.[15,45] As a result, when given in combination with MTX, the clearance of infliximab and adalimumab decreased and their serum concentrations increased.[15,38,40,44,45]

Basiliximab is a high-affinity chimeric IgG1-based mAb that binds to and blocks the interleukin 2 receptor α-chain (IL-2Rα) on the surface of activated T-lymphocytes and competitively inhibits IL-2-mediated activation of lymphocytes, a critical pathway in the cellular immune response involved in allograft rejection.[41] Basiliximab is often given in combination with triple immunosuppressive agents, including cyclosporine (CsA), corticosteroids, and either azathioprine or mycophenolate mofetil to prevent renal allograft rejection.[42] A cross-study comparison of basiliximab PK parameters in *de novo* renal allograft recipients showed that basiliximab clearance was 29 ± 14 mL/h when coadministered with azathioprine and 18 ± 8 mL/h with mycophenolate mofetil, and both were significantly lower compared to a clearance of 37 ± 15 mL/h from an earlier study of basiliximab with dual immunosuppressive agents of CsA and corticosteroids only.

Although the interstudy variability could potentially be a caveat of this observation, a plausible reason is that immunosuppressive agents such as azathioprine and mycophenolate mofetil may inhibit humoral immune responses and consequently decrease basiliximab clearance. Mycophenolate mofetil may have a stronger inhibition effect than azathioprine. The immunogenicity data in these studies further suggest that if the reduced clearance of basiliximab in the presence of azathioprine or mycophenolate mofetil is due to an inhibition of humoral immune responses, it appears that these responses are directed to a portion of basiliximab other than the IL-2Rα binding portion of basiliximab. Nonetheless, the PD effect of basiliximab as indicated by the range of the durations of IL-2Rα saturation when basiliximab was combined with triple immunosuppressive agents did not extend outside the range of the durations after the treatment of basiliximab in combination with dual immunosuppressive agents. Hence, regardless of the observed clearance difference, no dosing adjustment was necessary.[42]

2.3.2.2 Pharmacodynamic-Based Therapeutic Protein–Drug Interactions Owing to Changes in Target-Mediated Drug Disposition of Therapeutic Proteins Combinations that affect the TMDD pathway of some TPs may result in PD-based TP-DIs, when TMDD plays an important role in the clearance processes of these TPs at their therapeutic doses. Based on a theoretical simulation of a typical mAb with strong TMDD (i.e., TMDD accounts for >50% of total clearance in the therapeutically relevant dose range), an increase of baseline receptor expression level (e.g., 3-fold or 10-fold of baseline) leads to a significant increase of the observed total clearance; a decrease of baseline receptor expression level (e.g., 10%, 30% of baseline) leads to a significant decrease of the observed total clearance. The magnitudes of changes are higher at low dose levels, when TMDD accounts for higher percentage of total clearance (Figure 2-3*a*).

A clinical example is efalizumab in combination with immunosuppressive agents for preventing organ rejection after transplantation, is listed in Table 2-1. At clinical doses, TMDD of efalizumab after binding with CD11a receptors expressed on T-cells plays an important role in efalizumab clearance. Consequently, clearance of efalizumab is concentration and dose dependent and governed by CD11a expression level.[48] When efalizumab was given by subcutaneous administration in combination with oral administration of triple immunosuppressive agents (CsA, sirolimus, and prednisone) in transplant recipients, the clearance of efalizumab was 50% lower than the clearance observed in psoriatic patients. One of the reasons might be related to the reduction of CD11a+ T-cells in circulation by immunosuppressive agents given in combination in transplant recipients, resulting in reduced TMDD of efalizumab.[43,48]

However, in most clinical situations in which TPs are given at efficacious doses at which the TMDD pathway does not play a major role in their clearance, the nonspecific proteolytic degradation pathways, which are usually of high capacity and unsaturable, dominate the clearance of these TPs. They exhibit linear PK characteristics with relatively constant total clearance at therapeutic doses. For these TPs, the possibility of PD-based TP-DIs as a result of a target receptor expression level change by the combination drug is usually low. Based on a theoretical simulation of a typical mAb with negligible TMDD (i.e., TMDD accounts for ≤1% of total clearance in the therapeutically relevant dose range), an increase or a decrease of baseline receptor expression level (e.g., 3-fold, 10-fold, 10%, or 30% of baseline) does not have a significant impact on the observed total clearance (Figure 2-3*b*).

A clinical example is the combination of T-DM1 with pertuzumab, an ADC with a mAb.[49] Both the trastuzumab component of T-DM1 and pertuzumab bind to

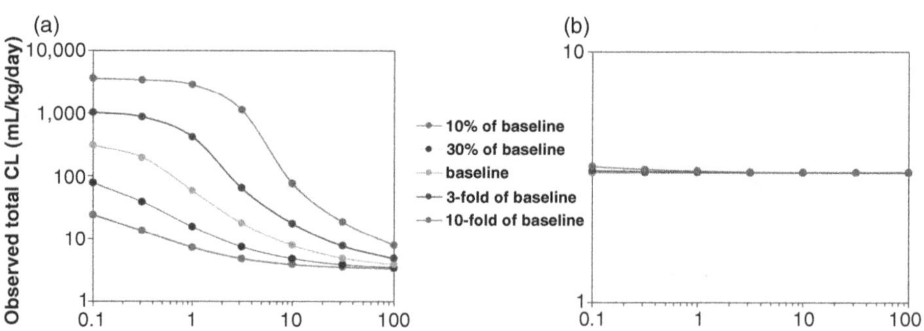

FIGURE 2-3 Effect of receptor expression level on the dose-total CL relationship for a typical mAb. Observed total CL = dose/area under the concentration–time profile. Simulation is performed using a standard TMDD model after a single dose of a typical mAb [46,47]. The baseline free receptor expression levels was altered to 3-fold, 10-fold, 10%, or 30% of baseline, and the observed total CL values were plotted versus dose levels. (a) TMDD accounts for >50% of total clearance in the therapeutically relevant dose range. (b) TMDD accounts for ≤1% of total clearance in the therapeutically relevant dose range. CL: clearance. (See the color version of this figure in Color Plates section.)

HER2 receptors. Therefore a potential for PD-based DIs might exist for this combination. However, pertuzumab and T-DM1 bind to distinct epitopes of HER2 simultaneously without steric hindrance.[50,51] Furthermore, both T-DM1 and pertuzumab have linear PK characteristics at their clinically efficacious dose ranges when given in combination,[12,22] indicating that HER2-mediated clearance is unlikely to play a major role in their clearance in these dose ranges. Hence a potential impact on HER2 biology by T-DM1 is unlikely to affect the PK characteristics of pertuzumab, and vice versa. Cross-study comparisons of PK parameters of T-DM1 and pertuzumab in the combination study and the historical single-agent studies demonstrate that there is no DI observed for this combination.[49]

2.3.3 Impact of Pharmacodynamic-Based Therapeutic Protein–Drug Interactions on the Exposure of Small Molecule Drugs Owing to Therapeutic Protein–Cytokine–Cytochrome P450 Modulation Effects

Because TPs and SMDs do not share common ADME mechanisms, TPs are not predicted to affect the hepatic, renal, or biliary elimination of SMDs directly. However, some immunomodulatory TPs may exert an indirect effect on the hepatic clearance pathways of SMDs through the TP-cytokine-CYP modulation effect, resulting in PD-based TP-DIs. As summarized in Figure 2-4, infection and inflammatory diseases may cause an elevation of systemic cytokine level, such as interferons (IFNs), interleukins, and tumor necrosis factors, which in turn inhibit the CYP

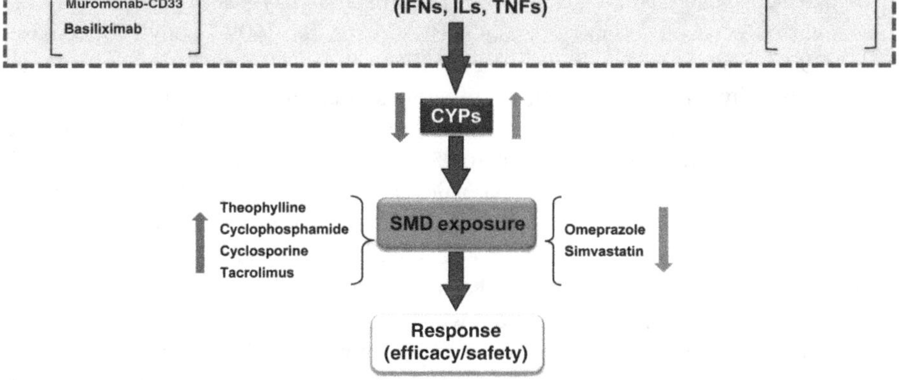

FIGURE 2-4 Theoretical mechanisms of PD-based TP-DIs that have an impact on the PK properties of SMDs through TP-cytokine-CYP modulation effect. CYP: cytochrome P450; IFN: interferon; IL: interleukin; SMD: small molecule drug; TNF: tumor necrosis factor; TP: therapeutic protein. (See the color version of this figure in Color Plates section.)

activity and consequently increase the exposure of SMDs that are CYP substrates.[52] Some immunomodulatory TPs, such as exogenous IFNα, muromanab-CD3, and basiliximab may upregulate certain cytokine levels, inhibit CYP activities, and increase the exposure of CYP substrates (e.g., theophylline, cyclophosphamide, CsA, and tacrolimus). In contrast, some immunomodulatory TPs such as tocilizumab may downregulate the pathologically elevated cytokine levels in the inflammation disease state, normalize CYP activities to a higher level similar to healthy individuals, and decrease the exposure of CYP substrates (e.g., omeprazole, simvastatin). In addition to these clinical observations, there is strong *in vitro* evidence suggesting the differential effects of cytokines on various CYP enzymes.[1] The TP-DIs induced by the TP-cytokine-CYP modulation effect may be clinically relevant for SMDs with a NTR. The mechanisms and various scenarios of this type of TP-DI are reviewed in detail in this section.

Pathophysiological changes in patients and animal models in infectious or inflammatory conditions are often associated with decreased production and activities of hepatic and intestinal CYP enzymes, which are important for the first-pass extraction and metabolism of most SMDs. Based on *in vitro* and *in vivo* data, proinflammatory cytokines such as ILs (IL-1β, IL-6), TNFs (TNFα), and IFNs (IFNγ, IFNα-2b) may be potentially important mediators of this CYP modulation effect. These cytokines may induce a decrease in transcription factor activity for CYP enzyme expression or a decrease in CYP enzyme stability. As a result, CYP activities decrease in patients with infectious and inflammatory disease with elevated proinflammatory cytokine levels. The patients usually have an increased exposure of the SMDs metabolized by CYP enzymes because of decreased clearance and/or increased bioavailability compared with healthy individuals.[52]

Direct administration of exogenous cytokines may produce an immunopathological state similar to infection or inflammation, downregulate CYP activities, and potentially increase the exposure of some SMDs given in combination.[52] For instance, a clinical study using the "Pittsburgh cocktail" approach in high-risk melanoma patients found that the high-dose IFNα-2b[53] therapy differentially impairs CYP-mediated metabolism, causing no effect on some enzymes (e.g., CYP2E1) but substantial effects on others (e.g., CYP1A2: median 60% activity decrease; CYP2C19, median 40% activity decrease).[54] IFNα-2b has also been used in combination with theophylline for the treatment of various diseases (e.g., chronic hepatitis B). The clearance of theophylline, a CYP1A2 substrate drug with a NTR, decreased after IFNα-2b treatment, resulting in a 100% increase in serum theophylline levels.[53] In a clinical study in multiple myeloma patients, it was found that the administration of IFNα before cyclophosphamide (CP) caused a decrease in CP clearance to 63%, a 137% longer half-life and a 137% higher peak plasma concentration compared to the results obtained when IFNα was administered 24 h after CP. Consequently, the active metabolite 4-hydroxycyclophosphamide, which is formed by CYP-mediated metabolism from the prodrug CP, had 45% lower exposure when IFNα was given before CP compared with that observed when IFNα was administered 24 h after CP. As a result, a significant decrease in leukocyte count was observed when IFNα was given 24 h after CP.[55] These examples (some of them are listed in Table 2-2) demonstrate the

TABLE 2-2 Examples and Clinical Relevance of TP-DIs Affecting the PK Properties of SMDs Caused by the TP-Cytokine-CYP Modulation Effect

Drug Combination	Type of TP and Indication	Possible Mechanisms for TP-DIs	Change in PK Properties	Clinical Relevance	References
IFNα-2b (Intron A) + theophylline	IFNα-2b: exogenous interferon Indication: various diseases (e.g., chronic hepatitis B)	Theophylline is a CYP1A2 substrate IFNα-2b inhibits CYP1A2.	IFNα-2b ↑ serum theophylline concentration by 100%.	Be cautious, considering theophylline is a NTR drug	53
IFNα + CP	IFNα: exogenous interferon Indication: multiple myeloma	CP is converted to its active metabolite 4-OHCP by CYPs; IFNα inhibits CYP activity.	Administration of IFNα before CP caused ↓ CP clearance by 37%, ↑ half-life by 37%, and ↑ peak plasma concentration by 37%, compared to the results obtained when IFNα was administered 24 h after CP.	Sequence-dependent toxicity profile: a significant decrease in leukocyte count when IFNα was given 24 h after CP.	55
Muromonab-CD3 + CsA	Muromonab-CD3: murine IgG2a mAb, targets CD3 on T-cells Indication: prevent acute transplant rejection	CsA is a CYP3A4 substrate. Muromonab-CD3 results in T-cell activation and subsequent release of multiple	↑ CsA trough by 100% at day 5 post-transplantation, compared to the control group	CsA doses were adjusted based on day 5 trough levels.	56,57

(continued)

TABLE 2.2 (*Continued*)

Drug Combination	Type of TP and Indication	Possible Mechanisms for TP-DIs	Change in PK Properties	Clinical Relevance	References
		cytokines immediately after injection, ↓ CYP3A4 activities.			
Basiliximab + tacrolimus Basiliximab + CsA	Basiliximab: chimeric mAb antagonist of IL-2Rα on activated T-lymphocytes Indication: prevent organ transplant rejection	Tacrolimus and CsA are CYP3A4 substrates Binding of basiliximab to the IL-2R on activated T-cells may allow circulating IL-2 binding to IL-2Rs on hepatic and intestinal cells, resulting in ↓ CYP3A4 enzyme activity.	↑ day 3 trough level of tacrolimus by 63% ↑ CsA levels in a pediatric patients receiving kidney allograft	Tacrolimus dose was decreased until the plasma concentration returned to desirable range. Recommend that the initial CsA dose should be limited to 400 mg/m^2 if used in combination with basiliximab.	41,58,59

Tocilizumab + simvastatin or omeprazole	Tocilizumab: anti-IL-6R mAb that binds specifically to both sIL-6R and mIL-6R and inhibits IL-6-mediated signaling through blocking these receptors Indication: RA	Simvastatin is a CYP3A4 substrate. Omeprazole is a CYP2C19 and CYP3A4 substrate. Tocilizumab may reverse IL-6-induced suppression of CYP activities in the disease state, leading to normalization of metabolism of drugs that are CYP substrates.	↓ simvastatin exposure by 57%; ↓ omeprazole exposure by 12–28%.	Effect may be clinically relevant for CYP substrates with a NTR, for which dose is individually adjusted.	60,61

CP: cyclophosphamide; CsA: cyclosporine; CYP: cytochrome P450; IFN: interferon; IL-2R: interleukin 2 receptor; IL-6: interleukin 6; IL-6R: IL-6 receptor; mAb: monoclonal antibody; mIL-6R: membrane-bound IL-6 receptor; NTR: narrow therapeutic range; 4-OHCP: 4-hydroxycyclophosphamide; PK: pharmacokinetic; RA: rheumatoid arthritis; sIL-6R: soluble IL-6 receptor; TP: therapeutic protein; TP-DIs: therapeutic protein-drug interactions.

effect of exogenous cytokines on the activity of CYPs in the clinical setting, which implies that endogenous cytokine levels may play an important role as modulators of hepatic metabolism.

Similarly, some mAbs may increase cytokine levels and activities by their PD effects and may lead to similar effects as the treatment by exogenous cytokines. As listed in Table 2-2, both muromonab-CD3 and basiliximab are mAbs for preventing rejection after organ transplantation.[41,56] Muromonab-CD3 is a murine IgG2a mAb targeting CD3 on T-cells to prevent acute organ rejection. Immediately after administration of muromonab-CD3, the activation of T-cells and release of numerous cytokines are observed followed by the blocking of all known T-cell functions. The initial T-cell activation and cytokine release may downregulate CYP enzyme activity and expression. CsA is a CYP3A4 substrate that may be combined with muromonab-CD3 in transplantation patients. It was found that muromonab-CD3 treatment caused a significant rise in CsA trough levels at day 5 post–renal transplantation compared to the group of patients receiving antilymphocyte globulin immunoprophylaxis therapy. As a result, CsA doses were adjusted based on trough levels obtained on day 5 post-transplantation.[57]

Clinically significant DIs were also reported for basiliximab in combination with CsA or tacrolimus. Both CsA and tacrolimus are metabolized by CYP3A4 and require therapeutic drug monitoring owing to their NTRs. In children who have undergone kidney transplant, basiliximab combination led to a higher trough concentration of CsA compared to the patients in the control group who were not on basiliximab. Consequently, it was recommended that the initial CsA dose should be limited to 400 mg/m^2 if used in combination with basiliximab in children with kidney transplants.[58]

Similarly, a 63% increase in tacrolimus trough levels was reported in basiliximab-treated adult patients on day 3 postrenal transplantation, when compared to the control group receiving antithymocyte globulin induction therapy without basiliximab treatment. As a result, the tacrolimus dose was decreased until the plasma concentration returned to a desirable range.[59] It was postulated that both of these clinically significant TP-DIs may be related to the effect of basiliximab on CYP activities via cytokine modulation. Interaction between IL-2 and its respective IL-2R on intestinal epithelial cells and hepatocytes may decrease the expression or activity of CYP3A4.[62] Binding of basiliximab to the IL-2R on activated T-cells may allow circulating IL-2 binding to IL-2Rs on hepatic and intestinal cells, resulting in a downregulation of CYP3A4 enzyme activity.

On the other hand, the treatment of some TPs results in a downregulation and normalization of elevated cytokine levels in the disease state through cytokine or cytokine receptor antagonism. Subsequent inhibition of cytokine-mediated downstream signaling transduction may restore the previously down-regulated CYP enzyme activity to a normal level and consequently normalize the exposure of SMDs that are CYP substrates.[52] As listed in Table 2-2, tocilizumab, an anti-IL-6R mAb for RA treatment, may have DIs based on this mechanism. In RA patients, IL-6 serum concentrations are elevated up to 50–60 pg/mL as compared to those in healthy subjects (~5 pg/mL). Tocilizumab binds specifically to both soluble and membrane-bound IL-6 receptors (sIL-6R and mIL-6R) and inhibits IL-6-mediated signaling through blocking these receptors. This ameliorates the inflammation status.

In vitro data suggest that IL-6 inhibits CYP activity, and tocilizumab prevents CYP3A4 suppression by IL-6 in human hepatocytes, indicating that tocilizumab may reverse IL-6-induced suppression of CYP activities in the disease state.[60] This was demonstrated in a clinical study in RA patients in which the tocilizumab combination resulted in up to a 57% decrease in the exposure of simvastatin, a CYP3A4 probe substrate, in comparison to simvastatin alone in the same individuals. In another study, tocilizumab resulted in a 12–28% decrease in exposure of omeprazole, a CYP3A4 and CYP2C19 substrate, by comparing the exposure in the same RA patients before and after tocilizumab dosing. These observations may have clinical relevance for patients who take tocilizumab with NTR drugs for which the doses need to be individually adjusted.[60,61] Some other approved cytokine modulators, such as canakinumab (anti-IL-1β mAb),[63] golimumab (anti-TNFα mAb),[64] rilonacept (anti-IL-1R fusion protein),[65] and ustekinumab (anti-IL-12/23 mAb)[66] may also potentially normalize CYP enzymes in patients. Although no formal DI studies have been conducted for these TPs, the potential risk of TP-DIs and the need to monitor the therapeutic effects or concentrations of drugs with a NTR when given in combination with these TPs are addressed in their package inserts.

Transporters play important roles for the ADME process of SMDs. Inflammatory disease states may also affect several important drug transporters, such as *p*-glycoproteins, multidrug resistance-associated proteins, and organic anion-transporting peptides.[67] Theoretically, the risk of transporter-mediated DIs for the combinations of TPs and SMDs exists. However, to date no clinically relevant TP-DIs caused by this mechanism have been reported.

2.3.4 Pharmacodynamic-Based Therapeutic Protein–Drug Interactions That Lead to Toxicity without Affecting Exposures

The major objective for using combination therapy is to improve efficacy by desirable PD interactions without increasing safety concerns. As discussed earlier, PD-based TP-DIs may result in exposure changes in either the TP or the SMD given in combination, which may have clinical relevance for drugs with a NTR. In addition, PD-based TP-DIs can also lead to toxicity without affecting the exposure of the drugs in combination. PD-based TP-DIs that result in undesirable toxicity are usually documented in the labels of these TPs to guide the prescribers to avoid certain combinations. An example is the combination of anakinra and etanercept. Anakinra is a recombinant IL-1R antagonist protein that competitively inhibits the binding of IL-1 to IL-1Rs.[68] Anakinra has a molecular weight of 17.3 kDa. Etanercept is a dimeric fusion protein consisting of the extracellular ligand-binding portion of the human 75 kDa tumor necrosis factor receptor linked to the Fc portion of human IgG1.[69] Both TPs were approved for RA treatment. Because anakinra is eliminated by renal excretion and etanercept is eliminated by proteolysis with the salvage pathway of FcRn-mediated recycling, there are no overlapping clearance pathways for them. In a phase III study in RA patients using concurrent therapy of anakinra with etanercept, the PK properties of both agents were unchanged by the combination. However, there was no significant improvement in efficacy for the combination therapy, and the risk of serious infection and neutropenia increased when compared to etanercept alone. PD-based DIs may potentially cause patients to be highly immunosuppressed and prone to infection, and an additive neutropenia effect also likely results from the use of this combination.[69,70]

Other examples are also recently reviewed.[30] For example, filgrastim, a human granulocyte colony–stimulating factor, during a period of 24 h before through 24 h after cytotoxic chemotherapy should be avoided because of the potential sensitivity of rapidly dividing myeloid cells to cytotoxic chemotherapy.[71] Aldesleukin,[72] a human recombinant IL-2 protein, may enhance the toxicity of drugs of nephrotoxic, myelotoxic, cardiotoxic, or hepatotoxic effects when given in combination. IFNα-2b[53] may cause increased myelotoxicity of myelosuppressive agents, such as zidovudine, when given in combination.

2.4 STRATEGIES TO ASSESS THE RISK OF THERAPEUTIC PROTEIN–DRUG INTERACTIONS IN CLINICAL DEVELOPMENT OF THERAPEUTIC PROTEINS

2.4.1 Challenges of Assessing Therapeutic Protein–Drug Interactions in Clinical Development

It is critical to understand the theoretical TP-DI mechanisms and build an individualized and iterative strategy to ensure the safe and effective use of TPs for combination treatment. The FDA guidance document for DI studies,[3] the European Medicines Agency (EMA) guideline on the clinical investigation of the PK characteristics of TPs,[73] and several recent review papers from the FDA highlighted the importance of evaluating TP-DIs.[1,2]

There are multiple challenges in assessing TP-DIs in clinical development of TPs, owing to the distinct PK and PD properties of TPs. Considering potential toxicity, most clinical trials involving TPs need to be conducted in the patient population instead of in healthy subjects. The PK characteristics of TPs may differ between patients and healthy subjects because of target expression levels. For immunomodulatory TPs, the CYP activity normalization by TP-cytokine-CYP interactions occurs only in patients with infection or inflammatory diseases. However, formal TP-DI studies in patients with life-threatening diseases (e.g., cancer, transplantation, severe autoimmune diseases) are rarely conducted owing to logistical constrains. Cross-over studies are difficult because of the long washout period required for TPs with long half-lives; a parallel design of two arms in a randomized study or a cross-over design for TP-DI assessment is not ethical if one of the drugs lacks efficacy when it is used alone. Concurrent medications also confound the DI evaluation of studied combinations[2,30] In addition to these challenges in clinical evaluation, *in vitro* and preclinical DI assessment strategies for TPs are very different from SMDs and are still evolving owing to the inherent differences in metabolic pathways between TPs and SMDs.[30] Therefore, more creative and less disruptive strategies are needed to assess the risk of TP-DIs in clinical development.

2.4.2 Question-Based Strategy of Therapeutic Protein–Drug Interaction Assessment During Clinical Development

As summarized in Figure 2-5, a question-based TP-DI risk assessment strategy is recommended.[30] This strategy recommends both theoretical TP-DI risk assessments based on PK and PD properties of the drugs in combination (as described earlier in this chapter) and PK evaluation in phase Ib/II studies to support the theoretical assessment. If a clinically relevant TP-DI signal is detected, further assessment and confirmation might be considered in phase II/III studies. All TP-DI relevant information–including the theoretical risk of TP-DIs based on the PK and PD properties of the drugs in combination, the PK and ADME data collected at various stages in the development of combination therapy, the potential impact of TP-DIs on the safety and/or efficacy outcomes, and whether the impact warrants clinical dose adjustment and/or therapeutic drug monitoring—is documented in the package insert to provide physicians with data that allow an informed decision about whether to adjust a drug dose when administered in combination or whether to avoid a particular concomitant medication. The TP-DI risk assessments need an individualized strategy for each molecule and are iterative based on emerging data.

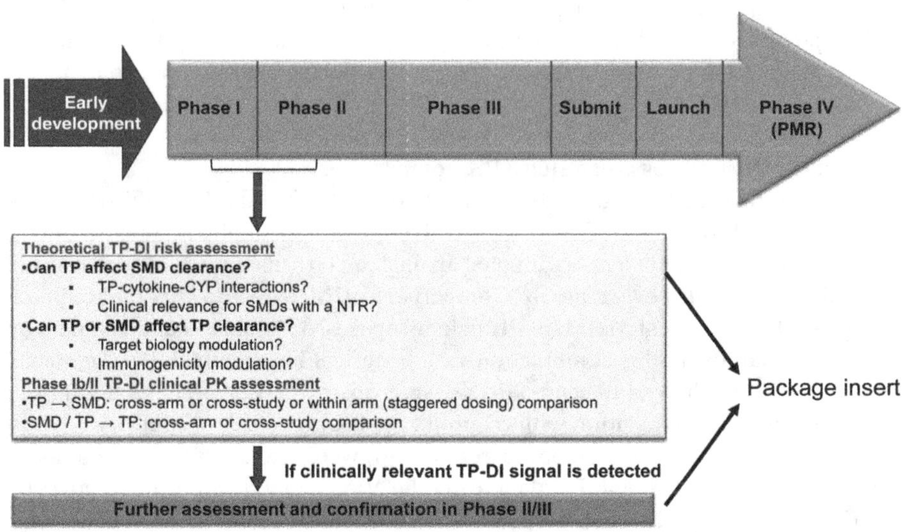

FIGURE 2-5 Question-based TP-DI risk assessment strategy during TP development. CYP: cytochrome P450; NTR: narrow therapeutic range; PK: pharmacokinetic; PMR: post-marketing requirement; SMD: small molecule drug; TP: therapeutic protein; TP-DI: therapeutic protein–drug interaction. (See the color version of this figure in Color Plates section.)

2.4.2.1 Theoretical Risk Assessments for Therapeutic Protein–Drug Interactions The PK and PD properties form the basis for the theoretical risk assessments of TP-DIs that may affect the exposure of combination drugs. To illustrate the general considerations for this theoretical assessment based on various mechanisms of TP-DIs, three tables provide examples for the combinations of a TP with a SMD or another TP in different therapeutic areas: a mAb combined with a SMD for cancer treatment (bevacizumab and irinotecan; Table 2-3), a mAb combined with a SMD for inflammatory disease treatment (tocilizumab and MTX or simvastatin; Table 2-4), and an ADC combined with a mAb for cancer treatment (T-DM1 and pertuzumab; Table 2-5).

2.4.2.2 Evaluation of TP–DI Risk in Combination Studies by Clinical PK Assessment In addition to the theoretical TP-DI risk assessments, PK comparison with single-agent PK properties is recommended in phase Ib/II combination studies. These assessments are particularly important for drugs with a NTR, such as ADCs and some SMDs.

PK analysis for TP-DI assessment can be conducted by various approaches such as noncompartmental analysis (NCA) of individual patient PK profiles and population PK approaches using nonlinear mixed-effect modeling.[82,83] Among population PK approaches, the indicator approach and the simulation-based visual predictive check (VPC) approach are commonly used. With the indicator approach, a dichotomous indicator with the value of 1 or 0, corresponding to the presence or absence of the combination drug, is used as a covariate for PK parameters. The confidence intervals of the ratios from the PK parameters between the presence and absence of the interacting drug can be calculated. This approach was used to assess the impact of multiple combination agents on the PK properties of ustekinumab[84] and the impact of pertuzumab on the PK properties of T-DM1.[49] In addition, the VPC approach to PK data from combination studies based on simulations using a historical population PK model is also frequently used for cross-study PK comparisons.

Cross-arm and/or cross-study PK comparisons are often used for TP-DI assessment in phase Ib/II studies of combination therapy. For example, in a phase I combination study, MetMAb, a mAb directed against the tyrosine kinase Met, was given in combination with bevacizumab. Comparison of MetMAb in this combination with phase I single-agent MetMAb PK data using the NCA method and comparison of bevacizumab PK in this combination with historical bevacizumab PK data using the VPC method showed that no TP-DIs were observed. For a phase II study of MetMAb given in combination with erlotinib, the PK properties of MetMAb, erlotinib, and OSI-420 (a major metabolite of erlotinib) were also not altered in comparison to historical single-agent PK data using the VPC method and/or in comparison to PK parameters of the placebo plus the erlotinib arm in the same study.[85] In a phase II study of T-DM1 in combination with pertuzumab, the PK properties of pertuzumab and T-DM1-related analytes, including T-DM1, total trastuzumab, and DM1, were not changed compared to their historical single-agent data based on a cross-study comparison using both the NCA and population PK approach.[49]

TABLE 2-3 Theoretical TP-DI Risk Assessments for the Combination of Bevacizumab and Irinotecan in Cancer Patients [13,31,74–76]

		Bevacizumab	Irinotecan
PK characteristics		Mainly cleared by nonspecific proteolytic degradation at clinically efficacious doses; linear PK property; $t_{1/2} = \sim 20$ days; $CL = 0.2–0.3$ L/day; interindividual CL variability $= \sim 26\%$.	A prodrug metabolized to its active form, SN-38, by carboxyesterases; SN-38 is inactivated by UGT1A and CYP3A4; irinotecan $t_{1/2} = 6–12$ h; SN-38 $t_{1/2} = 10–20$ h. There is large interindividual PK variability for both irinotecan and its metabolite SN-38. UGT1A gene polymorphism may play a role in PK variability.
Risk of PK-based TP-DIs		Low: No overlapping clearance pathways for bevacizumab and irinotecan	
PD characteristics	MOA	Binds to VEGF and prevents the interaction of VEGF to its receptors, leading to antiangiogenesis effect	SN-38 binds to topoisomerase I–DNA complex, preventing religation of a single-strand break induced by topoisomerase I, leading to DNA damage and cytotoxicity.
	TMDD	Not a major clearance route at clinically efficacious dose ranges	NA
	Immunogenicity	Low	NA
	Cytokine modulator	No	No
	Immuno suppressive agent	No	Yes, due to the neutropenia effect
	Therapeutic range	Moderate	Narrow

(continued)

TABLE 2.3 (*Continued*)

	Bevacizumab	Irinotecan
Risk of PD-based TP-DIs	Risk of impact on bevacizumab exposure is low. TMDD does not play a major role in total CL at clinically efficacious dose ranges. Bevacizumab has relatively low immunogenicity; there is no documented impact of immunogenicity on bevacizumab CL.	Risk of impact on irinotecan exposure is low. UGT1A and CYP activities are unlikely to be impacted by combination with bevacizumab
Risk of clinically relevant TP-DIs	Low	

CL: clearance; CYP: cytochrome P450; MOA: mechanism of action; NA: not applicable; PD: pharmacodynamic; PK: pharmacokinetic; TMDD: target-mediated drug disposition; TP-DIs: therapeutic protein–drug interactions; UGT1A: uridine diphosphate glucuronosyltransferase 1A; VEGF: vascular endothelial growth factor.

For single-arm studies of TP-SMD combinations, a staggered dosing approach in the first treatment cycle may be applied to assess the impact of the TP on the PK properties of the SMD. In this approach, since the short half-lives of most SMDs determine a short wash-out period, a SMD may be administered alone in the first treatment cycle to allow PK assessment in the absence of the TP. In subsequent treatment cycles, the TP and the SMD are administered concurrently, and PK properties of the SMD in the presence of the TP can be assessed. This approach was used to evaluate the effect of T-DM1 on the PK properties of taxane,[36,37] the effect of bevacizumab on the PK of irinotecan,[31] and the effect of tocilizumab on the PK properties of simvastatin.[60]

A confirmatory TP-DI substudy in phase II/III might be needed when theoretical analysis and TP-DI assessment in phase Ib/II studies suggest a clinically relevant TP-DI potential. For example, it is hypothesized that tocilizumab may restore CYP activity in RA patients and therefore normalize the required doses of combination drugs of CYP substrates. To evaluate this hypothesis, a dedicated TP-DI study using simvastatin as the CYP3A4 probe substrate was conducted.[60]

In some situations, postmarketing studies to evaluate TP-DI might have to be conducted when the registration trial did not adequately address the potential for clinically relevant TP-DIs. For example, a dedicated TP-DI study for the combination of bevacizumab and irinotecan (as part of the FOLFIRI regimen combining irinotecan, fluorouracil, and leucovorin) was conducted as a postmarketing

TABLE 2-4 Theoretical TP-DI Risk Assessments for the Combination of Tocilizumab and Other SMDs (e.g., Methotrexate or Simvastatin) in Rheumatoid Arthritis Patients [60,61,77,78]

		Tocilizumab	MTX	Simvastatin
PK characteristics		Nonlinear PK behavior at clinically efficacious doses; the total CL is the sum of linear and nonlinear CL; linear CL = 0.3 L/day; the concentration-dependent apparent $t_{½}$ = up to 11 days for 4 mg/kg and up to 13 days for 8 mg/kg at steady-state given on an every-4-weeks regimen.	Undergoes hepatic and intracellular metabolism to polyglutamated forms, which can be converted back to MTX by hydrolase enzymes; $t_{½}$ = 3–10 h; 80–90% of the administered dose is excreted unchanged in the urine within 24 h. MTX clearance rates vary widely and are generally decreased at higher doses. Not a CYP substrate	A lactone that is readily hydrolyzed *in vivo* to the corresponding β-hydroxyacid form, mediated mainly by CYP3A4, and extensively metabolized by the liver.
Risk of PK-based TP-DIs		Low: No overlapping CL pathways for tocilizumab and MTX or tocilizumab and simvastatin		
PD characteristics	MOA	Tocilizumab binds specifically to both soluble and membrane-bound IL-6R and inhibits IL-6-mediated signaling through these receptors; used for treatment of RA.	MTX inhibits dihydrofolic acid reductase, thus interferes with DNA synthesis, repair, and cellular replication; combination with tocilizumab is used for treatment of moderate to severe active RA in adult patients.	β-hydroxyacid is a potent inhibitor of HMG-CoA reductase, the enzyme that catalyzes the conversion of HMG-CoA to mevalonate, an early and rate-limiting step in the biosynthetic pathway for cholesterol.
	TMDD	Plays an important role in total CL at clinically efficacious dose ranges	NA	NA

(*continued*)

TABLE 2.4 (*Continued*)

		Tocilizumab	MTX	Simvastatin
	Immunogenicity	Low (~2%)	NA	NA
	Cytokine modulator	Yes	No	No
	Immuno-suppressive agent	Yes	Yes	No
	Therapeutic range	Moderate	Narrow	Moderate
Risk of PD-based TP-DIs		Risk of impact on tocilizumab exposure is low. MTX is unlikely to modulate IL-6R-mediated CL of tocilizumab. Tocilizumab has low immunogenicity, and there is no documented impact of immunogenicity on tocilizumab CL.	Risk of impact on MTX exposure is low. In RA patients, IL-6 concentrations are elevated, and CYP activities are suppressed. Tocilizumab may restore the previously down-regulated CYP enzyme activity in RA patients to normal levels and consequently increase the metabolism of SMDs that are CYP substrates, through its inhibition of IL-6-mediated signaling. However, CYPs don't play a major role in MTX metabolism. Risk of impact on simvastatin exposure is high: Simvastatin is mainly metabolized by CYP3A4.	
Risk of clinically relevant TP-DIs		Impact on exposure of CYP (e.g., CYP3A4) substrates: dose increase may be needed to maintain its therapeutic effect when given in combination with tocilizumab. The effect of tocilizumab on CYPs may be persistent for several weeks after stopping treatment.		

CL: clearance; CYP: cytochrome P450; HMG-CoA: 3-hydroxy-3-methylglutaryl-coenzyme A; IL-6: interleukin 6; IL-6R: interleukin-6 receptors; MOA: mechanism of action; MTX: methotrexate; NA: not applicable; PD: pharmacodynamic; PK: pharmacokinetic; RA: rheumatoid arthritis; SMDs: small molecule drugs; TMDD: target-mediated drug disposition; TP-DIs: therapeutic protein–drug interactions.

TABLE 2-5 Theoretical TP-DI Risk Assessments for the Combination of T-DM1 and Pertuzumab in HER2+ Metastatic Breast Cancer Patients [12,22,49,79–81]

		T-DM1-Related Analytes		
		T-DM1	DM1	Pertuzumab
PK characteristics		Mainly cleared by nonspecific proteolytic degradation at clinically efficacious doses; deconjugation may also clear T-DM1; linear PK property; CL = ~0.7 L/day; $t_{1/2}$ = 3–4 days; interindividual CL variability = ~21%.	Downstream catabolite of T-DM1, metabolized by CYP3A4 and CYP3A5; excreted mainly by biliary secretion; mean maximal concentration at clinically efficacious dose: 5–6 ng/mL	Mainly cleared by nonspecific proteolytic degradation at clinically efficacious doses; linear PK property; CL = ~0.2 L/day; terminal $t_{1/2}$ = 15–28 days; interindividual CL variability = ~31%.
Risk of PK-based TP-DIs		Low: nonspecific proteolytic degradation pathways play a major role in the clearance of T-DM1and pertuzumab at clinically efficacious doses. The combination is unlikely to saturate these pathways. No overlapping clearance pathway of DM1 and pertuzumab. Low levels of unconjugated DM1 in systemic circulation and DM1 is not a CYP inhibitor or inducer and not a *p*-gp inhibitor at clinically relevant concentrations. So DM1 is not expected to be a perpetrator of DIs.		
PD characteristics	MOA	T-DM1 binds to an epitope in subdomain IV of HER2, specifically delivers DM1 to HER2+ cancer cells, and induces cancer cell killing by antimitotic effect of DM1. Trastuzumab component of T-DM1 inhibits HER2-mediated intracellular signaling and induces ADCC of HER2+ cancer cells.		Pertuzumab binds to an epitope in subdomain II of HER2 and inhibits ligand-activated heterodimerization of HER2 with other HERs (e.g., HER1, HER3, and HER4) and subsequent intracellular signaling, leading to growth arrest and apoptosis of HER2+ cancer cells.

(*continued*)

TABLE 2.5 (*Continued*)

	T-DM1-Related Analytes		
	T-DM1	DM1	Pertuzumab
TMDD	Not a major CL route at clinically efficacious dose ranges	NA	Not a major CL route at clinically efficacious dose ranges
Immunogenicity	Low	NA	Low
Cytokine modulator	No	No	No
Immuno-suppressive agent	No	No	No
Therapeutic range	Narrow	Narrow	Wide
Risk of PD-based TP-DIs that impact on exposure	Low: T-DM1 and pertuzumab bind to distinct epitopes of HER2 simultaneously without steric hindrance. TMDD does not play a major role in the CL of T-DM1and pertuzumab at clinically efficacious doses. T-DM1and pertuzumab have low immunogenicity rates; there is no identified impact of immunogenicity on their CL; immunogenicity and CYP activities are unlikely to be affected by the combination.		
Risk of clinically relevant TP-DIs	Low		

ADCC: antibody-dependent cellular cytotoxicity; CL: clearance; CYP: cytochrome P450; DIs: drug interactions; DM1: cytotoxic component of T-DM1; HER: human epidermal growth factor receptor; MOA: mechanism of action; NA: not available; PD: pharmacodynamic; *p*-gp: *p*-glycoprotein; PK: pharmacokinetic; T-DM1: trastuzumab emtansine; TMDD: target-mediated drug disposition; TP-DIs: therapeutic protein–drug interactions.

requirement, because a phase III study suggested a 33% increase of SN-38 (active metabolite of irinotecan) exposure and an increased incidence of grade 3 or 4 diarrhea and leukopenia in the bevacizumab and FOLFIRI combination arm compared to the FOLFIRI alone arm.[74] Whether there was a clinically relevant TP-DI potential was unknown. There are no theoretical mechanisms supporting this based on the PK and PD properties of bevacizumab and irinotecan, as assessed in Table 2-3. Unfortunately, the PK substudy in this phase III trial used limited sampling for irinotecan and thus was not definitive given irinotecan's large intersubject PK variability. The study also lacked an analysis of the uridine diphosphate glucuronosyltransferase 1A (UGT1A) genotype, which has been thought to affect the metabolic pathway of irinotecan. As a result, a controlled and fully powered clinical trial to fulfill the postmarketing requirement was conducted to investigate whether bevacizumab affected the PK properties of irinotecan and SN-38 using a staggered dosing design in cycle 1 and concurrent dosing in later cycles for bevacizumab and FOLFIRI. An exploratory analysis of UGT1A genotyping was also performed. This confirmatory study found that the PK characteristics of irinotecan and its active metabolite SN-38 were not altered by the bevacizumab combination. Pharmacogenetic analysis further confirmed that UGT1A1 gene polymorphisms were associated with variability in irinotecan and SN-38 exposure, which might explain the cross-arm exposure difference of SN-38 observed in the phase III study.[31]

2.5 SUMMARY

The PK properties of TPs are distinct from SMDs. TPs with a large molecular weight usually have a relatively small volume of distribution that is confined to plasma with limited penetration to the extravascular space. Unlike SMDs, most TPs are mainly catabolized via nonspecific proteolysis pathways with no involvement of CYP enzymes. FcRn-mediated recycling prolongs the half-lives of TPs that bind to FcRn, such as mAbs and some Fc fusion proteins. Some TPs may exhibit dose- and concentration-dependent PK due to a TMDD pathway. The clearance of TPs may also be affected by immunogenicity. Biliary secretion and renal excretion, the common routes of excretion for SMDs, are not common for TPs. The evolving class of ADCs belong to a more complex type of TPs made up of both a mAb and a small molecule cytotoxic agent and cleared by mechanisms relevant to both TPs and SMDs, as well as the deconjugation process determined by linker chemistry.

Owing to differences in the PK and PD properties of TPs and SMDs, the plausible mechanisms and relative risks for TP-DIs are distinct from SMD-DIs. PK-based DIs are not common for TPs, considering that the nonspecific proteolysis pathways are usually not saturable at the therapeutic concentrations of most TPs and there are usually no overlapping clearance pathways for TPs and SMDs. PD-based DIs induced by the PD effects of the combination drugs are possible for TPs. The major mechanisms of PD-based TP-DIs include modulation of immunogenicity-mediated clearance of TPs, modulation of TMDD, and modulation of CYP enzyme activity and expression through TP-cytokine-CYP interactions. To date, only modest changes in exposure owing to TP-DIs have been observed with little relevance in the clinical setting. However, the risk of potential TP-DIs involving SMDs with a NTR may be clinically relevant. Because most drugs with a NTR are closely monitored by therapeutic drug monitoring, potential changes in exposure can typically be adjusted as needed to maintain efficacious concentrations and minimize the occurrence of adverse events. ADCs are likely to have similar mechanisms of PD-based TP-DIs, with the additional possibility of DIs between the small molecule cytotoxic component of the ADC and the combination drug.

The distinct PK and PD properties of TPs pose multiple challenges for assessing TP-DIs. Consequently, novel strategies that are different from the traditional DI assessment strategies for SMDs are required. A question-based TP-DI risk assessment approach is recommended during clinical development. This approach includes both theoretical risk assessments based on PK and PD properties of TPs and SMDs in combination and clinical PK evaluation in support of the theoretical assessment.

ACKNOWLEDGMENTS

The authors acknowledge Dr. Yanan Zheng for her assistance with the preparation of Figure 2-3. The preparation of this manuscript was supported by Genentech Inc.

REFERENCES

1. Lee JI, Zhang L, Men AY, et al. CYP-mediated therapeutic protein-drug interactions: clinical findings, proposed mechanisms and regulatory implications. *Clin Pharmacokinet* **49**(5), 295–310 (2010).
2. Huang SM, Zhao H, Lee JI, et al. Therapeutic protein-drug interactions and implications for drug development. *Clin Pharmacol Ther* **87**(4), 497–503 (2010).
3. FDA. Guidance for industry: drug interaction studies–study design, data analysis, and implications for dosing and labeling. 2006. Available at www.fda.gov/downloads/Drugs/GuidanceComplianceRegulatoryInformation/Guidances/ucm072101.pdf (retrieved October 2012).
4. Tabrizi MA, Tseng CM, Roskos, LK. Elimination mechanisms of therapeutic monoclonal antibodies. *Drug Discov Today* **11**(1–2), 81–88 (2006).
5. Mould DR, Green B. Pharmacokinetics and pharmacodynamics of monoclonal antibodies: concepts and lessons for drug development. *BioDrugs* **24**(1), 23–39 (2010).
6. Meibohm B. *Pharmacokinetics and pharmacodynamics of biotech drugs: principles and case studies in drug development*. Wiley-VCH, San Francisco, 2006.
7. Porter CJ, Edward, GA, Charman SA. Lymphatic transport of proteins after S.C. injection: implications of animal model selection. *Adv Drug Deliv Rev* **50**(1–2), 157–71 (2001).
8. Deng R, Iyer S, Theil FP, et al. Projecting human pharmacokinetics of therapeutic antibodies from nonclinical data: what have we learned? *MAbs* **3**(1), 61–66 (2010).

9. Ng CM, Joshi A, Dedrick RL, et al. Pharmacokinetic-pharmacodynamic-efficacy analysis of efalizumab in patients with moderate to severe psoriasis. *Pharm Res* **22**(7), 1088–100 (2005).

10. Ma P, Yang BB, Wang YM, et al. Population pharmacokinetic analysis of panitumumab in patients with advanced solid tumors. *J Clin Pharmacol* **49**(10), 1142–156 (2009).

11. Dowell JA, Korth-Bradley J, Liu H, et al. Pharmacokinetics of gemtuzumab ozogamicin, an antibody-targeted chemotherapy agent for the treatment of patients with acute myeloid leukemia in first relapse. *J Clin Pharmacol* **41**(11), 1206–14 (2001).

12. Ng CM, Lum BL, Gimenez V, et al. Rationale for fixed dosing of pertuzumab in cancer patients based on population pharmacokinetic analysis. *Pharm Res* **23**(6), 1275–84 (2006).

13. Lu JF, Bruno R, Eppler S, et al. Clinical pharmacokinetics of bevacizumab in patients with solid tumors. *Cancer Chemother Pharmacol* **62**(5), 779–86 (2008).

14. Bruno R, Washington CB, Lu JF, et al. Population pharmacokinetics of trastuzumab in patients with HER2+ metastatic breast cancer. *Cancer Chemother Pharmacol* **56**(4), 361–69 (2005).

15. Bartelds GM, Wijbrandts CA, Nurmohamed MT, et al. Clinical response to adalimumab: relationship to anti-adalimumab antibodies and serum adalimumab concentrations in rheumatoid arthritis. *Ann Rheum Dis* **66**(7), 921–26 (2007).

16. Haraoui B, Cameron L, Ouellet M, White B. Anti-infliximab antibodies in patients with rheumatoid arthritis who require higher doses of infliximab to achieve or maintain a clinical response. *J Rheumatol* **33**(1), 31–36 (2006).

17. Anderson NL, Anderson NG. The human plasma proteome: history, character, and diagnostic prospects. [Review]. *Mol Cell Proteomics* **1**(11), 845–67 (2002).

18. Chanson P, Timsit J, Harris AG. Clinical pharmacokinetics of octreotide. Therapeutic applications in patients with pituitary tumours. *Clin Pharmacokinet* **25**(5), 375–91 (1993).

19. Alley SC, Okeley NM, Senter PD. Antibody-drug conjugates: targeted drug delivery for cancer. *Curr Opin Chem Biol* **14**(4), 529–37 (2010).

20. Prabhu S, Boswell CA, Leipold D, et al. Antibody delivery of drugs and radionuclides: factors influencing clinical pharmacology. *Therapeutic Delivery* **2**(6), 769–91 (2011).

21. Shen B-Q, Bumbaca D, Saad O, et al. Metabolic fate and pharmacokinetic characterization of trastuzumab emtansine (T-DM1): an emphasis on preclinical and clinical catabolism *Clin Pharmacol Ther* **89**(S1), S90 (2011).

22. Girish S, Gupta M, Wang B, et al. Clinical pharmacology of trastuzumab emtansine (T-DM1): an antibody-drug conjugate in development for the treatment of HER2-positive cancer. *Cancer Chemother Pharmacol* **69**(5), 1229–40 (2012).

23. Seattle Genetics, Inc. Package insert for Adcetris (brentuximab vedotin, SGN-35). Bothell, WA, 2011.

24. Younes A, Bartlett NL, Leonard JP. Brentuximab vedotin (SGN-35) for relapsed CD30-positive lymphomas. *N Engl J Med* **363**(19), 1812–21 (2010).

25. European Medicines Agency. Guideline on the investigation of drug interactions. London, 2010. Available at www.ema.europa.eu/docs/en_GB/document_library/Scientific_guideline/2010/05/WC500090112.pdf.

26. Pegram M, Chan D, Dickmann R. Phase II combined biological therapy targeting the HER2 proto-oncogene and vascular endothelial growth factor using trastuzumab (T) and

bevacizumab (B) as first line therapy of HER2 amplified breast cancer. *Breast Cancer Res Treat* **100** (Suppl 1), S28 (2006).

27. Baselga J, Pfister D, Cooper MR, et al. Phase I studies of anti-epidermal growth factor receptor chimeric antibody C225 alone and in combination with cisplatin. *J Clin Oncol* **18**(4), 904–14 (2000).
28. Bourhis J, Rivera F, Mesia R, et al. Phase I/II study of cetuximab in combination with cisplatin or carboplatin and fluorouracil in patients with recurrent or metastatic squamous cell carcinoma of the head and neck. *J Clin Oncol* **24**(18), 2866–72 (2006).
29. Folprecht G, Lutz MP, Schoffski P, et al. Cetuximab and irinotecan/5-fluorouracil/folinic acid is a safe combination for the first-line treatment of patients with epidermal growth factor receptor expressing metastatic colorectal carcinoma. *Ann Oncol* **17**(3), 450–56 (2006).
30. Girish S, Martin SW, Peterson MC, et al. AAPS Workshop report: strategies to address therapeutic protein-drug interactions during clinical development. *AAPS J* **13**(2), 405–16 (2011).
31. Denlinger CS, Blanchard R, Xu L, et al. Pharmacokinetic analysis of irinotecan plus bevacizumab in patients with advanced solid tumors. *Cancer Chemother Pharmacol* **65**(1), 97–105 (2009).
32. Mahmood, I, Green MD. Drug interaction studies of therapeutic proteins or monoclonal antibodies. *J Clin Pharmacol* **47**(12), 1540–54 (2007).
33. Seitz K, Zhou H. Pharmacokinetic drug-drug interaction potentials for therapeutic monoclonal antibodies: reality check. *J Clin Pharmacol* **47**(9), 1104–18 (2007).
34. Zhou H, Davis HM. Risk-based strategy for the assessment of pharmacokinetic drug-drug interactions for therapeutic monoclonal antibodies. *Drug Discov Today* **14**(17–18), 891–98 (2009).
35. Zhou H, Mascelli MA. Mechanisms of monoclonal antibody-drug interactions. *Annu Rev Pharmacol Toxicol* **51**, 359–72 (2011).
36. Lu D, Krop I, Modi S, et al. Pharmacokinetics (PK) of trastuzumab-DM1 (T-DM1) and paclitaxel (T) in patients with HER2-positive locally advanced or metastatic breast cancer (MBC) Previously treated with a trastuzumab-containing regimen. Paper given at the San Antonio Breast Cancer Symposium, 2010, P3-14-22.
37. Lu D, Modi S, Elias A, et al. Pharmacokinetics (PK) of Trastuzumab emtansine and paclitaxel or docetaxel in patients with HER2-positive MBC previously treated with a trastuzumab-containing regimen. Paper given at the San Antonio Breast Cancer Symposium, 2011.
38. Maini RN, Breedveld FC, Kalden JR, et al. Therapeutic efficacy of multiple intravenous infusions of anti-tumor necrosis factor alpha monoclonal antibody combined with low-dose weekly methotrexate in rheumatoid arthritis. *Arthritis Rheum* **41**(9), 1552–63 (1998).
39. Janssen Biotech, Inc. Package insert for Remicade (infliximab). Horsham, PA, 2011.
40. Abbott, Laboratories. Package insert for Humira (adalimumab). North Chicago, 2002.
41. Novartis Pharmaceuticals, Corp. Package insert for Simulect (basiliximab). East Hanover, NJ, 2003.
42. Kovarik JM, Pescovitz MD, Sollinger HW, et al. Differential influence of azathioprine and mycophenolate mofetil on the disposition of basiliximab in renal transplant patients. *Clin Transplant* **15**(2), 123–30 (2001).

43. Vincenti F, Mendez R, Pescovitz M, et al. A phase I/II randomized open-label multicenter trial of efalizumab, a humanized anti-CD11a, anti-LFA-1 in renal transplantation. *Am J Transplant* **7**(7), 1770–77 (2007).

44. Emi Aikawa N, de Carvalho JF, Artur Almeida Silva C, Bonfa E. Immunogenicity of anti-TNF-alpha agents in autoimmune diseases. *Clin Rev Allergy Immunol* **38**(2–3) 82–89 (2010).

45. Vermeire S, Noman M, Van Assche G. et al. Effectiveness of concomitant immunosuppressive therapy in suppressing the formation of antibodies to infliximab in Crohn's disease. *Gut* **56**(9), 1226–31 (2007).

46. Ng CM. Stefanich E, Anand BS, et al. Pharmacokinetics/pharmacodynamics of nondepleting anti-CD4 monoclonal antibody (TRX1) in healthy human volunteers. *Pharm Res* **23**(1), 95–103 (2006).

47. Zheng Y, Scheerens H, Davis JC Jr, et al. Translational pharmacokinetics and pharmacodynamics of an FcRn-variant anti-CD4 monoclonal antibody from preclinical model to phase I study. [Clinical Trial, Phase I Research Support, Non-U. S. Gov't]. *Clin Pharmacol Ther* **89**(2), 283–90 (2011).

48. Joshi A, Bauer R, Kuebler P, et al. An overview of the pharmacokinetics and pharmacodynamics of efalizumab: a monoclonal antibody approved for use in psoriasis. *J Clin Pharmacol* **46**(1), 10–20 (2006).

49. Lu D, Burris HA 3rd, Wang B, et al. Drug interaction potential of trastuzumab emtansine (T-DM1) combined with pertuzumab in patients with HER2-positive metastatic breast cancer. *Curr Drug Metab* **13**(7), 911–22 (2012).

50. Franklin MC, Carey KD, Vajdos FF, et al. Insights into ErbB signaling from the structure of the ErbB2-pertuzumab complex. *Cancer Cell* **5**(4), 317–28 (2004).

51. Scheuer W, Friess T, Burtscher H, et al. Strongly enhanced antitumor activity of trastuzumab and pertuzumab combination treatment on HER2-positive human xenograft tumor models. *Cancer Res* **69**(4), 9330–36 (2009).

52. Morgan ET. Impact of infectious and inflammatory disease on cytochrome P450-mediated drug metabolism and pharmacokinetics. *Clin Pharmacol Ther* **85**(4), 434–38 (2009).

53. Schering, Corp. Package insert for Intron A (interferon-alfa-2b). Whitehouse Station, NJ, 2008.

54. Islam M, Frye RF, Richards TJ, et al. Differential effect of IFNalpha-2b on the cytochrome P450 enzyme system: a potential basis of IFN toxicity and its modulation by other drugs. *Clin Cancer Res* **8**(8), 2480–87 (2002).

55. Hassan M, Nilsson C, Olsson H, et al. The influence of interferon-alpha on the pharmacokinetics of cyclophosphamide and its 4-hydroxy metabolite in patients with multiple myeloma. *Eur J Haematol* **63**(3), 163–70 (1999).

56. Janssen, Pharmaceuticals. Package insert for Orthoclone (muromonab-CD3). Titusville, NJ, 1992.

57. Vasquez EM, Pollak R. OKT3 therapy increases cyclosporine blood levels. *Clin Transplant* **11**(1), 38–41 (1997).

58. Strehlau J, Pape L, Offner G, et al. Interleukin-2 receptor antibody-induced alterations of ciclosporin dose requirements in paediatric transplant recipients. *Lancet* **356**(9238), 1327–28 (2000).

59. Sifontis NM, Benedetti E, Vasquez, EM. Clinically significant drug interaction between basiliximab and tacrolimus in renal transplant recipients. *Transplant Proc* **34**(5), 1730–32 (2002).
60. Schmitt C, Kuhn B, Zhang X, et al. Disease-drug-drug interaction involving tocilizumab and simvastatin in patients with rheumatoid arthritis. *Clin Pharmacol Ther* **89**(5), 735–40 (2011).
61. Genentech, Inc. Package insert for Actemra (tocilizumab). South San Francisco, 2011.
62. Elkahwaji J, Robin MA, Berson A, et al. Decrease in hepatic cytochrome P450 after interleukin-2 immunotherapy. *Biochem Pharmacol* **57**(8), 951–54 (1999).
63. Novartis Europharm, Ltd. Package insert for Ilaris (canakinumab). West Sussex, UK, 2009.
64. Centocor Ortho Biotech Inc. Package insert for Simponi (golimumab). Horsham, PA, 2010.
65. Regeneron Pharmaceuticals, Inc. Package insert for Arcalyst (rilonacept) Tarrytown, NY, 2008.
66. Centocor Ortho Biotech, Inc. Package insert for Stelara (ustekinumab). Horsham, PA, 2009.
67. Petrovic V, Teng S, Piquette-Miller M. Regulation of drug transporters during infection and inflammation. *Mol Interv* **7**(2), 99–111 (2007).
68. Swedish Orphan Biovitrum AB. Package insert for Kineret (anakinra). Stockholm, 2011.
69. Immunex Corp. Package insert for Enbrel (etanercept). Thousand Oaks, CA, 2009.
70. Genovese MC, Cohen S, Moreland L, et al. Combination therapy with etanercept and anakinra in the treatment of patients with rheumatoid arthritis who have been treated unsuccessfully with methotrexate. *Arthritis Rheum* **50**(5), 1412–19 (2004).
71. Amgen Inc. Package insert for Neupogen (filgrastim). Thousand Oaks, CA, 2010.
72. Novartis Pharmaceuticals Corp. Package insert for Proleukin (aldesleukin) East Hanover, NJ, 2008.
73. European Medicines Agency. Guideline on the clinical investigation of the pharmacokinetics of therapeutic proteins. London, 2007. Available at www.ema.europa.eu/docs/en_GB/document_library/Scientific_guideline/2009/09/WC500003029.pdf.
74. Hurwitz H, Fehrenbacher L, Novotny W, et al. Bevacizumab plus irinotecan, fluorouracil, and leucovorin for metastatic colorectal cancer. *N Engl J Med* **350**(23), 2335–42 (2004).
75. Genentech, Inc. Package insert for Avastin (bevacizumab). South San Francisco, 2009.
76. Sagent Pharmaceuticals, Inc. Package insert for Camptosar (irinotecan). Schaumburg, IL, 2008.
77. Mayne Pharma (USA) Inc. Package insert for Trexall (methotrexate). Paramus, NJ, 2005.
78. Merck & Co, Inc. Package insert for Zocor (simvastatin). Whitehouse Station, NJ, 2011.
79. Gupta M, Lorusso PM, Wang B, et al. Clinical implications of pathophysiological and demographic covariates on the population pharmacokinetics of trastuzumab emtansine, a HER2-targeted antibody-drug conjugate, in patients with HER2-positive metastatic breast cancer. *J Clin Pharmacol* **52**, 691–703 (2012).
80. Krop IE, Beeram M, Modi S, et al. Phase I study of trastuzumab-DM1, an HER2 antibody-drug conjugate, given every 3 weeks to patients with HER2-positive metastatic breast cancer. *J Clin Oncol* **28**(16), 2698–704 (2010).

81. Lewis Phillips GD, Li G, Dugger DL, et al. Targeting HER2-positive breast cancer with trastuzumab-DM1, an antibody-cytotoxic drug conjugate. *Cancer Res* **68**(22), 9280–90 (2008).

82. Sheiner LB, Beal S, Rosenberg B, Marathe VV. Forecasting individual pharmacokinetics. *Clin Pharmacol Ther* **26**(3), 294–305 (1979).

83. Sheiner LB, Rosenberg B, Marathe VV. Estimation of population characteristics of pharmacokinetic parameters from routine clinical data. *J Pharmacokinet Biopharm* **5**(5), 445–79 (1977).

84. Zhu Y, Hu C, Lu M, et al. Population pharmacokinetic modeling of ustekinumab, a human monoclonal antibody targeting IL-12/23p40, in patients with moderate to severe plaque psoriasis. *J Clin Pharmacol* **49**(2), 162–75 (2009).

85. Bai S, Xin Y, Jin D, et al. Population pharmacokinetic analysis from phase I and phase II studies of the humanized monovalent antibody, MetMAb, in patients with advanced solid tumors. *J Clin Oncol* **29** (Suppl.) 2571 (2011).

CHAPTER 3

DRUG INTERACTION ASSESSMENT STRATEGIES: SMALL MOLECULES VERSUS THERAPEUTIC PROTEINS

SHANNON DALLAS, CARLO SENSENHAUSER,
SOUVIK CHATTOPADHYAY, and JOSE SILVA

3.1 INTRODUCTION

Paramount to every small molecule drug (SMD) development program is the assessment of drug interaction (DI) liabilities involving metabolic enzymes or drug transporters. Generally, these assessments are undertaken to understand how one SMD (perpetrator) may affect the absorption, distribution, metabolism, and excretion (ADME) of another SMD (victim) and therefore determine the potential for drug–drug interactions (DDIs). The importance of DDI assessment during SMD drug development (generally in the context of pharmacokinetics-based interactions) has been highlighted in recent guidance documents and papers published by regulatory agencies worldwide, including the FDA[1–3] and the European Medicines Agency (EMA).[4] Many preclinical studies (mostly *in vitro* but also *in vivo*) have been developed over the last few decades to understand how a particular SMD may have a potential DDI risk and whether such a risk will translate into the clinic. In contrast, for therapeutic proteins (TPs), or biologics, recent concern has centered more around the ability of a TP to affect the ADME of a coadministered SMD (i.e., as a perpetrator only), although exceptions do exist. These types of TP-DI studies historically were not part of standard Biologics License Application (BLA) submissions. Table 3-1 gives a few examples of approved TPs for which DI assessment was performed and reported in the compound label.[5–10]

DDI risk assessment for a SMD often starts with well-established *in vitro* studies using human-derived models (e.g., human hepatocytes and microsomes). Inhibition (reversible, time-dependent, or mechanism-based) and induction studies of the

Drug–Drug Interactions for Therapeutic Biologics, First Edition.
Edited by Honghui Zhou and Bernd Meibohm.

TABLE 3-1 Examples of TPs That Have Drug Interaction Information Listed on the Label

Drug	Drug Interactions Indicated on Label
Adalimumab (Humira); TNF blocker	MTX: Adalimumab has been studied in RA patients taking concomitant MTX. Although MTX reduced the apparent adalimumab clearance, the data do not suggest the need for dose adjustment of either drug.
Tocilizumab (Actemra); IL-6 receptor inhibitor	MTX: Population pharmacokinetic analyses of RA patients did not detect any effect of MTX, NSAIDs, or corticosteroids on tocilizumab clearance. Other drugs for treatment of concomitant administration of a single dose of 10 mg/kg tocilizumab with 10–25 mg MTX once weekly had no clinically significant effect on MTX exposure.
	CYP450 enzymes: *In vitro* studies showed that tocilizumab has the potential to affect expression of multiple CYP enzymes, including CYP1A2, CYP2B6, CYP2C9, CYP2C19, CYP2D6, and CYP3A4. Its effects on CYP2C8 or transporters is unknown. *In vivo* studies with omeprazole (metabolized by CYP2C19 and CYP3A4) and simvastatin (metabolized by CYP3A4) showed up to a 28% and 57% decrease, respectively, in exposure 1 week after a single dose of tocilizumab. The effect of tocilizumab on CYP enzymes may be clinically relevant for CYP450 substrates with a narrow therapeutic index, for whichthe dose is individually adjusted.
Trastuzumab (Herceptin); HER2/neu receptor antagonist	Chemotherapeutic drugs: In a study involving metastatic breast cancer patients, the mean serum trough concentration of trastuzumab (4 mg/kg initial dose of trastuzumab followed by 2 mg/kg weekly) was consistently elevated approximately 1.5-fold when administered in combination with paclitaxel (175 mg/m^2 over 3 hours every 21 days for at least six cycles) as compared to trough concentrations of trastuzumab when administered in combination with an anthracycline and cyclophosphamide (doxorubicin 60 mg/m^2 or epirubicin 75 mg/m^2 plus 600 mg/m^2 cyclophosphamide every 21 days for six cycles). In other pharmacokinetic studies, where trastuzumab was administered in combination with paclitaxel, docetaxel, or doxorubicin, trastuzumab did not alter the plasma concentrations of these chemotherapeutic agents or the metabolites that were analyzed.
	In another drug interaction study involving gastric cancer patients, the pharmacokinetics of cisplatin (80 mg/m^2 on day 1), capecitabine (1000 mg/m^2 orally twice a day on days 1–14), and their metabolites were not altered when administered in combination with trastuzumab (initial dose of 8 mg/kg administered on day 1 prior to chemotherapy, followed by 6 mg/kg every 21 days until disease progression).

TABLE 3.1 (*Continued*)

Drug	Drug Interactions Indicated on Label
Belatacept (Nulojix); fusion protein (Fc portion of human IgG1 linked to CTLA-4	MMF: In a pharmacokinetic study the plasma concentrations of MPA were measured in 41 patients who received fixed MMF doses of 500–1500 mg twice daily with either 5 mg/kg of belatacept or cyclosporine. The mean dose-normalized MPA $C_{\max}$ and AUC_{0-12} were approximately 20% and 40% higher, respectively, with belatacept coadministration than with cyclosporine coadministration.
Peginterferon α-2a (Pegasys)	Theophylline: Treatment with peginterferon α-2a once weekly for 4 weeks in healthy subjects was associated with an inhibition of CYP1A2 and a 25% increase in theophylline AUC. Methadone: In a pharmacokinetic study of HCV patients concomitantly receiving methadone, treatment with peginterferon α-2a once weekly for 4 weeks was associated with methadone levels that were 10–15% higher than at baseline. The clinical significance of this finding is unknown.
Interferon γ-1b (Actimmune)	CYP450s: Preclinical studies in rodents using species-specific interferon-γ have demonstrated a decrease in hepatic microsomal CYP450 concentrations. This could potentially lead to a depression of the hepatic metabolism of certain drugs that use this degradative pathway.

CYP450: cytochrome P450; HCV: hepatitis C virus; HER2: human epidermal growth receptor 2; IgG: immunoglobulin G; IL-6: interleukin 6; MPA: mycophenolic acid; MMF: mycophenolate mofetil; MTX: methotrexate; NSAID: nonsteroidal anti-inflammatory drug; RA: rheumatoid arthritis; TNF: tumor necrosis factor.

major cytochrome P450s, inhibition studies of non-CYP pathways (e.g., uridine 5′-diphospho-glucuronosyltransferases), and metabolic enzyme phenotyping are well established and run routinely. Although the liver is often the first organ examined, studies can be expanded to other organ systems (e.g., kidney or intestine) depending on a compound's characteristics (e.g., physiochemical properties or pharmacokinetics profile), mode and site of action, molecular target, clearance mechanisms, and clinical population. The potential role of drug transporters in compound disposition (as substrates, inhibitors, inducers, or suppressors) is assessed by numerous *in vitro* assays, which are also readily available. However, the relevance of individual transporters for potential DDIs is currently less well understood compared to metabolic enzymes. The *in vitro* kinetic data generated from these various assays (e.g., kinetic constants) can then be used to predict DDI risk clinically, using mechanistic modeling and simulation, such as physiologically based pharmacokinetic (PBPK) modeling. If a potential risk is identified, clinical DDI studies are conducted to characterize the magnitude of the interaction and predict how this might affect the safety and use of a potential new drug entity with comedications. PBPK modeling can provide simulations broadly across multiple populations, including special populations, such as pediatrics, geriatrics, and in renal or hepatic insufficiency. In addition, PBPK modeling is emerging as a viable alternative to actual clinical DDI trials, in some cases. When submitting a new drug for approval, these studies would be a standard part of the drug application process. More recently, the possibility of some TPs, such as monoclonal antibodies and fusion proteins, to cause clinically relevant DIs (either directly or indirectly) has received renewed attention. The FDA and EMA have included language in their drug development guidance documents about how to assess the risk of a TP-DI.[3,11]

This chapter is not intended to give an expansive and detailed review of all of the methods and assays used for DDI assessment during SMD development because this has already been accomplished by others in the field.[12] Rather, general concepts and methods will be discussed for some of the typical assays undertaken in SMD development, including whether these methods can be, should be, or already have been applied in the context of TP-DI assessments. This chapter serves as a starting point for understanding the complexity and difficulties in assessing TP-DIs. The reader will be guided to other chapters within this volume that discuss these concepts more extensively and to specific case studies when appropriate.

3.2 DRUG-METABOLIZING ENZYMES

3.2.1 Cytochrome P450 Enzymes

Metabolic clearance generally involves biotransformation reactions catalyzed by a wide variety of drug-metabolizing enzymes (DME). Biotransformations are typically classified as either phase I or phase II reactions. Generally, phase I metabolic reactions aim at inactivating a drug through the introduction of a more polar functional group (—OH, —COOH, —NH_2, —SH, etc.) by oxidation, dealkylation, reduction, hydrolysis, cyclization, etc. Various classes of enzymes can be involved in phase I metabolism: cytochrome P450 (CYP), flavin-containing monooxygenases, amidases, carboxylesterases, and epoxide hydrolases. Because as much as 60% of currently marketed drugs are metabolized by one or several CYP isoforms,[13,14] focus has been directed at DDIs involving CYPs, although similar kinetic approaches can be applied to other classes of phase I enzymes.

Phase II metabolism, on the other hand, leads to formation of conjugates of the initial substrate—either the parent compound through direct conjugation or a phase I metabolite through indirect conjugation via uridine 5′-diphospho-glucuronosyltransferases (UGTs), glutathione-*S*-transferases, sulfotransferases, *N*-acetyltransferases, and methyltransferases—resulting in higher molecular weight compounds with improved water solubility, which increases renal or biliary clearance.[15,16] Phase II metabolism also provides a mechanism to inactivate potentially toxic electrophilic products of phase I metabolism. Although clinically relevant DDIs involving phase II metabolism have not been reported as frequently as with the CYPs, bioactivation through conjugation can occur, for instance through the formation of acyl glucuronides or thioesters resulting from conjugation of carboxylic acid moieties. Their inherent electrophilicity, deriving from the spontaneous tendency to undergo hydrolysis and intramolecular rearrangements (e.g., acyl migration) makes them susceptible to protein adduct formation, and they have been implicated in observed adverse effects.[17] One notable example of phase II metabolism-derived DDI is the inactivation of CYP2C8 by gemfibrozil glucuronide, which will be briefly discussed later in the chapter.

The CYP superfamily of enzymes is widespread throughout nature and is present in all types of living organisms, ranging from bacteria to plants and animals. Of the more than 11,000 CYP proteins identified to date,[18] only a few dozen isoforms are of interest from a drug metabolism perspective and in terms of potential for adverse DDIs. Their nomenclature is based on the extent of amino acid sequence homology. Only the CYP1, CYP2, and CYP3 subfamilies are known to be primarily involved in the metabolism and bioactivation of pharmaceutical SMDs.[19] CYPs are proteins that contain a heme molecule that is noncovalently bound in the active site. The heme binds molecular oxygen and allows the catalytic cycle to carry out a biotransformation reaction by receiving protons and electrons from a cofactor, NADPH, and a closely associated oxidoreductase protein, producing 1 mol of water as a byproduct. CYP proteins, together with other phase I and some phase II enzymes are situated in the lipid bilayer of the endoplasmic reticulum, where they can efficiently catalyze the conversion of lipophilic xenobiotics to their oxidized metabolites and conjugates.

Other DMEs are found in the cytosolic region of the cell. Most DMEs share the characteristic of having broad substrate selectivity, being able to accommodate structurally diverse substrates in the active sites, albeit at the expense of turnover rates, which are typically much lower than those observed for enzymes involved in endogenous cell regulation and biosynthesis. This lack of substrate specificity, leading to the possibility of the same substrate being metabolized by different CYPs or different CYPs metabolizing the same substrate to produce different metabolites at different turnover rates, can ultimately lead to observed adverse DDIs. Characterization of the kinetic parameters of these reactions from *in vitro* assays can provide a valuable predictive tool during SMD development when incorporated with current modeling and simulation methods, such as PBPK.

3.2.2 Inhibition

A drug binding to a CYP enzyme typically displays hyperbolic saturation (Michaelis-Menten) kinetics, meaning that as the concentration of substrate is increased the first-order rate of product formation (often referred to as rate of substrate turnover or velocity), reaches a maximal rate of efficiency (V_{max}). Under these conditions, the enzyme present is completely in the enzyme–substrate (ES) complex form and is therefore saturated. Rates higher than V_{max} are not possible; therefore, increasing substrate concentration beyond a certain point will not have any effect on substrate turnover, and zero-order kinetics are attained. In cases that deviate from classical Michaelis-Menten kinetics, such as substrate inhibition or autoactivation, changes in V_{max} as substrate concentration increases can be observed). In addition, the substrate concentration that yields 50% of the maximal turnover rate, the Michaelis-Menten constant (K_m) represents the affinity of a substrate for the enzyme catalyzing its metabolic conversion. The ratio V_{max}/K_m, or intrinsic clearance (CL_{int}), expressed in units of volume over time, provides a measure of the overall enzymatic capacity for a specific substrate. When this basic kinetic model is modified by the introduction of a second compound, which can either compete with the substrate for the active site of the same enzyme or act as an effector molecule, binding to an allosteric site of the enzyme, the rate of substrate turnover may be either inhibited or activated. In these cases, metabolic conversion of the substrate, the victim drug, will be driven not only by its concentration but also by the concentration of the perpetrator drug and the relative affinities of the two compounds to the enzyme. If a victim drug does not have alternate routes of elimination, inhibition of the enzyme will affect the exposure and pharmacokinetic profile more significantly. In addition to the *in vitro* determination of the kinetic parameters of inhibition, consideration of the fraction metabolized (f_m) by the inhibited enzyme and by other CYP and non-CYP enzymes, contributions by the gut to first-pass metabolism, the extent of biliary and renal clearance of the unchanged drug, and plasma protein binding all have to be taken into account as inputs in the prediction of changes in exposure (AUC) of a victim drug resulting from coadministration of a perpetrator drug.

In SMD development, several models, ranging from whole cell (hepatocytes) to subcellular fractions (microsomes; S9) and single enzyme expression systems, are routinely used in assays designed to evaluate the effect of a drug on CYP activity. For the determination of kinetic parameters, however, microsomes are the system of choice[20] because they provide kinetic data independent of other cellular and transport mechanisms present in whole cell systems. It is also widely recognized that Michaelis-Menten kinetic models require certain assumptions based on the steady-state, rapid equilibrium approach: (a) the amount of enzyme is negligible compared to the amount of substrate; (b) all of the enzyme can be accounted for as either free enzyme (E) or ES and enzyme-inhibitor (EI) complexes, all of which are in equilibrium for the duration of the assay; and (c) the amount of product formed is minimal compared to the substrate concentration, so the reaction remains within the linear range and substrate depletion is not rate limiting and does not affect assay kinetics. Consistent with these assumptions, microsomal assays should be run at low protein concentrations (typically <0.5 mg/mL, provided product formation is sufficiently detectable and linear within the time frame it is being measured), with constant conditions (cofactor, pH, temperature, and ionic strength) throughout the reaction.[20] Recent advances in methodologies, highlighted in the newest FDA DDI guidance[3] have included the identification and standardization of CYP-specific probe substrates and inhibitors, which specifically address the need to accurately monitor the activity of a CYP isoform without significant contributions from other isoforms. The widespread availability of metabolite reference standards and their deuterated analogs also allow for more accurate and sensitive quantification methods using current high-performance liquid chromatography tandem mass spectrometry (HPLC/MS/MS) analytical techniques.

3.2.2.1 Reversible Inhibition Reversible inhibition results from noncovalent binding of an inhibitor to the enzyme in such a way that access of a substrate to the

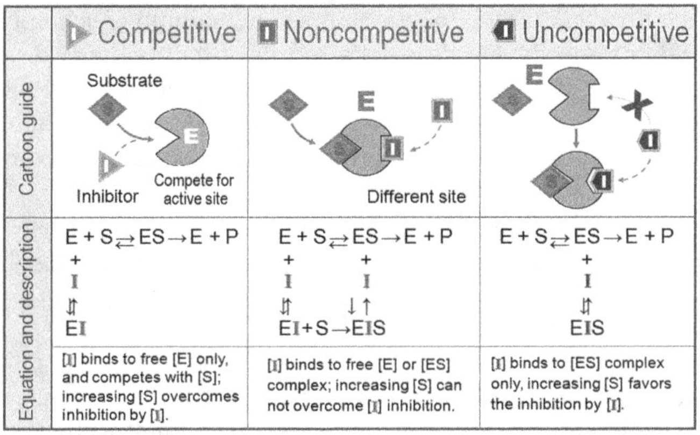

FIGURE 3-1 Enzyme-substrate-inhibitor interactions and kinetic equilibria for the three major models of reversible inhibition. E: enzyme; ES: enzyme–substrate; I: inhibitor; P: product; S: substrate. (Reprinted with permission from Rong-Huay Juang, National Taiwan University.[70]) (See the color version of this figure in Color Plates section.)

active site is prevented, either by directly competing for the active site, or by causing conformational changes to the enzyme (caused by allosteric effects) that affect the efficiency (turnover rate) of the reaction. The inhibitor could be a molecule that is structurally similar to the substrate but that is not necessarily chemically modified by the enzyme it inhibits or is modified at a substantially lower rate. It could also be a molecule that binds to a site different from the active site (allosteric or effector site), causing changes that prevent the substrate from binding. Therefore, from a kinetics standpoint, reversible inhibition models vary based on whether the inhibitor binds only to the free enzyme or also to the ES complex and whether product can be formed from the ES complex only or also from the enzyme-substrate-inhibitor (ESI) complex, albeit possibly at a reduced rate. Briefly, these models are summarized as follows (Figure 3-1):

- *Competitive inhibition.* Both substrate and inhibitor compete for the active site of the enzyme, and the equilibrium can be driven one way or the other by increasing the concentration of either compounds. As such, the $V_{\max}$ is not affected, although more substrate is needed to compensate for the presence of the inhibitor and therefore apparent K_m increases. Competitive inhibition can also be observed as a result of binding to an allosteric site on the enzyme, provided that it binds only to the free enzyme.
- *Mixed inhibition.* The inhibitor can bind to both E and the ES complex, to an allosteric site different from the binding site, causing conformational changes and thus reducing the efficiency of the enzyme and the affinity of the substrate for the enzyme. However, the equilibrium can be partially driven by increases in substrate concentration; therefore, apparent K_m increases.

- *Noncompetitive inhibition.* A subset of mixed inhibition, the inhibitor binds to a different enzyme site but does not affect the affinity of the substrate binding to the active site, although it cannot be chemically modified by the enzyme. It reduces the efficiency of the enzyme, therefore affecting the maximal turnover rate (V_{max} decreases), but because the affinity of the substrate for the enzyme does not change, apparent K_m is not affected.
- *Uncompetitive inhibition.* The least common model with CYP enzymes, in which the inhibitor binds only to the ES complex, reducing its efficiency but increasing the apparent affinity of the substrate for the enzyme. In this case, both V_{max} and K_m decrease.

Experimental assessment of the potential to inhibit a specific enzyme typically begins by determining if incubation of the test compound in the presence of an established probe substrate causes loss of enzymatic activity as represented by a decrease in formation of the metabolite of the probe substrate. If the appropriate range of test compound concentrations is selected and, most important, if the probe substrate is incubated at a concentration that is equal or below its determined K_m value, the IC_{50} value of the inhibitor–that is, the concentration of inhibitor that lowers enzymatic basal activity levels by 50% can be determined. IC_{50} values provide an approximate estimation of the potential of a compound to act as an inhibitor, but they are extrinsic constants and depend on experimental conditions and substrate concentrations.

The EI complex dissociation constant, K_i, is an intrinsic kinetic parameter and more appropriate in estimating potential DDIs. It can be estimated from determined IC_{50} values, provided that substrate concentration is equal to its K_m value and that the mode of inhibition is consistent with either the competitive model (in which case $IC_{50} = 2 \times K_i$) or the noncompetitive model ($IC_{50} = K_i$).[21] Experimentally, it is determined by incubating several substrate concentrations (preferably bracketing the substrate K_m value) in the presence of a range of inhibitor concentrations (preferably bracketing 0.2 to 5 times the expected K_i value)[22] and plotting the enzyme velocities (v, turnover rates) as functions of inhibitor ([I]) and substrate ([S]) concentrations. Graphical representations of the data, such as Michaelis-Menten (v versus [S]), Lineweaver-Burk ($1/v$ versus 1/[S]), Dixon ($1/v$ versus [I]) or Eadie-Hofstee plots (v versus v/[S]), can also provide insight on the appropriate inhibition mechanism (Figure 3-2).

More complex models have been extensively observed with enzymes exhibiting kinetic profiles that deviate from the traditional Michaelis-Menten model. CYP3A4, the most abundant and primary isoform involved in the metabolism of xenobiotics, has been shown to commonly exhibit non-Michaelis-Menten kinetics, due to homotropic cooperative effects resulting from the presence of multiple binding sites[23,24] or to flexible binding sites that can accommodate multiple substrates simultaneously.[25] These models can be represented as nonhyperbolic kinetic profiles—for example, in the case of substrate inhibition, as with midazolam, velocity decreases at higher substrate concentrations after an initial hyperbolic phase. In the case of autoactivation, as with testosterone, a sigmoidal profile of velocity as a function of

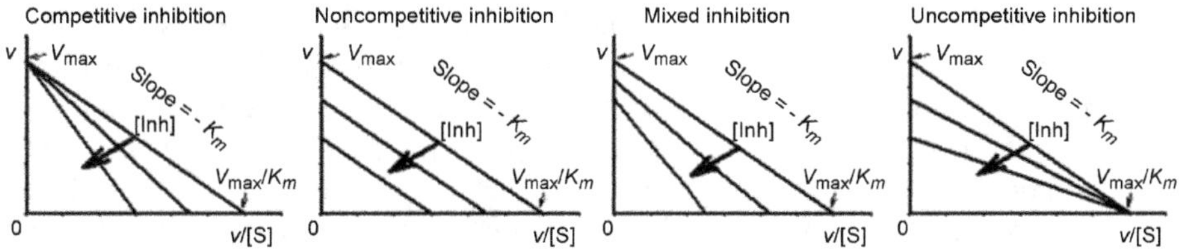

FIGURE 3-2 Enzyme inhibition models. The different reversible inhibition models (competitive, noncompetitive, mixed, and uncompetitive) can be differentiated by using Eadie-Hofstee plots. The *y*-intercept corresponds to $V_{\max}$, the slope corresponds to $-K_m$ and the *x*-intercept corresponds to $V_{\max}/K_m$, or the intrinsic clearance (CL_{int}). Inh: inhibition.

substrate concentration is observed.[26,27] It is common practice, therefore, to screen interactions between a potential CYP3A4 inhibitor with more than one CYP3A4-specific substrate, to assess binding to the different domains (midazolam, testosterone, and nifedipine are used as the prototypical substrates).[28]

As inhibition constants provide an indication whether exposure of an enzyme to a perpetrator drug may have a concomitant effect on the pharmacokinetic profile of a victim drug, their value is essential in estimating the potential for *in vivo* DDI by a SMD. Therapeutic proteins, though unlikely to have similar kinetic intracellular interactions with CYP enzymes, could trigger different indirect mechanisms affecting the basal activity levels of an enzyme, prompting as a result, similar pharmacokinetic interactions (see section 3.2.4).

3.2.2.2 Time-Dependent Inhibition When an inhibitor or substrate is converted enzymatically to a reactive species, it may inactivate the enzyme in a time- and concentration-dependent fashion.[29] Recently, emphasis has been placed on differentiating between time-dependent inhibition (TDI) and a particular subset defined as mechanism-based inactivation (MBI). Several mechanisms can trigger TDI of an enzyme. They can be typically identified after a preincubation period in the presence of a metabolizing system and are therefore NADPH dependent. Some of these mechanisms may not necessarily require bioactivation and may result, for instance, from the formation of species with a higher inhibitory affinity than the parent compound. A compound causing MBI, on the other hand, is one that binds catalytically to the target enzyme and is metabolically converted to a reactive species. This reactive metabolite can then bind covalently and irreversibly to a moiety of the active site of the apoprotein or in a quasi-irreversible fashion, if a reactive species forms a metabolic intermediate complex (MIC) with the heme group of the CYP enzyme. For example, alkyl and aromatic amines can be bioactivated to form a nitroso intermediate species, which coordinates tightly with the heme but in a non-covalent way. This interaction can be reversed almost completely by oxidizing agents, such as potassium ferricyanide, as in the case with troleandomycin.[30]

Greater concern from a DDI perspective is generated when these reactive intermediates interact covalently or quasi-irreversibly to CYP3A4. Often, DDIs triggered by these irreversible mechanisms can be more unfavorable compared to DDIs owing to reversible inhibition because the inactivated enzyme has to be replaced by newly synthesized protein (*de novo* synthesis). As in the case with reversible inhibition, the potency of inactivation can be assessed by determining the kinetic parameters of inactivation, k_{inact}, defined as the maximal rate of inactivation and K_I, indicating the potency or affinity of the inactivator for the enzyme, defined as the concentration that causes 50% of the maximal rate of inactivation. An additional parameter provides a measure of efficiency of the inactivation and is defined as the number of inactivator (substrate) molecules that are turned over per molecule of enzyme that is inactivated; it is called the apparent partition ratio (APR).[31–33]

APR screens can be carried out easily and effectively by incubating the test compound at serially diluted concentrations, with and without NADPH, in a primary reaction mixture. After 60 min, an aliquot of the reaction mixture is transferred to a secondary reaction that contains an excess of CYP-specific probe substrate and additional NADPH, to assay for residual enzymatic activity. A dilution in the transfer step, typically 10- to 20-fold and the excess of probe substrate, ensure that any observed loss in residual activity is the result of inactivation of the enzyme during the primary incubation and not due to possible reversible interactions with the unchanged inactivator or any generated metabolite transferred to the secondary reaction mixture. The residual activity data are normalized to the maximal activity observed in the absence of inactivator and are plotted as a function of the molar ratio of inactivator to CYP enzyme. Any separation between the curves corresponding to the +NADPH and −NADPH primary incubation samples indicate that the observed loss of activity appears to be time, concentration, and NADPH dependent, identifying a TDI and possible MBI. The APR value can be determined from the linear range of the activity curve as a function of molar ratio, which can be extrapolated to the intercept of the x-axis, as shown by the interaction of mifepristone a known 3A4 MBI (Figure 3-3). In contrast, ketoconazole, a strong reversible inhibitor exhibits the opposite behavior (greater loss of activity without NADPH), as the potent reversible inhibition interacts with the probe substrate.

Although microsomes have traditionally been the system of choice for TDI screens and kinetic determinations, hepatocytes have recently gained increasing value as a better predictive model that incorporates mechanisms and transport not present with microsomes. A study of several known CYP3A4 inactivators resulted in generally lower potencies, either lower affinities or lower rates of inactivation than those typically calculated from microsomal data. However the hepatocyte-derived lower potencies appeared to correlate better with observed clinical data when simulated with PBPK.[34]

Determination of the kinetic parameters of inactivation, k_{inact} and K_I, is an extension of the APR assay. Again, serially diluted test compound concentrations are incubated in a primary reaction containing NADPH, but activity is assayed at several time points. The apparent rates of inactivation generated from the semilog time course can be plotted as a function of the inactivator to yield the customary Michaelis-Menten-like hyperbolic plot, from which the kinetic parameters of inactivation can be determined by nonlinear regression (Figure 3-4). The concept of screening for potential TDIs or MBIs has become pivotal in SMD development, as evidenced also by the postmarket withdrawals of several previously approved drugs found to

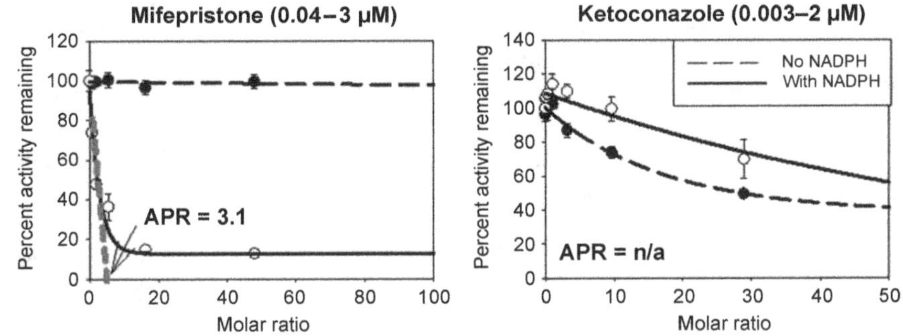

FIGURE 3-3 APR curves for mifepristone, a known CYP3A4 TDI, and ketoconazole, a potent CYP3A4 reversible inhibitor. Serially diluted concentrations of either compound were incubated with an NADPH (solid lines) or without an NADPH (dashed lines) regenerating system for 60min. Residual CYP3A4 activity was assayed by transferring an aliquot of each initial incubation to secondary incubations containing midazolam (CYP3A4 probe substrate) and NADPH. Residual activity was normalized to the activity in a vehicle control and was graphed as a function of inactivator to CYP3A4 molar ratio. The concentration- and NADPH-dependent loss of activity was consistent with a TDI. Ketoconazole, as a potent but reversible inhibitor, does not cause NADPH-dependent loss of activity and exhibits stronger loss of activity in the absence of NADPH. APR: apparent partition ratio; TDI: time-dependent inhibitor.

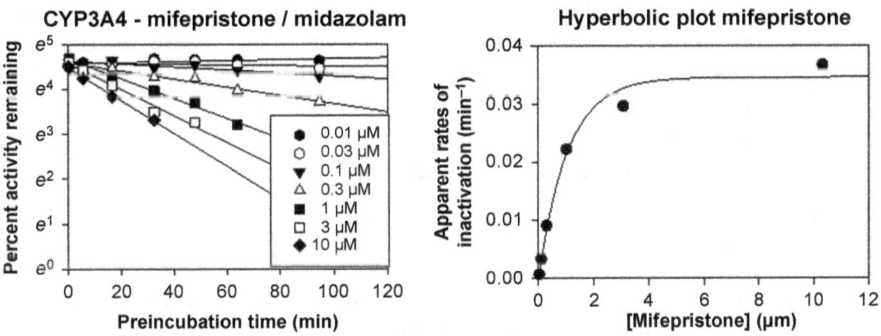

FIGURE 3-4 Inactivation time course of CYP3A4 by mifepristone and hyperbolic: plot of apparent first-order rates of inactivation as a function of mifepristone concentrations. Mifepristone was serially diluted (0.01 to 30 μM) and incubated in the presence of an NADPH-regenerating system. Aliquots were removed for each concentration level at specific time points (0, 5, 15, 30, 45, 60 and 90 min) and transferred to a secondary reaction mixture containing 10 μM midazolam (5× K_m) to assay for residual activity. Samples were normalized for each time point to vehicle incubates and plotted as the natural log of residual activity as a function of time. Apparent first-order rates of inactivation were derived from the slopes of the decay curve for each concentration and were plotted as a function of inactivator concentration to yield a hyperbolic inactivation plot. The maximal rate of inactivation, k_{inact} ($= 0.035\ \text{min}^{-1}$), was determined from the zero-order kinetic plateau of the hyperbolic curve and the inactivation constant, K_{I} ($= 0.74$ μM) was calculated as the inactivator concentration yielding 50% of the maximal inactivation rate.

be TDIs and the cause of unexpected drug-induced liver injuries and idiosyncratic toxicities.[35,36] Because of the different nature of the interactions between TPs and DMEs, TDI in the classical sense is unlikely, although interactions affecting signaling pathways leading to time-dependent effects in the expression or activity levels of DMEs are certainly possible. In these cases, investigating the TP as a potential perpetrator would assess the likelihood of potential DDIs.

3.2.2.3 Inhibition of Non-CYP Enzymes A number of additional classes of phase I enzymes play significant roles in drug metabolism. Recurrence of DDI is less frequent, and as a result, enzymes such as the flavin-containing monooxygenase (FMOs), epoxide hydrolases, and carboxylesterases have not been the source of consistent scrutiny. Their significant role in drug design is to provide alternate routes of elimination of drugs to mitigate the possibilities of CYP-related DDI. Phase II enzymes also do not tend to be involved in clinically relevant DDI. Glucuronidation and other phase II reactions tend to have higher capacity, faster turnover rates than phase I reactions and consequently do not contribute to marked changes in AUC and other PK parameters. Several significant exceptions should however be noted: the UGT1A1-catalyzed glucuronidation of bilirubin, the rate-limiting step in the clearance of bilirubin,[37] and the UGT2B7-catalyzed conversion of gemfibrozil to the acyl glucuronide conjugate, which was found to be a more potent reversible inhibitor of CYP2C8 than the parent drug.[38] Subsequently, the gemfibrozil acyl glucuronide conjugate was also identified as a strong time-dependent inhibitor of the same isoform through an uncommon CYP2C8-catalyzed oxidation of the conjugated form (phase I hydroxylation following a phase II conjugation), generating a reactive intermediate (possibly a substituted hydroquinone), which can then covalently bind to and inactivate the CYP enzyme.[39] As with the CYPs, DDIs arising from direct interactions of TPs with non-CYP enzymes such as UGTs are unlikely. However, indirect interactions via transcriptional or post-transcriptional regulation cannot be ruled excluded.

3.2.3 Reaction Phenotyping

Reaction phenotyping in SMD development is designed to establish if a drug candidate has the potential to be a victim drug and consequently have a higher *in vivo* DDI risk. Determining the CYP isoforms, non-CYP enzymes, and phase II reaction pathways that are involved as routes of metabolic clearance of a new molecular entity will allow the identification of potential metabolic liabilities. Phenotyping studies should therefore be designed to provide a complete scheme of the biotransformation landscape of a drug candidate. Commercially available single recombinant enzyme expression systems from bacterial, insect, or mammalian cells have facilitated determining these metabolic pathways. These assays can be routinely used to confirm involvement of specific enzymes in the catalytic conversion of a SMD. Studies are usually designed in a sequential approach as follows:

- General metabolic stability is initially determined in either microsomal or intact cell systems.

- An initial review of the major metabolites detected can subsequently trigger the selection of the possible biotransformation pathways to be investigated.
- Determination of the kinetic parameters (K_m, V_{max}) of the major metabolic pathways is carried out in microsomes by monitoring rates of metabolite formation at varying substrate concentrations. Establishing the rates of conversion and the affinity for a specific pathway will provide a better understanding of the potential for saturation of an enzyme and possible DDI consequences. This should be done in relation to the expected systemic concentration levels based on anticipated dosing regimens.
- Contributions of a specific pathway to the overall metabolic clearance are then investigated by simulated DDI scenarios; varying levels of known chemical inhibitors of the targeted pathway or inhibitory monoclonal antibodies of the specific enzyme are incubated with the substrate to monitor and quantify relative changes in the metabolic profile.

Reaction phenotyping as applied to enzyme substrates in SMD development is presumably not applicable to TP as the likelihood of similar enzyme–substrate interactions is remote. But it can be used as an important tool for defining the risk of TP-DI, with the protein acting as a potential perpetrator.

3.2.4 Induction and Suppression

Induction potential of compounds toward CYP1A2, 2B6, and 3A4 via the aryl hydrocarbon (AhR), constitutive androstane (CAR), and pregnane X (PXR) nuclear receptors (NRs), respectively, is routinely determined during SMD development. Significant CYP induction can result in therapeutic failure from clinical exposure of a compound outside the therapeutic window, if the respective enzymes are responsible for a significant portion of its overall metabolism and clearance. Similarly, exposure of a comedication that is also metabolized by the induced enzymes can be decreased significantly in a similar manner. Over repeated doses, autoinduction can occur by which the exposure of the drugs will decrease over time and result in nonlinear PK.[40] In this case, subtherapeutic concentrations of the drugs may result, which can lead to reduced efficacy at the site of target—that is, therapeutic failure. In addition, if metabolism via the induced CYP enzyme results in toxic or pharmacologically active compounds being produced, exaggerated pharmacological effects may be seen resulting in direct and/or indirect toxicity.

Multiple well-characterized models are available for examining CYP induction, either in a discovery screening setting (e.g., NR assays, hepatocyte cells lines) or later during drug development (e.g., human hepatocytes). A complete list of *in vitro* and *in vivo* methods used to study CYP induction, along with respective advantages and disadvantages is available.[41,42] Nuclear receptor luminescence assays such as PXR and AhR are well suited for a discovery setting because they can be easily automated and provide a first readout on potential enzyme induction. In general, there appears to be good agreement between EC_{50} values generated in the nuclear receptor assays compared to those generated in human hepatocytes.[41] Use of cryopreserved or freshly isolated human hepatocytes plated on simple multiwell plates (2D cultures) is considered the gold standard *in vitro* assay for examining potential induction of CYP enzymes via nuclear receptors.[3] It is well known that upon isolation and culture, basal levels of CYP enzymes decrease dramatically in hepatocytes owing to rapid cell dedifferentiation. Therefore these models are well suited for detecting a potential increase in enzyme expression after incubation for a predetermined time frame. Typically, hepatocytes are plated on collagen-coated microwell plates, allowed to attach, and then cultured for hours to several days before treatment with known inducers or test compounds for an additional 48–72 h. In this simple system, the cells cannot survive beyond 5 days. Use of a matrigel or collagen overlay to generate sandwich cultures has been used to extend the life of hepatocyte cultures for an additional 1–2 days. At the end of the dosing period, gene expression and enzyme activities can be assessed.[43] Kinetic constants (i.e., EC_{50}, E_{max}) generated from these studies can then be used for PBPK modeling to predict clinical effects. Although the *in vitro* to *in vivo* correlation (IVIVC) using primary hepatocytes has been relatively good, the absolute fold induction tends to be overpredicted by the *in vitro* assay.[41]

In addition to primary hepatocytes, hepatocyte-derived cell lines such as Fa2N-4, HepG2, BC2, and HepRG have been used previously for CYP induction or other mechanistic studies.[44,45] Ease of use, unlimited availability, and ability for manipulation via SiRNA or transfection has made these models attractive. However, at this time they are not considered a suitable replacement for primary human hepatocyte models because they either are not fully equivalent to primary hepatocytes in terms of enzyme expression or have not been fully validated.[46,47] For example, the Fa2N-4 cell line does not express functional levels of the CAR receptor and therefore cannot be used to assess CYP2B6.[48] Induction of other metabolic enzymes has been studied during SMD development, but more as a case-by-case basis. UGT1A1 induction can be examined in 2D hepatocytes alongside the CYPs. However, in this system induction of other UGT isoforms is generally modest.

Preclinical *in vivo* models (e.g., rodents, dogs, primates) to understand CYP induction risk are generally used in a more mechanistic manner during SMD development, for example, to understand nonlinear PK or toxicity observed during chronic dose studies. Even for SMD, extrapolating enzyme changes in preclinical species to humans is difficult due to species differences in enzyme expression and activity.[49] Recent advances in transgenic technologies have led to the availability of humanized liver mice, which may hold promise for bridging the gap between *in vitro* human and clinical studies, however this remains to be proven. Chapter 6 provides an overview of the potential use of animal models for predicting DDI by TP. Induction potential of metabolic enzymes in the liver by TP (*in vitro* or *in vivo*) has not been previously reported. It is interesting that exposure of nonhepatic cell types, including peripheral blood mononuclear and brain parenchymal cells to proinflammatory cytokines, does result in CYP induction[50,51] The relevance of induction of metabolic enzymes in these cells, which have little metabolic capability, is currently unknown.

Generally, suppression of CYP enzyme expression (gene or activity) has not been of particular importance during SMD development. Suppression of CYP enzymes by biologics such as immunomodulators (e.g., interferon-γ; IFN-γ) appears to be a more likely liability. The concern for enzyme suppression in the context of TP is as follows: a patient with an inflammatory disease has increased circulatory cytokines, such as interleukin 6 (IL-6) and tumor necrosis factor α (TNF-α). These cytokines have the potential capacity to decrease CYP enzyme levels through global suppression of NRs via interaction with a shared heterodimer subunit RXR and the master regulator NF-κB.[52,53] In these patients doses of narrow therapeutic index compounds such as theophylline are titrated according to the patient's metabolic capacity to achieve effective and safe drug exposures. If a patient is then treated with an anti-IL-6 or anti-TNF-α antibody, a reversal of the reduced basal CYP activity levels might occur, resulting in therapeutic failure of the coadministered SMD, via increased drug metabolism and decreased exposure and/or increased toxicity (see Chapter 11 on Tocilizumab example). In contrast, for TPs that are direct immunomodulators (e.g., IFN-γ), suppression of CYP levels might result in decreased clearance of coadministered drugs, which could result in increased drug exposure and toxicity.

Because the mechanism of CYP suppression by cytokines is likely via NRs, 2D hepatocyte cultures have been used by several groups to understand how particular cytokines, either alone or in combination, may affect the expression and/or activity of the major CYPs.[54–57] However, studies on the suppression of non-CYP enzymes by cytokines in these same models are currently lacking. The advantages, disadvantages, and open questions regarding this methodology as well as other *in vitro* methods in the context of TP-DI are addressed in detail in Chapters 5 and 13. At this time, the predictability of simple hepatocyte assays has been questioned by multiple groups.[55,58] This lack of confidence in predictability likely explains the current regulatory opinion regarding *in vitro* assessments of TP-DI (see Chapter 10). Nonetheless, further understanding of the molecular mechanisms underlying cytokine-mediated CYP suppression may help bridge the gap of IVIVC for TP-DI. For example, the suppression of CYP3A4 by IL-6 is known to occur in a time-dependent manner based on both mRNA and activity measurements[57]. Furthermore, interaction of IL-6 with its cell surface receptor (a required step to start the cascade toward CYP suppression) results in what one could consider a quasi-irreversible activation of molecular pathways. Once activated, a subsequent invagination and recycling of the receptor occurs. Based on these observations, Kenny[59] has suggested modeling the suppression of CYP mRNA using kinetic methodology similar to SMD TDI assessments (also time dependent and irreversible). By measuring the suppression of mRNA over time and using multiple concentrations of IL-6, K_I, and k_{inact} constants can be determined and used for subsequent PBPK modeling of TP-DI risk for comedications. General strategies around model-based methods and TP-DIs are described in Chapter 4. While still early, these types of analyses show promise.

3.3 TRANSPORTERS

In general, compounds can cross biological membranes in several different ways: through paracellular or transcellular routes or via transporter proteins embedded in

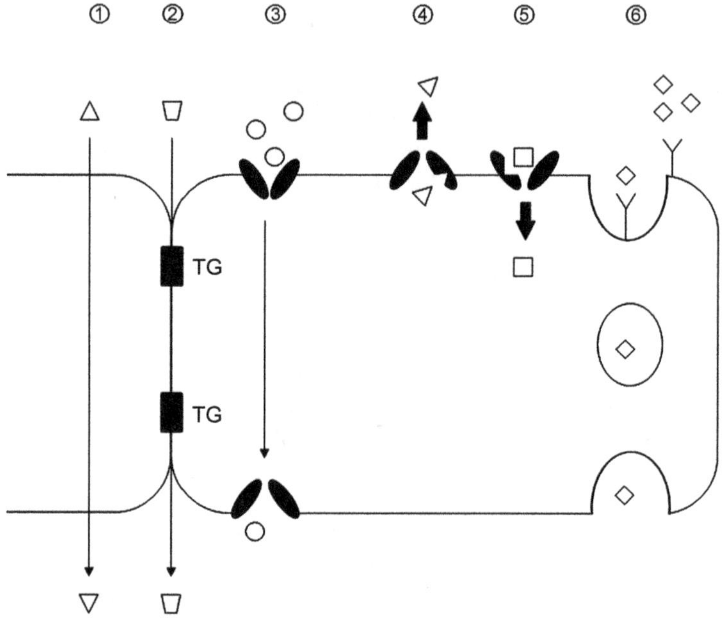

FIGURE 3-5 Routes of translocation for pharmaceutical compounds through cellular membranes. 1. Lipid soluble compounds can permeate through the epithelium by simple passive diffusion (transcellular route). 2. Small molecular weight compounds such as CO_2 can permeate through the tight junctions (TG) between cells (paracellular route). 3. Facilitative transport (e.g., uptake of amino acids) occurs via a transport carrier down the concentration gradient of the solute. Transport can also occur via a carrier but against the concentration gradient of the solute. 4. In the case of efflux transporters such as P-glycoprotein, ATP-hydrolysis is the driving force for this transport. 5. Uptake carriers, such as the organic anion, and organic cation transporters can also transport substances against a concentration gradient. The driving force depends on the particular transporter but can be pH dependent or electrogenic, for example. 6. Some hormones are transported through the epithelium via receptor-mediated transcytosis. The ligand binds to a cell surface receptor, which causes an invagination in the membrane. This invagination is internalized and carried to the opposite membrane, where the ligand is then released (Reprinted with permission from Ref. 62.)

the membrane (Figure 3-5). From a pharmaceutical perspective, active efflux and uptake transporters (routes 4 and 5 in the figure) are considered most relevant with respect to SMD. Two superfamilies of transporters are particularly important: the solute carrier or SLC family and the ATP binding cassette or ABC superfamily. Within these families several dozen drug transporters have been identified, and their characteristics (substrates, inhibitors, clinical relevance) have been reviewed in detail elsewhere.[60] Recently seven of these drug transporters have been highlighted to have particular clinical relevance for DDI and have been described in publications by the FDA or consortia, such as the International Drug Transporter (ITC) consortium.[3,61] These transporters include P-glycoprotein (P-gp), breast cancer resistance protein (BCRP), organic anion transporter polypeptide 1B1 and 1B3

(OATP1B1/1B3), organic cation transporter 2 (OCT2), and organic anion transporters 1 and 3 (OAT1/3).

3.3.1 Substrate and Inhibitor Studies

Drug transporters are found in multiple organs throughout the body and play an important role in absorption, distribution, and overall drug clearance (metabolism and excretion).[62] Multiple organ-specific *in vitro* transporter models exist that are of varying value for understanding compound transport characteristics. These include 2D suspension hepatocytes (liver uptake), 2D hepatocyte overlay, or "sandwich" cultures (liver efflux), brain endothelial cell models (brain efflux/influx), and Caco-2 cells (intestinal permeability). These cellular systems are easy to use, readily available, contain multiple tissue-relevant transporters (along with metabolizing enzymes in some cases), and can be useful in determining whether a potential DDI liability may exist. For example, uptake of a compound into hepatocytes, which is both saturable and inhibited by statins (e.g., rosuvastatin), may indicate an OATP1B1 liability. This potential DDI risk can then be followed up mechanistically using single transporter transfected/transduced cell lines. Madin-Darby canine kidney (MDCKII), human embryonic kidney (HEK), and Chinese hamster ovary (CHO) cells are most commonly used. These polarized cells can be grown on permeable inserts to study compound flux (from the apical-to-basal and basal-to-apical directions), grown on impermeable plates to access apically located transporters only, or turned into inside-out vesicles. The latter model is useful for compounds that are poorly permeable, such as conjugated drug metabolites. As with the CYPs, these *in vitro* systems can be used to generate kinetic values for DDI risk assessment using PBPK modeling and simulation. Michaelis-Menten constants such as K_m and $J_{\max}$ describe the affinity and maximal rate, respectively, of a compound for a particular transporter. Inhibition studies that generate IC_{50} values can assess the risk of inhibiting transport resulting in a DDI, either as a victim or a perpetrator. Compared to the CYPs, the prediction of clinically relevant DDI involving transporters is still in its infancy. Generally, the ability of a compound to interact with a given drug transporter (as a substrate or inhibitor) is based largely on physiochemical properties (log P, MW, solubility, etc.). Therefore, it is unlikely that large TPs such as monocloncal antibodies or fusion proteins will be direct substrates or inhibitors of classical drug transporters. However, drug-transporter function can also be affected indirectly through perturbations of lipid bilayer integrity and composition, alterations in required transporter-driving forces (e.g. Na^+/K^+ ATPase, pH gradient) or alterations in proper membrane trafficking.[63,64] Whether these mechanisms will be of importance in the context of TP-DI is yet to be determined.

3.3.2 Induction and Suppression

The clinical significance of induction of transporters, in contrast to CYPs, is poorly understood, and difficult to replicate *in vitro*. Although regulation of transporters occurs via nuclear receptors such as PXR and CAR, the duplicative cross-talk that occurs between these NRs and others such as the farnesoid X receptor and AhR is still yet to be fully explored. At present, there are no clinical examples of a TP that has caused an increase in transporter expression or function directly. However, similar to the CYPs, there is a theoretical concern that suppression of transporter expression may occur, particularly for immunosuppressant compounds. The consequence of this suppression depends largely on the characteristics of the victim SMD. For example, for an anionic drug cleared exclusively via the kidney, a suppression of OAT1 or OAT3 could decrease renal excretion and therefore increase exposure and potential drug toxicity.[65] On the other hand, for a cationic compound, this suppression may be of no consequence. Unlike the CYPs, for which the major focus of possible suppression has been centered mainly on the liver, in the case of transporters changes in function or expression in multiple organs may affect the overall ADME of a coadministered compound, including the gut (absorption), kidney (elimination), and brain (central nervous system distribution).

Thus far, work around drug transporter suppression in the liver via cytokines has been undertaken in preclinical species such as rodents[66,67] and, more recently, in primary human hepatocytes using methodologies similar to those used for CYPs.[68,69] Known species differences in transporter expression and affinities may complicate interpretation of rodent generated data. In addition, sandwich hepatocyte cultures will be more appropriate for examining canalicular transporters, such as P-gp and BCRP, because the correct orientation of those transporters is lost in simple 2D hepatocyte cultures. Because the underlying mechanism for transporter suppression may be similar to that of the CYPs, presumably one could apply similar K_I/k_{inact}-type PBPK analyses for TP-DI prediction. However, the inherent complexity of drug transporter regulation and the propensity for multifaceted interactions involving metabolism/transporter interplay might make these seemingly simplistic models difficult to interpret. A significant amount of research is certainly needed in this area.

3.4 CONCLUSION

In the SMD arena, a robust toolbox of *in vitro* and preclinical models have been developed to derisk many DDI liabilities and decrease the number of clinical DDI studies required for a given compound. Nonetheless, even for small molecules the predictability of some DDI assessments (e.g., transporter induction and transporter-metabolism interplay) remains difficult owing to a lack of well-validated models and a poor understanding of the molecular mechanisms. Although it seems logical to apply decades of knowledge and experience from the small molecule world to TP-DI assessment, it is not surprising that the inherent differences between these two types of compounds have made this challenging. As our understanding of the underlying mechanisms driving TP-mediated interactions increases and as technology platforms evolve and improve (both *in silico* and *in vitro*), accurate prediction of clinically relevant TP-DIs will, we hope, become possible. In the meantime, there is much work to do to make this a practical reality.

REFERENCES

1. Huang SM, Zhao H, Lee JI, et al. Therapeutic protein-drug interactions and implications for drug development. *Clin Pharmacol Ther* **87**(4), 497–503 (2010).
2. Lee JI, Zhang L, Men AY, et al. CYP-mediated therapeutic protein-drug interactions: clinical findings, proposed mechanisms and regulatory implications. *Clin Pharmacokinet* **49**(5), 295–310 (2010).
3. US Food and Drug Administration, US Department of Health and Human Services, Center for Drug Evaluation and Research, Center for Biologics Evaluation and Research (CBER). Guidance for industry: drug interaction studies–study design, data analysis, and implications for dosing and labeling. Silver Spring, MD, 2012. Available at www.fda.gov/downloads/Drugs/GuidanceComplianceRegulatoryInformation/Guidances/UCM292362.pdf.
4. European Medicines Agency. Guideline on the investigation of drug interactions. London, 2010. Available at www.ema.europa.eu/docs/en_GB/document_library/Scientific_guideline/2010/05/WC500090112.pdf.
5. Abbott Laboratories. *Packet insert for Humira (Adalimumab)*. North Chicago, 2012.
6. Genentech, Inc. *Packet insert for Actemra (Tocilizumab)*. South San Francisco, 2011.
7. Genentech, Inc. *Packet insert for Herceptin (Trastuzumab)*. South San Francisco, 2011.
8. Bristol-Myers Squibb. *Packet insert for Nulojix (Belatacept)*. Princeton, NJ, 2011.
9. Hoffmann-La Roche. *Packet insert for Pegasys (Peginterferon α-2a)*. Nutley, NJ, 2011.
10. InterMune, Inc. *Packet insert for Actimmune (Interferon γ-1b)*. Brisbane, CA, 2007.
11. European Medicines Agency. Guideline on the clinical investigation of the pharmacokinetics of therapeutic proteins. London, 2009. Available at www.emea.europa.eu/docs/en_GB/document_library/Scientific_guideline/2009/09/WC500003029.pdf.
12. Pang KS, Rodrigues AD, Peter RM, eds. *Enzyme- and transporter-based drug-drug interactions: progress and future challenges*. New York, Springer, 2010.
13. Venkatakrishnan K, Von Moltke LL, Greenblatt DJ. Human drug metabolism and the cytochromes P450: application and relevance of in vitro models. *J Clin Pharmacol* **41**(11), 1149–179 (2001).
14. Williams JA, Hyland R, Jones BC, et al. Drug-drug interactions for UDP-glucuronosyltransferase substrates: a pharmacokinetic explanation for typically observed low exposure (AUCi/AUC) ratios. *Drug Metab Dispos* **32**(11), 1201–08 (2004).
15. Smith DA, Jones BC, Walker DK. Design of drugs involving the concepts and theories of drug metabolism and pharmacokinetics. *Med Res Rev* **16**(3), 243–66 (1966).
16. Thompson TN. Optimization of metabolic stability as a goal of modern drug design. *Med Res Rev* **21**(5), 412–49 (2001).
17. Regan SL, Maggs JL, Hammond TG, et al. Acyl glucuronides: the good, the bad and the ugly. *Biopharm Drug Dispos* **31**(7), 367–95 (2010).
18. Nelson DR. The cytochrome P450 homepage. *Hum Genomics* **4**(1), 59–65 (2009).
19. Clarke SE, Jones BC. Human cytochrome P450 and their role in metabolism-based drug-drug interactions. In Rodrigues AD, ed. *Drug-drug interactions*. 2nd ed. New York, Informa Healthcare USA, 2008: 53–85.
20. Bjornsson TD, Callaghan JT, Einolf HJ, et al. The conduct of in vitro and in vivo drug-drug interaction studies: a Pharmaceutical Research and Manufacturers of America (PhRMA) perspective. *Drug Metab Dispos* **31**(7), 815–32 (2003).

21. Segel IH. Simple inhibition systems. In Segel IH, ed. *Enzyme kinetics: behavior and analysis of rapid equilibrium and steady-state enzyme systems.* New York, Wiley Interscience, 1975: 100–60.
22. Houston JB, Galetin A. In vitro techniques to study drug-drug interactions of drug metabolism: cytochrome P450. In Pang KS, Rodrigues AD, Peter RM, eds. *Enzyme- and transporter-based drug-drug interactions: progress and future challenges.* New York, Springer; 2010: 169–215.
23. Galetin A, Clarke SE, Houston JB. Quinidine and haloperidol as modifiers of CYP3A4 activity: multisite kinetic model approach. *Drug Metab Dispos* **30**(12), 1512–22 (2002).
24. Galetin A, Clarke SE, Houston JB. Multisite kinetic analysis of interactions between prototypical CYP3A4 subgroup substrates: midazolam, testosterone, and nifedipine. *Drug Metab Dispos* **31**(9), 1108–16 (2003).
25. Atkins WM. Non-Michaelis-Menten kinetics in cytochrome P450-catalyzed reactions. *Annu Rev Pharmacol Toxicol* **45**, 291–310 (2005).
26. Clarke SE. In vitro assessment of human cytochrome P450. *Xenobiotica* **28**(12), 1167–202 (1998).
27. Houston JB, Kenworthy KE. In vitro-in vivo scaling of CYP kinetic data not consistent with the classical Michaelis-Menten model. *Drug Metab Dispos* **28**(3), 246–54 (2000).
28. Houston JB, Kenworthy KE, Galetin A. Typical and atypical enzyme kinetics. In Fisher M, Lee J, Obach RS, eds. *Drug metabolizing enzymes: cytochrome P450 and other enzymes in drug discovery and development.* Lausanne, Switzerland, Fontis Media, 2003, 211–54.
29. Silverman RB. Mechanism-based enzyme inactivators. *Methods Enzymol* **249**, 240–83 (1995).
30. Masubuchi Y, Horie T. Toxicological significance of mechanism-based inactivation of cytochrome P450 enzymes by drugs. *Crit Rev Toxicol* **37**(5), 389–412 (2007).
31. O'Donnell JP, Dalvie DK, Kalgutkar AS, Obach RS. Mechanism-based inactivation of human recombinant P450 2C9 by the nonsteroidal anti-inflammatory drug suprofen. *Drug Metab Dispos* **31**(11), 1369–77 (2003).
32. Yang J, Jamei M, Yeo KR, et al. Kinetic values for mechanism-based enzyme inhibition: assessing the bias introduced by the conventional experimental protocol. *Eur J Pharm Sci* **26**(3–4), 334–40 (2005).
33. Lim HK, Duczak N Jr, Brougham L, et al. Automated screening with confirmation of mechanism-based inactivation of CYP3A4, CYP2C9, CYP2C19, CYP2D6, and CYP1A2 in pooled human liver microsomes. *Drug Metab Dispos* **33**(8), 1211–19 (2005).
34. Chen Y, Liu L, Monshouwer M, Fretland AJ. Determination of time-dependent inactivation of CYP3A4 in cryopreserved human hepatocytes and assessment of human drug-drug interactions. *Drug Metab Dispos* **39**(11), 2085–92 (2011)
35. Robin MA, Maratrat M, Le Roy M, et al. Antigenic targets in tienilic acid hepatitis. Both cytochrome P450 2C11 and 2C11-tienilic acid adducts are transported to the plasma membrane of rat hepatocytes and recognized by human sera. *J Clin Invest* **98**(6), 1471–80 (1996).

36. Kalgutkar AS, Obach RS, Maurer TS. Mechanism-based inactivation of cytochrome P450 enzymes: chemical mechanisms, structure-activity relationships and relationship to clinical drug-drug interactions and idiosyncratic adverse drug reactions. *Curr Drug Metab* **8**(5), 407–47 (2007).

37. Kang TW, Kim HJ, Ju H, et al. Genome-wide association of serum bilirubin levels in Korean population. *Hum Mol Genet* **9**(18), 3672–78 (2010).

38. Shitara Y, Hirano M, Sato H, Sugiyama Y. Gemfibrozil and its glucuronide inhibit the organic anion transporting polypeptide 2 (OATP2/OATP1B1:SLC21A6)-mediated hepatic uptake and CYP2C8-mediated metabolism of cerivastatin: analysis of the mechanism of the clinically relevant drug-drug interaction between cerivastatin and gemfibrozil. *J Pharmacol Exp Ther* **311**(1), 228–36 (2004).

39. Ogilvie BW, Zhang D, Li W, et al. Glucuronidation converts gemfibrozil to a potent, metabolism-dependent inhibitor of CYP2C8: implications for drug-drug interactions. *Drug Metab Dispos* **34**(1), 191–97 (2006).

40. Jana S, Paliwal J. Molecular mechanisms of cytochrome p450 induction: potential for drug-drug interactions. *Curr Protein Pept Sci* **8**(6), 619–28 (2007).

41. Hewitt NJ, Lecluyse EL, Ferguson SS. Induction of hepatic cytochrome P450 enzymes: methods, mechanisms, recommendations, and in vitro-in vivo correlations. *Xenobiotica* **37**(10–11), 1196–224 (2007).

42. McGinnity DF, Zhang G, Kenny JR, et al. Evaluation of multiple in vitro systems for assessment of CYP3A4 induction in drug discovery: human hepatocytes, pregnane X receptor reporter gene, and Fa2N-4 and HepaRG cells. *Drug Metab Dispos* **37**(6), 1259–68 (2009).

43. Roymans D, Annaert P, Van Houdt J, et al. Expression and induction potential of cytochromes P450 in human cryopreserved hepatocytes. *Drug Metab Dispos* **33**(7), 1004–16.

44. Gomez-Lechon MJ, Donato T, Jover R, et al. Expression and induction of a large set of drug-metabolizing enzymes by the highly differentiated human hepatoma cell line BC2. *Eur J Biochem* **268**(5), 1448–59 (2001).

45. Westerink WM, Schoonen WG. Cytochrome P450 enzyme levels in HepG2 cells and cryopreserved primary human hepatocytes and their induction in HepG2 cells. *Toxicol In Vitro* **21**(8), 1581–91 (2007).

46. Chu V, Einolf HJ, Evers R, et al. In vitro and in vivo induction of cytochrome p450: a survey of the current practices and recommendations: a Pharmaceutical Research and Manufacturers of America perspective. *Drug Metab Dispos* **37**(7), 1339–54 (2009).

47. Gerets HH, Tilmant K, Gerin B, et al. Characterization of primary human hepatocytes, HepG2 cells, and HepaRG cells at the mRNA level and CYP activity in response to inducers and their predictivity for the detection of human hepatotoxins. *Cell Biol Toxicol* **28**(2), 69–87 (2012).

48. Hariparsad N, Carr BA, Evers R, Chu X. Comparison of immortalized Fa2N-4 cells and human hepatocytes as in vitro models for cytochrome P450 induction. *Drug Metab Dispos* **36**(6), 1046–55 (2008).

49. Kraynov E, Dowty ME, Fahmi OA, Fields O. Animal models for evaluation of drug-drug interaction potential of biotherapeutics. *Curr Drug Metab* **13**(7), 947–950 (2012).

50. Tindberg N, Baldwin HA, Cross AJ, Ingelman-Sundberg M. Induction of cytochrome P450 2E1 expression in rat and gerbil astrocytes by inflammatory factors and ischemic injury. *Mol Pharmacol* **50**(5), 1065–72 (1996).
51. Liptrott NJ, Penny M, Bray PG, et al. The impact of cytokines on the expression of drug transporters, cytochrome P450 enzymes and chemokine receptors in human PBMC. *Br J Pharmacol* **156**(3), 497–508 (2009).
52. Zordoky BN, El-Kadi AO. Role of NF-kappaB in the regulation of cytochrome P450 enzymes. *Curr Drug Metab* **10**(2), 164–78 (2009).
53. Yang J, Hao C, Yang D, et al. Pregnane X receptor is required for interleukin-6-mediated down-regulation of cytochrome P450 3A4 in human hepatocytes. *Toxicol Lett* **197**(3), 219–26 (2010).
54. Aitken AE, Morgan ET. Gene-specific effects of inflammatory cytokines on cytochrome P450 2C, 2B6 and 3A4 mRNA levels in human hepatocytes. *Drug Metab Dispos* **35**(9), 1687–93 (2007).
55. Dickmann LJ, Patel SK, Rock DA, et al. Effects of interleukin-6 (IL-6) and an anti-IL-6 monoclonal antibody on drug-metabolizing enzymes in human hepatocyte culture. *Drug Metab Dispos* **39**(8), 1415–22 (2011).
56. Dallas S, Sensenhauser C, Batheja A, et al. De-risking bio-therapeutics for possible drug interactions using cryopreserved human hepatocytes. *Curr Drug Metab* **13**(7), 923–929 (2012).
57. Dickmann LJ, Patel SK, Wienkers LC, Slatter JG. Effects of Interleukin 1beta (IL-1beta) and IL-1beta/interleukin 6 (IL-6) Combinations on drug metabolizing enzymes in human hepatocyte culture. *Curr Drug Metab* **13**(7), 930–937 (2012).
58. Girish S, Martin SW, Peterson MC, et al. AAPS workshop report: strategies to address therapeutic protein-drug interactions during clinical development. *AAPS J* **13**(3), 405–16 (2011).
59. Kenny J.Cytokine mediated regulation of human cytochrome P450s—can we translate in vitro evaluation to clinical relevance? Paper presented at the AAPS/National Biotechnology Conference, San Francisco, CA, May 17, 2011.
60. You G, Morris ME, eds. *Drug transporters: molecular characterization and role in drug disposition*. Hoboken, NJ, Wiley, 2007.
61. Giacomini KM, Huang SM, Tweedie DJ, et al. Membrane transporters in drug development. *Nat Rev Drug Discov* **9**(3), 215–36 (2010).
62. Dallas S, Ward P. Drug Transporters and ADME. In Swarbrick J, ed. *Encyclopedia of Pharmaceutical Technology*. 4th ed. New York, Informa Healthcare, in press 2012.
63. Meyer dos Santos S, Weber CC, Franke C, et al. Cholesterol: coupling between membrane microenvironment and ABC transporter activity. *Biochem Biophys Res Commun* **354**(1), 216–21 (2007).
64. Akhtar N, Ahad A, Khar RK, et al. The emerging role of P-glycoprotein inhibitors in drug delivery: a patent review. *Expert Opin Ther Pat* **21**(4), 561–76 (2011).
65. Hagos Y, Wolff NA. Assessment of the role of renal organic anion transporters in drug-induced nephrotoxicity. *Toxins (Basel)* **2**(8), 2055–82 (2010).
66. Hartmann G, Cheung AK, Piquette-Miller M. Inflammatory cytokines, but not bile acids, regulate expression of murine hepatic anion transporters in endotoxemia. *J Pharmacol Exp Ther* **303**(1), 273–81 (2002).

67. Geier A, Dietrich CG, Voigt S, et al. Cytokine-dependent regulation of hepatic organic anion transporter gene transactivators in mouse liver. *Am J Physiol Gastrointest Liver Physiol* **289**(5), G831–41 (2005).

68. Diao L, Li N, Brayman TG, Hotz KJ, Lai Y. Regulation of MRP2/ABCC2 and BSEP/ABCB11 expression in sandwich cultured human and rat hepatocytes exposed to inflammatory cytokines TNF-α, IL-6, and IL-1β. *J Biol Chem* **285**(41), 31185–92 (2010).

69. Le Vee M, Jouan E, Stieger B, et al. Regulation of drug transporter expression by oncostatin M in human hepatocytes. *Biochem Pharmacol* **82**(3), 304–11 (2011).

70. Juang RH. Enz04 inhibition. Paper presented at the National Taiwan University, 2004. Available at http://juang.bst.ntu.edu.tw/BCbasics.

CHAPTER 4

MODEL-INDEPENDENT AND MODEL-BASED METHODS TO ASSESS DRUG–DRUG INTERACTIONS FOR THERAPEUTIC PROTEINS

DI WU and JEFFREY S. BARRETT

4.1 INTRODUCTION

Therapeutic proteins (TPs) are playing a greater role in pharmacotherapy due to their high potency and mechanism of action. The main therapeutic protein classes include monoclonal antibodies, interferons, interleukins, hormones such as growth hormone and insulin, colony-stimulating factors, and blood clotting factors. TP therapies have been applied in a variety of disease areas, such as oncology, autoimmune disorders, infectious diseases, and hormone deficiencies.

The value of the global market for TPs was estimated at $93 billion in 2010, with a compound annual growth rate (CAGR) of 16.4% between 2002 and 2010. The market is forecasted to grow at a CAGR of 6.2% between 2010 and 2017, to reach $141.5 billion in 2017.[1] Therefore, TPs are expected to occupy a significant portion of the commercial portfolios of many large pharmaceutical companies. Likewise, their presence on drug formularies and in the marketplace in general necessitates a greater understanding for both the lay and the scientific communities.

By the year 2050, 34.6% of the world population will be aged more than 65 years.[2] Polypharmacy is frequently encountered when treating senior patients, resulting in a high drug–drug interaction (DDI) potential. Unexpected DDIs can lead to reduced efficacy or toxicity. Because the approval of human recombinant insulin as the first biologic by the FDA in 1982,[3] TPs have been used clinically. With the increasing use of TPs, TP-drug interactions (TP-DIs) are more prevalent, and public awareness has increased.

Drug–Drug Interactions for Therapeutic Biologics, First Edition.
Edited by Honghui Zhou and Bernd Meibohm.

4.2 TP-DIs

Because TP-DI research is still a new and emerging area, limited data are available from the drug development of TPs. The interactions of TPs with small molecule drugs (SMDs) and other TPs[4] are likewise based on sparse experience. The basic mechanisms behind TP-DIs certainly need to be further investigated.

4.2.1 Therapeutic Target and Window

Monoclonal antibodies (mAbs) are one of the most important classes of TPs. They have been approved in a variety of indications, but most are developed to treat cancer and autoimmune diseases. A distinguishing feature of the class is that mAbs can specifically bind to therapeutic targets (i.e., target cells or proteins). Binding may then stimulate a patient's immune system to attack those targets. The therapeutic targets can be identified as either soluble or cellular. The mAbs targeting cellular antigens exhibit complex and nonlinear PK, whereas mAbs that bind to soluble antigens usually have linear PK over a wide dose range with clearance usually assessed as dose independent. In most cases, the clearance of cellular antigen-binding mAbs decreases with increasing dose in a low dose range, stabilizing to a constant value after the target-mediated, nonlinear elimination pathway is saturated. Thus it is important to select appropriate dose(s) for a DDI study of a mAb that binds to cellular antigen.[3]

Similar to small molecules, TPs with a wide therapeutic range would not present a safety concern in regard to DDI. A safety margin of 10 or greater is empirically considered a wide therapeutic window for a TP.[3]

4.3 *IN VITRO* AND *IN VIVO* APPROACHES FOR EVALUATING TP-DI

In vitro and *in vivo* approaches for evaluating DDI for TPs can be grouped into model-independent and model-based methods. The strategy to design and conduct a TP-DI study should be based on the clearance mechanisms and biological mechanisms of action.[5] Multiple doses of small molecules are sometimes required for testing TP-DIs because small molecules usually have much shorter half-lives than those of TPs.[6] Compared with small molecule DDI evaluation, a scientifically based and operationally feasible strategy and design should be incorporated into TP-DI exploration and investigation. Selection of interacting drugs should be based on clinical probability of concomitant utilization, the therapeutic range of the small molecule drug, and the pharmacokinetic (PK) and pharmacodynamic (PD) characteristics of the drugs.

The FDA published draft guidance for DDIs for industry, including a section on TP-DI, in February 2012.[7] This draft guidance provides instructions for *in vitro* and *in vivo* studies on TP-DIs. However, case-by-case-based TP-DI studies are commonly designed and evaluated. Risk-based strategies for TP-DI evaluation are extensively executed in the pharmaceutical industry in the United States.[3,5]

Model-independent methods include *in vitro* screening, dedicated DDI studies, and "cocktail" DDI studies. Model-based methods include population PK/PD-based DDI studies.

4.3.1 *In Vitro* Screening

Currently, human-cultured hepatocyte assay is considered as the most commonly used preclinical approach to assess the effects of TPs on CYP enzymes and transporters using clinically relevant concentrations of TPs or physiologically relevant concentration of cytokine/cytokine modulators.[3,8,9] This method allows for direct assessment of the effect of TPs on CYP isozymes and drug transporters. It has been suggested that preclinical evaluations should not be required if a TP drug candidate is one of the cytokines or cytokine modulators with reported effects on cytochrome P450 (CYP) isozymes and drug transporters,[10] or if it is a mAb directed against one of those cytokines and/or their receptors. This class of TP drug candidates should be directly assessed in a clinical DDI study.[8,10] However, if the TP drug candidate is one of the cytokines or cytokine modulators with no reported effects on CYP isozymes and transporters,[10] or it is a mAb directed against one of those cytokines and/or their receptors, using a human cultured hepatocyte assay is recommended for testing the TP-DI propensity with a minimum of three different hepatocyte donors.[3,8] No preclinical studies are generally required if a TP drug candidate is not cytokine related.

An *in vitro* cytokine release assay has been used to evaluate cytokine release potentials from a TP drug candidate. Because healthy human blood is applied, the assay is limited to the safety assessment of TP immunomodulatory capacity and results from the assay are difficult to translate to the *in vivo* setting where TP targets are located in the body outside the circulation.[8]

The *in vitro* approach needs to be optimized further to ensure both qualitative and quantitative evaluation of TP-DIs. At this point, all results from *in vitro* screening should be viewed qualitatively. Furthermore, *in vitro–in vivo* correlation for TP-DIs has not been established and needs further investigation. In addition, *in silico* translational prediction built on a good understanding of the basic mechanisms underlying TP-DIs is still under development.[5]

A good attempt has been made to predict the human PK of a mAb exhibiting target-mediated drug disposition (TMDD).[11] System parameters were obtained from *in vitro* experiments or literature, such as binding affinity (k_{on} and k_{off}), internalization of the drug–target complex (k_{int}), target degradation rate (k_{deg}), and target abundance (R_o), incorporated into a TMDD model and developed in a population-based approach to predict human PK using monkey PK data as well. The modeling and simulation results showed that prediction of human PK was within one- to two-fold of the observed data from a phase I study. This well-designed study illustrates the excellent combination of *in vitro* experimental methods with population-based modeling to predict clinical data in a clinically relevant range.

Animal models are rarely used in the assessment of TP-DIs at present. One limitation of animal models used in small molecule drug development is the fact of highly species-specific immune tolerance. Because tolerance is highly species specific and good correlations between animals and humans are hardly established, animal models have so far been rarely used in drug development of TPs and evaluation of TP-DIs.[12]

4.3.2 Clinical DDI Investigations

Clinical DDI studies are recommended to be conducted in patients, not healthy subjects, for several reasons. Firstly, PK differences of TPs may exist between healthy subjects and patients, and secondly, there is the potential of toxicity of some TPs and small molecule drugs, making it unethical to conduct the studies in healthy subjects. More important, CYP isozyme levels may have been down-regulated in chronic inflammatory diseases, such as rheumatoid arthritis and some infectious diseases. If TP-DIs are mediated through CYP isozymes, anticytokine effects of TPs would normalize the down-regulated CYP enzyme activities or expressions and thereby alter the clearance of coadministered small molecule drugs.[5,10]

4.3.2.1 Dedicated Clinical DDI Studies A dedicated clinical DDI study is often applied for the reliable evaluation of TP-DIs. Common study designs for dedicated clinical DDI studies include parallel, crossover (TPs with short half-lives), and single-sequence crossover (TPs with long half-lives; small molecule drug PK during run-in period compared with small molecule drug PK after single or multiple doses of TP).[8,13]

In a pharmacokinetic study of repeated subcutaneous and intravenous administration of golimumab in patients with rheumatoid arthritis, the effect of concomitant administration of methotrexate (MTX) was examined in an open-label, randomized study.[14] A parallel design was used in the clinical DDI study of golimumab. Golimumab pharmacokinetics was described using noncompartmental analysis (NCA). In the group receiving subcutaneous golimumab administration, the geometric means of golimumab CL/F in patients cotreated with MTX and patients not cotreated with MTX were 13.9 and 21.2 mL/d/kg, respectively. Compared with the patients not cotreated with MTX, the mean C_{max} and AUC_{0-28d} at steady state were 24% and 37% higher, respectively, in patients cotreated with MTX. In this example, NCA is used efficiently to evaluate TP-DI by comparing C_{max}, AUC, and CL/F in the two scenarios. NCA requires a sufficient number of samples collected especially in the terminal phase and around C_{max}.

Another dedicated DDI study investigated the effects of concomitant tocilizumab on MTX PK in patients with rheumatoid arthritis. The study showed that the 90% CIs for mean ratios in systemic exposure (AUC_{last} and C_{max}) of a combination therapy of tocilizumab and MTX and a monotherapy of either MTX or its metabolite 7-hydroxymethotrexate were all within the 80–125% bounds. C-reactive protein, a PD marker, returned to baseline within a week after tocilizumab injection. Therefore, tocilizumab and MTX can be administered concurrently without any concern of DDI in this population.[15]

In addition, a cross-study comparison approach can be used in TP-DI evaluations involving PK data comparison between healthy subjects and patients or historical data. Basiliximab clearance was significantly lower when co-treated with azathioprine (29 ± 14 mL/h) or mycophenolate mofetil (18 ± 8 mL/h) when compared with its established clearance of 37 ± 15 mL/h. Population-based nonlinear mixed-effect models were used to derive all the PK parameters used in the comparison.[16] Another example of using historical data for TP-DI study comparisons is a study of trastuzumab emtansine and pertuzumab.[17] However, results generated from such comparisons should be confirmed in specific DDI studies in the same patient population.[4] As another approach, a crossover design can be used to evaluate the effect of a coadministered drug on a TP. TP PK after cotreatment with another TP or a small molecule drug in patients can be compared with TP PK obtained in a first-in-patient TP monotherapy study. The most rigorous approach for assessing TP-DI is to implement a within-patient comparison. Careful interpretation should be given on the possible interference of antidrug antibodies resulting from the crossover design.[8]

4.3.2.2 Cocktail Clinical DDI Study The "cocktail" clinical DDI study is designed to screen the effect of a TP on multiple CYP isozymes via the use of corresponding prototypical CYP probes. Compared to administration of only a single probe in a dedicated DDI study, multiple CYP probes are simultaneously administered in a cocktail DDI study to identify several potential CYP isoenzymes involved in TP-DIs. The cocktail used should be safe and use validated probe drugs for specific enzymes. Ideally, only one enzyme should be involved in the elimination of the probe drug. But if the second enzyme is metabolizing the same probe drug, its con tribution to the total clearance of the very probe drug should be less than 10%. The probe drugs used should be validated by the applicant or published in the literature. In cocktail DDI studies, doses used for probe drugs should be validated both individually and in combination to ensure no interference in the pharmacokinetic interaction(s) among them. In any DDI studies, dose and duration of the investigational drug should be sufficient to estimate the maximum inhibition and/or induction achieved at a clinically relevant dose.[7,18,19] Though cocktail DDI studies are sometimes perceived as less optimal than studies testing DDIs individually, they are an efficient approach to gather more information in a single study.[3] Principles and different designs of cocktail DDI studies have been reviewed.[20,21] (see also Chapter 7).

Few TP cocktail DDI studies have been published at present. It is still questionable if the design and experimental conditions used for TPs would be the same or similar as those used for a small-molecule cocktail study. One example in cocktail TP-DI studies has been reported in a study assessing the effect of IFNα-2b on CYP isozymes.[22] High-dose IFNα-2b (HDI) is used as a standard adjuvant therapy for the treatment of high-risk melanoma. The potential effect of HDI on CYP isozymes was evaluated in 17 patients with high-risk melanoma. A Pittsburgh cocktail was used in this study with a single-sequence crossover design. A CYP isozyme activity was measured using the index derived by quantifying a selectively isozyme probe drug (and/or its metabolite) which were given at certain points of time. The results showed that HDI had substantial effects on some CYP isozymes (e.g., CYP1A2 by 60% decrease) and no effect on other isozymes (e.g., CYP2E1). A significant association has been found between the degree of CYP inhibition and the occurrence of side effects in HDI therapy, indicating a pathophysiological basis of the HDI-induced toxicity.

TP-DI studies are preferentially conducted early in the drug development process (e.g., during phase Ib/II) because of the relatively small study size needed. In addition, early assessment of DDIs by rigorous scientific design can provide safety information and development strategies to ensure the success of late-stage drug development.

4.3.3 Model-Based Method: Population PK/PD–based DDI Study

Population PK/PD modeling plays a significant role in drug development of TPs. Population PK/PD modeling can be a screening tool to assess and identify TP-DIs initially, which can then be confirmed in subsequent formal DDI studies if deemed necessary. It can also be used to explore whether a TP is a perpetrator or a victim for a TP-DI if both TP and small molecule drug are sampled simultaneously.[3]

The structure and mechanism of action of mAbs are closely linked to the systemic exposure and pharmacological response of mAbs. PK/PD modeling on mAbs has been used to elucidate the mechanisms of action and to optimize dose regimens.[23–25]

Both NCA and population PK approaches have their advantages and limitations. Population PK modeling aims to develop models to describe drug disposition and obtain insights on mechanisms of drug actions when linked with PD as well as assess covariate effects. NCA allows the estimation of terminal half-lives and exposure statistics. Population PK modeling can incorporate data more efficiently (e.g., sparse or rich data as well as data from different occasions) and provide uncertainty measures for the estimated PK parameters. Certainly NCA requires less effort to perform and generate results in a much faster manner. However, population PK modeling approach provides advantages over NCA when sparse samples are present and covariate effects or mechanisms are needed to explore in the target population.

In the DDI study of MTX and infliximab (IFX) in ankylosing spondylitis patients, population PK modeling was used to estimate population parameters and evaluate the MTX effect on final parameters.[26] A parallel design was used in this investigation. MTX cotreatment did not influence any population PK parameter when examined as a covariate in the established IFX population model. Furthermore, no significant difference of AUC_{18} and AUE (area under the curve of PD marker) was observed between the IFX and the IFX + MTX groups.

4.4 BIOANALYTICAL CONSIDERATIONS

Most TP concentrations are quantified by ligand-binding assays (LBAs) in designated biological matrices. Specificity and selectivity issues are the bioanalytical challenges when using LBA. Specificity and selectivity of an assay for a TP of interest should be assessed during method validation of LBA.[8,27] LBA capture and detection reagents are crucial to determine assay specificity for free, partially bound, fully bound, or total antibodies as well as specificity for free and total targets. Existing metabolites and inactive species in the biological matrices might interfere with the determination of the TP concentrations and lead to overestimation of the assay results. It is critical to understand the mechanisms of the assays for a better interpretation of study results in regard to data reliability and limitations. The design of LBA bioanalytical methods (free or total concentrations) should be conducted according to the aims of the applications.[28]

In a recent study of identifying the discrepancy between enzyme-linking immunosorbent assay (ELISA) and liquid chromatography tandem mass spectrometry (LC-MS/MS) assay results of a PEGylated scaffold protein in postdose monkey plasma samples, the presence of antidrug antibodies specific to the engineered antigen-binding region of this scaffold protein affected the ability of drug to bind to the target antigen used in the ELISA assay.[29] Thus the ELISA method measured only the active circulating PEGylated scaffold protein, whereas the LC-MS/MS method measured the total circulating drug. LC-MS/MS results exerted an approximately 1.53-fold higher total exposure (AUC_{tot}) than the ELISA results. The bioanalysis of TPs can be compromised by drug-binding endogenous components or ADAs in the incurred samples. Therefore, good understanding of the principles of different bioanalytical methods is essential to select the appropriate approaches for specific TPs.

4.5 CONCLUSION

Individualized and iterative approaches for evaluation of TP-DIs should clearly be implemented in the TP drug development program. With a thorough understanding of the mechanisms of actions of TPs, TP-DI studies can be scientifically assessed, designed, and conducted. Thus a risk based strategy is expected to guide the appro priate design and execution of TP-DIs. The risk-based strategy takes in account an evaluation of therapeutic target (cellular or soluble antigens), therapeutic window (wide or narrow), and mechanisms of DDI (*in vitro* and *in vivo*) of a particular TP of interest. Once the risk factors are evaluated and identified, the appropriate approach can be selected to design the DDI study (parallel, single-sequence crossover, or crossover) and analyze the data (NCA and/or population PK-based modeling).[3]

Model-independent and model-dependent approaches are the two main methods used to evaluate DDI between a TP and a small-molecule drug as well as between two TPs, and to predict DDI in different scenarios within model defined populations and scopes. Further investigations into the likely mechanisms involved in TP-based DDIs are needed. A critical next step is the early identification of DDI potential for TPs as well as a more granular understanding of those likely to be clinically relevant.

REFERENCES

1. GBI, Research. Therapeutic proteins market to 2017—high demand for monoclonal antibodies will drive the market. GBI Research 2011.
2. De Luca d'Alessandro E, Bonacci S, Giraldi G. Aging populations: the health and quality of life of the elderly. *Clin Ter* **162**, e13-8 (2011).
3. Zhou H, Davis HM. Risk-based strategy for the assessment of pharmacokinetic drug-drug interactions for therapeutic monoclonal antibodies. *Drug Discov Today* **14**, 891–98 (2009).
4. Pang KS, Peter RM, Rodrigues AD. Drug-drug interactions: what have we learn and where are we going? In Pang KS, Peter RM, Rodrigues AD, eds *Enzyme- and transporter-based drug-drug interaction*. Springer, New York, 2010: 701–22.
5. Girish S, Martin SW, Peterson MC, et al. AAPS workshop report: strategies to address therapeutic protein-drug interactions during clinical development. *AAPS J* **13**, 405–16 (2011).
6. Seitz K, Zhou H. Pharmacokinetic drug-drug interaction potentials for therapeutic monoclonal antibodies: reality check. *J Clin Pharmacol* **47**, 1104–18 (2007).
7. US Food and Drug Administration, US Department of Health and Human Services, Center for Drug Evaluation and Research, Center for Biologics Evaluation and Research (CBER). Draft guidance for industry: drug interaction studies—study design, data analysis, implications for dosing, and labeling. Silver Spring, MD, 2012. Available at www.fda.gov/downloads/Drugs/GuidanceComplianceRegulatoryInformation/Guidances/UCM292362.pdf.
8. Kraynov E, Martin SW, Hurst S, et al. How current understanding of clearance mechanisms and pharmacodynamics of therapeutic proteins can be applied for evaluation of their drug-drug interaction potential. *Drug Metab Dispos* **39**, 1779–83 (2011).
9. Zhou H, Mascelli MA. Mechanisms of monoclonal antibody-drug interactions. *Annu Rev Pharmacol Toxicol* **51**, 359–72 (2011).
10. Huang SM, Zhao H, Lee JI, et al. Therapeutic protein-drug interactions and implications for drug development. *Clin Pharmacol Ther* **87**, 497–503 (2010).
11. Luu KT, Bergqvist S, Chen E, et al. A model-based approach to predicting the human pharmacokinetics of amonoclonal antibody exhibiting target-mediated drug dispostion. *J Pharmacol Exp Ther* **341**, 702–8 (2012).
12. Baumann A. Early development of therapeutic biologics-pharmacokinetics. *Current Drug Metab* **7**, 15–21 (2006).
13. Vincenti F, Mendez R, Pescovitz M, et al. A phase I/II randomized open-label multicenter trial of efalizumab, a humanized anti-CD11a, anti-LFA-1 in renal transplantation. *Am J Transplant* **7**, 1770–77 (2007).
14. Zhuang Y, Xu Z, Frederick B, et al. Golimumab pharmacokinetics after repeated subcutaneous and intravenous administration in patients with rheumatoid arthritis and effect of concomitant methotrexate: an open-label, randomized study. *Clin Ther* **34**, 77–90 (2012).
15. Schmitt C, Kuhn B, Zhang X, et al. Tocilixumab has no clinically relevant effects on methotrexate pharmacokinetics in patients with rheumatoid arthritis. *Int J Clin Pharmacol Ther* **50**, 218–23 (2012).

16. Kovarik JM, Pescovitz MD, Sollinger HW, et al. Differential influence of azathioprine and mycophenolate mofetil on the disposition of basiliximab in renal transplant patients. *Clin Transplant* **15**, 123–30 (2001).
17. Lu D, Burris HA, Wang B, et al. Drug interaction potential of trastuzumab emtansine combined with pertuzumab in patients with HER2-positivie metastatic breast cancer. *Curr Drug Metab* **13**, 911–22 (2012).
18. Huang S-M, Temple R, Throckmorton DC, et al. Drug interaction studies: study design, data analysis, and implications for dosing and labeling. *Clin Pharmacol Ther* **81**, 298–304 (2007).
19. Turpault S, Brian W, Van Horn R, et al. Pharmacokinetics assessment of a five-probe cocktail for CYPs 1A2, 2C9, 2C19, 2D6, and 3A. *Br J Clin Pharmacol* **68**, 928–35 (2009).
20. Zhou H, Tong Z, McLeod JF. "Cocktail" approaches and strategies in drug development: valuable tool or flawed science? *J Clin Pharmacol* **44**, 120–34 (2004).
21. Fuhr U, Jetter A, Kirchheiner J. Appropriate phenotyping procedures for drug metabolizing enzymes and transporters in humans and their simutaneous use in the "cocktail" approach. *Clin Pharmacol Ther* **81**, 270–83 (2007).
22. Islam M, Frye RF, Richards TJ, et al. Differential effect of IFNα-2b on the cytochrome P450 enzyme system: a potential basis of IFN toxicity and its modulation by other drugs. *Clin Cancer Res* **8**, 2480–87 (2002).
23. Mould DR, Sweeney KR. The pharmacokinetics and pharmacodynamics of monoclonal antibodies—mechanistic modeling applied to drug development. *Curr Opin Drug Discov Devel* **10**, 84–96 (2007).
24. Mould DR, Green B. Pharmacokinetics and pharmacodynamics of monoclonal antibodies: concepts and lessons for drug development. *BioDrugs* **24**, 23–39 (2010).
25. Mould D, Frame B. Population pharmacokinetic-pharmacodynamic modeling of biological agents: when modeling meets reality. *J Clin Pharmacol* **50**, 91S–100S (2010).
26. Ternant D, Mulleman D, Lauferon F, et al. Influence of methotrexate on infliximab pharmacokinetics and pharmacodynamics in ankylosing spondylitis. *Br J Clin Pharmacol* **73**, 55–65 (2011).
27. Hamuro LL, Kishnani NS. Metabolism of biologics: biotherapeutic proteins. *Bioanalysis* **4**, 189–95 (2012).
28. Garofolo F, Rocci Jr, ML Dumont I, et al. White paper on recent issues in bioanalysis and regulatory findings from audits and inspections. *Bioanalysis* **3**, 2081–96 (2011).
29. Wang SJ, Wu ST, Gokemeijer J, et al. Attribution of the discrepancy between ELISA and LC-MS/MS assay results of a PEGylated scaffold protein in post-dose monkey plasma samples due to the presence of anti-drug antibodies. *Anal Nioanal Chem* **402**, 1229–39 (2012).

CHAPTER 5

UTILITY OF *IN VITRO* METHODS IN DRUG–DRUG INTERACTION ASSESSMENT AND PREDICTION FOR THERAPEUTIC BIOLOGICS

THERESA NGUYEN, NARENDRA KISHNANI, and RAYMOND EVERS

5.1 INTRODUCTION

Therapeutic proteins (TPs) are increasingly being coadministered with small molecule (SM) drugs and/or other biologics such as monoclonal antibodies (mAbs). Historically, interactions between TPs and SMs have not been considered very likely because mechanisms involved in the disposition of both modalities are fundamentally different. In the case of SMs, various drug-metabolizing enzymes and transporters expressed primarily in the intestine, liver, and kidney are involved in absorption, distribution, and elimination. For mAb, non-ligand and ligand-mediated clearance pathways are involved. In addition, binding to the neonatal Fc receptor (FcRn) functions as a salvage pathway that protects IgG molecules from lysosome-mediated degradation. In general, these differences will make it less likely that drug–drug interactions (DDIs) will occur between TPs and SMs. However, the situation may be different for TPs involved in pathways that modulate the immune system because it has been demonstrated that increased cytokine exposure can result in reduced levels of cytochrome P450 (CYP) enzymes and drug transporters.

Theoretically, the effects of cytokines on CYPs can therefore mediate DDIs between TPs and SMs when immunomodulating biologics (e.g., cytokines and cytokine modulators) are used in the treatment of cancer, infections, or inflammation. The scenarios that could lead to suppression or desuppression of enzymes have been outlined by Morgan[1] (Figure 5-1): When an individual, previously stabilized on a SM drug regimen becomes afflicted with an infection or inflammatory

Drug–Drug Interactions for Therapeutic Biologics, First Edition.
Edited by Honghui Zhou and Bernd Meibohm.

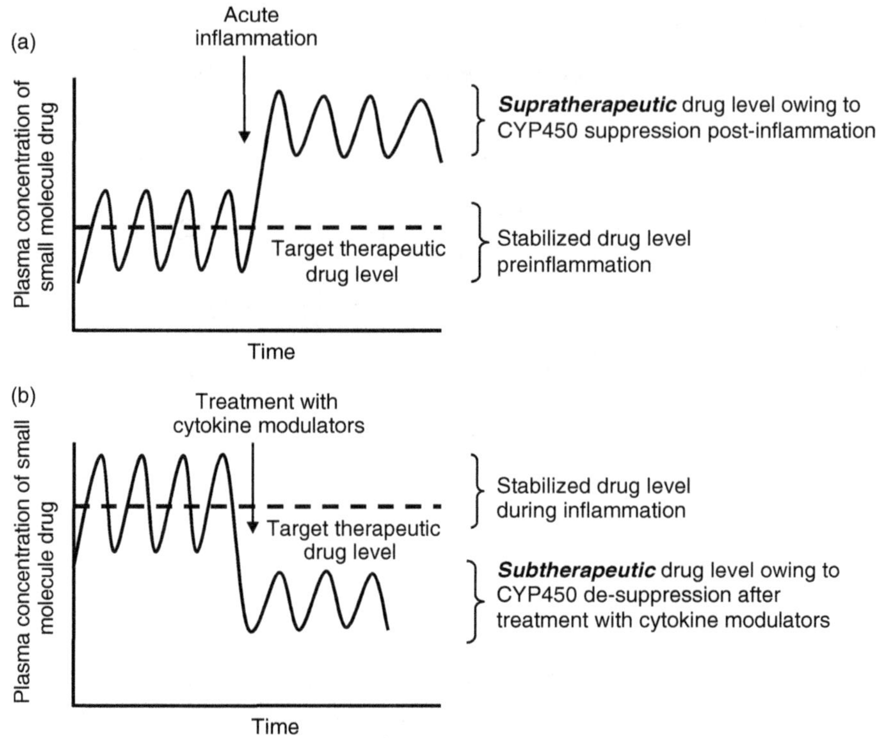

FIGURE 5-1 Suppression and desuppression of CYP450 enzymes owing to cytokines and cytokine modulators. (a) An individual previously stabilized on a small molecule drug regimen becomes afflicted with an infection or inflammatory response. As a result, the associated elevated levels of proinflammatory cytokines (e.g., tumor necrosis factor α, interleukin 1β, and interferon γ) may decrease levels of drug-metabolizing enzymes and potentially lead to a decline in drug clearance and supratherapeutic small molecule drug exposure. (b) In contrast, in patients with a preexisting inflammatory disease accompanied with increased levels of the proinflammatory cytokines, treatment with an anticytokine drug could result in normalization of previously suppressed drug-metabolizing enzymes levels and potentially increase drug clearance, leading to subtherapeutic exposure of small molecule drugs.

response, the associated elevated levels of proinflammatory cytokines (e.g., tumor necrosis factor α, interleukin 1β, interleukin 6, and interferon γ) may decrease the levels of drug-metabolizing enzymes and potentially lead to a decline in drug clearance and supratherapeutic SM drug exposure.[1] In contrast, in patients with a preexisting inflammatory disease accompanied with increased levels of the proinflammatory cytokines, treatment with an anticytokine drug could result in normalization of previously suppressed drug-metabolizing enzyme levels and potentially increase of drug clearance, leading to subtherapeutic exposure of SM drugs.

A classic example of metabolizing enzyme suppression owing to infection or inflammation is the administration of theophylline (a CYP1A2 substrate) to patients with influenza, resulting in toxic accumulation of drug as a result of reduced levels of CYP1A2.[2] Reduced clearance (10–50%) of theophylline was also found in patients infected with the hepatitis B or C virus who were treated with interferon α (IFN-α) or IFN-β.[3] Recently, an example of desuppression of CYP3A4 activity was described in patients with rheumatoid arthritis (RA) treated with the anti-IL-6 receptor mAb tocilizumab.[4] It was observed that administration of simvastatin to RA patients resulted in higher plasma drug concentrations than those reported in healthy subjects. Conversely, simvastatin exposure approached normal levels after treatment with tocilizumab. Because clearance by CYP3A4 contributes substantially to the clearance of simvastatin, treatment with tocilizumab may have normalized the inflammation-mediated reduction in CYP3A4 activity.

Currently, drug–drug interactions reported for TPs as perpetrators and SM drugs as victims in general have been small, and a generalization of the extent of interactions is not feasible at this point owing to the limited availability of clinical data. Excellent reviews on the clinical examples available have been published[5–8] and therefore will not be discussed further in this chapter.

Due to the long half-life of commonly used TPs such as mAbs and the difficulty in many cases to recruit patients for trial enrollment, the conduct of well-designed clinical interaction studies is generally costly and difficult. Therefore, there is a strong desire to develop *in vitro* approaches allowing predictions of whether significant interactions between TPs and SMs are likely. Goals of this chapter are to provide an overview of the current literature on the effects of proinflammatory cytokines on absorption, distribution, metabolism, and excretion (ADME) genes *in vitro* and to provide a perspective on the current challenges and future opportunities to establish improved *in vitro* systems.

5.2 MECHANISMS INVOLVED IN SUPPRESSION OF DRUG-METABOLIZING ENZYMES

The data presented in Table 5-1 suggest variability in the effect of individual cytokines on CYP enzymes in hepatocytes. For example, interleukin 6 (IL-6) and IFN in general caused *pan*-suppression across most CYP isoforms analyzed. In contrast, IL-1 appears to affect only a subset of the analyzed CYP isoforms. This suggests that there may be more than one mechanism in hepatocytes that could contribute to cytokine-mediated modulation of CYP expression levels. Potentially, this could have implications for DDI potential because CYP regulation may be affected by different types of inflammatory disease.[1] The general mechanisms by which cytokines interact with metabolic enzymes and transporters are illustrated in Figure 5-2. Several reviews on the molecular mechanisms identified to date that could lead to suppression of drug-metabolizing enzymes and transporters are available.[1,24–27] The major mechanisms resulting in gene suppression are summarized below.

The transcriptional activation of CYP3A4 as illustrated in Figure 5-3 is known to be mediated by the nuclear receptor pregnane X receptor (PXR), a ligand-dependent transcription factor.[28] PXR forms a complex with its obligate partner retinoid X

TABLE 5-1 Effects of Cytokines on Cytochrome P450 Enzymes

Cytokines	CYP450 Enzymes					
	1A1	1A2	2A6	2B6	2C8	2C9	2C11
IL-1β	↓r	↓r					
		↓a,r			↓r (2C)		
				↓r		↓r	
				↔r↓p	↓r	↔r↓p	
IL-1							
IL-2	↓a,r			↓r(2B1/2)			↓r
IL-4		↔r			↔r (2C)		
IL-6	↔r	↔r					
		↓a,r			↓r (2C)		
		↓a,p		↓a			
				↓r	↓r	↓r	
				↓p,r	↓r	↓p,r	
	↓r	↓a,r		↓r	↓r	↓r	
Oncostatin M		↓a,p,r	↓a	↓a			
TNF-α		↓a,r			↓r (2C)		
				↔r↓p	↓r	↔r↓p	
IFN-γ		↓a,r			↔ r (2C)		
		↓a,p	↓a	↓a			
		↓a,p	↓a	↓a			
				↓p,r	↓r	↔r↓p	
IFN-α							
IFN-α2b		↓p,r		↓r			
Cytokine mix[a]				↓p,r			

Upregulation (↑), downregulation (↓), or no effect (↔) on activity (a), protein levels (p), and/or mRNA (r) expression; IL: interleukin; IFN: interferon; KC: Kupffer cell; NPC: nonparenchymal cells; TNF: tumor necrosis factor.

[a]TNFα, IL-1β, and IFNγ.

2C18	2C19	2D1	2E1	3A4	Model	Reference
					Rat hepatocytes	9
			↓r	↓a,r (3A)	Human hepatocytes	10
				↓r	Human hepatocytes	11
↔r	↔r			↓p,r	Human hepatocytes	12
				↓a	Human hepatocytes	13
		↓r		↓a,r (3A)	Rat hepatocytes	14
				↓a	Human hepatocytes–KC co-culture	13
			↑r	↔ a,r (3A)	Human hepatocytes	10
					Rat hepatocytes	9
			↓r	↓ a,r (3A)	Human hepatocytes	10
				↓a	Human hepatocytes	15
				↓r	Human hepatocytes	16
				↔ a,r	Human hepatocytes	17
				↓a (3A)	Human hepatocytes	13
↔ r	↓r			↓p,r	Human hepatocytes	12
	↓r			↓a,r	Human hepatocytes	18
				↓a,p,r	Human hepatocytes	15
			↓r	↓a,r (3A)	Human hepatocytes	10
↔ r	↔ r			↓r ↔p	Human hepatocytes	12
				↓r	Human hepatocytes	19
			↓r	↔r↓a (3A)	Human hepatocytes	10
				↓a,p,r	Human hepatocytes	20
				↓a,p	Human hepatocytes	15
↔ r	↔ r			↓p,r	Human hepatocytes	12
				↓a,p,r	Human HepG2	21
			↓r	↑p,r	Human hepatocytes–NPC co-culture	22
			↓r ↔ p	↓p,r	Human hepatocytes	23

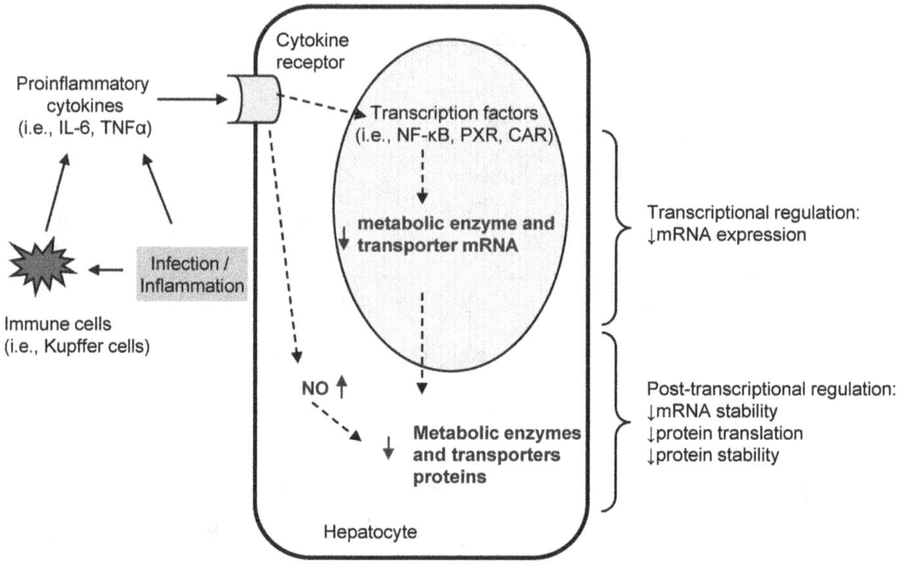

FIGURE 5-2 Proposed general mechanisms of cytokine regulation of metabolic enzymes and transporters. At the transcriptional level, cytokines can modulate gene expression of metabolic enzymes and transporters through interaction with various nuclear transcription factors—such as NF-kB, PXR, and CAR. At the post-transcriptional level, decreased enzyme and transporter proteins may be the result of reduction in mRNA stability owing to oxidative stress, suppression of protein translation, and/or increase in enzyme degradation due to elevated levels of NO associated with inflammation. Dotted lines: multiple intermediate steps may be involved; solid lines: a more direct effect. CAR: constitutive androstane receptor; IL: interleukin; NO: nitric oxide; PXR: pregnane X receptor; TNF: tumor necrosis factor. (See the color version of this figure in Color Plates section.)

receptor (RXR), and this complex in turn binds to the consensus DNA sequence elements in target genes (Figure 5-3a). PXR can be activated after binding to agonists such as rifampicin and play an important role in the induction of CYP gene transcription such as CYP3A4. In addition, it has been demonstrated that, at least in mice, PXR also plays a role in the basal transcription of Cyp3a.[29] Gu and coworkers[19] showed in human primary hepatocytes that TNF-α (2 ng/mL, 24 h) inhibited rifampicin-induced CYP3A4 mRNA levels by more than 50%. It was suggested that suppression of CYP3A4 may be mediated through the interference of the PXR-RXR-regulated pathway, and not by reduction in PXR gene transcription, because mRNA levels were unchanged in human primary hepatocytes. The researchers showed that activation of NF-κB in HepG2 cells was directly responsible for the suppression of the PXR-regulated gene expression and the presence of SRIκBα, an NF-κB super repressor, was able to reverse the suppression. In addition, it was demonstrated that NF-κB directly interacts with RXR to create a steric hindrance and disrupts the interaction of the RXR-PXR heterodimer to the PXR response element. The result is a suppression of gene transcription (Figure 5-3b).

The constitutive androstane receptor (CAR) is also involved in the transcriptional control of a number of drug-metabolizing enzymes and transporters, such as

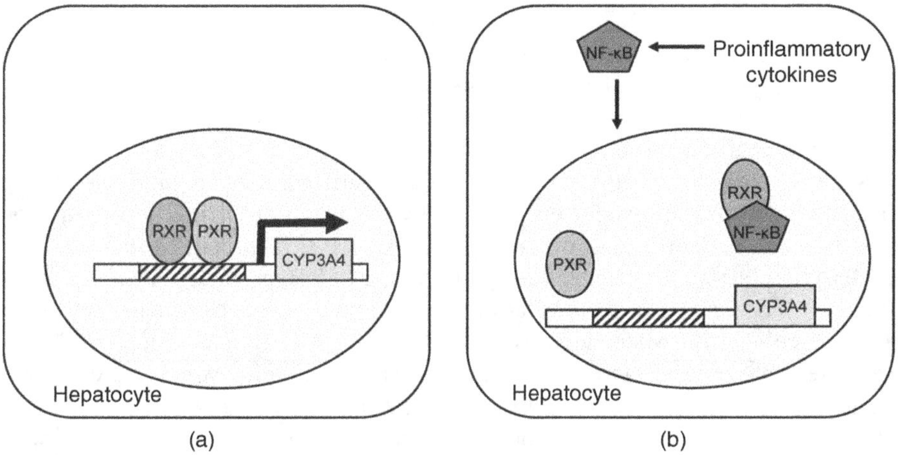

FIGURE 5-3 Regulation of CYP3A4 by pregnane X receptor (PXR) and its obligate partner retinoid X receptor (RXR). (a) Upon heterodimerization of PXR with RXR and co-activators, the complex binds to the consensus PXR response elements in target genes, leading to gene transcription. (b) The presence of proinflammatory cytokines will activate NF-κB, which in turn will bind to RXR and disrupt interaction of the RXR-PXR heterodimer to the PXR response element. The result is a suppression of gene transcription. It is hypothesized that the same mechanism could result in suppression of CAR-mediated transcription activity (CAR: constitutive androstane). (See the color version of this figure in Color Plates section.)

CYP3A4, CYP2B6, and UGT1A1.[30] Similar to PXR, CAR-mediated gene transcription also requires the formation of the CAR-RXR heterodimer complex. It is hypothesized that the same mechanism by which NF-κB interacts with the PXR-RXR complex (Figure 5-3b) may also explain the interaction of NF-κB with the CAR-RXR complex to suppress of CAR-mediated transcription activity. In addition, NF-κB may also regulate CAR-mediated transcription of drug-metabolizing enzymes and transporters through interaction with the glucocorticoid receptor (GR). Assenat and coworkers[11] demonstrated that IL-1β activation of NF-κB can inhibit the transcriptional activation of GR, resulting in diminished CAR mRNA expression and leading to reduced CAR-mediated gene transcription (i.e., CYP3A4). A similar mechanism has been proposed for the downregulation of the aryl hydrocarbon receptor (AhR), a nuclear receptor involved in the regulation of CYP1A transcription after treatment with lipopolysaccharide (LPS) or TNF-α.[31]

IFN-α activation of gene transcription is mediated through the Janus kinase–signal transducer and activator of transcription (JAK-STAT) pathway. In the hepatoma cell line HepG2, it was demonstrated that transfection with STAT1 downregulated transcription of a CYP3A4 promoter containing reporter construct, whereas overexpression of a dominant negative STAT1 partially reversed the down-regulated CYP3A4 expression.[21] Based on these results, it was concluded that components of the JAK-STAT pathway are either directly or indirectly involved in the downregulation of CYP3A4 by IFN-α. These results are in contrast to earlier data in rodents which suggested that IFN regulation of CYP3A4 is through the induction of xanthine oxidase, which can lead to elevated reactive oxygen intermediates and result in CYP3A4 protein degradation.[32] Subsequent work, however, indicated that this would not be a plausible mechanism and that downregulation is more likely to occur at the transcriptional level.[33]

Reactive oxygen species can also mediate transcriptional regulation of drug transporters by activating NF-κB. Ros and coworkers[34] noted that TNF-α induction of Mdr1b can be facilitated through various pathways, including NF-κB and p53, transcription factors for which binding sites in the Mdr1b promoter have been identified.[35,36] The authors observed that in both rat primary hepatocytes and hepatoma H-4-II-E cells, a time-dependent stimulation of the Mdr1b expression by TNF-α was detectable. Furthermore, in rat hepatoma H-4-II-E cells expressing dominant negative p53 (rendering nonfunctional proteins), TNF-α upregulation of Mdr1b was unaffected, suggesting that induction was independent of functional p53. In addition, in transient transfection assays, Ros' group also showed that Mdr1b upregulation correlated with activation of the promoter and that mutation of the NF-κB site in the Mdr1b promoter prevented its induction by TNF-α.[34]

Collectively, these results from both rat primary hepatocytes and hepatoma H-4-II-E cells suggest that induction of the rat Mdr1b gene by TNF-α is a result of activated NF-κB signaling (by elevated reactive oxygen species) and independent of p53.

Although the majority of evidence suggests that cytokines exert their effect on drug-metabolizing enzymes at the transcriptional level, work by Morgan's group has provided evidence that post-transcriptional mechanisms can be involved as well.[1] For instance, in human and rat hepatocytes, the presence of either LPS or IL-1 has been shown to induce nitric oxide (NO) synthase, resulting in elevated NO levels, which can lead to CYP2B protein degradation.[23,37] Lee and coworkers demonstrated that in rat hepatocytes IL-1 can stimulate NO-dependent CYP2B1 degradation via increased ubiquitination and subsequent proteasome-mediated protein degradation.[37] Donato and coworkers demonstrated that in human primary hepatocytes within 24 h of IFN-γ treatment (300 U/mL), an increase in NO release (200% over untreated control) was observed in parallel with the inhibition of CYP1A2 activity and protein levels.[20] Similarly, after the addition of *S*-nitroso-*N*-acetylpenicillamine (SNAP), which spontaneously generates NO in the cell culture medium, the concentration of nitrite increased and a corresponding concentration-dependent inhibition of CYP1A2 activity was observed; the highest inhibition (70% compared to untreated cells) reached at 2 mM SNAP. Conversely, the addition of up to 1 mM of N^G-monomethyl-L-arginine, a competitive inhibitor of NO biosynthesis, lead to reduction in IFN-γ-induced NO formation, and partial reversal of IFN-γ inhibition of basal and 3-methylcholanthrene-induced CYP1A2 activity.

5.3 *IN VITRO* ASSAYS

The present landscape of available *in vitro* methodologies for examining the interactions between cytokines and cytokine modulators and the array of metabolic

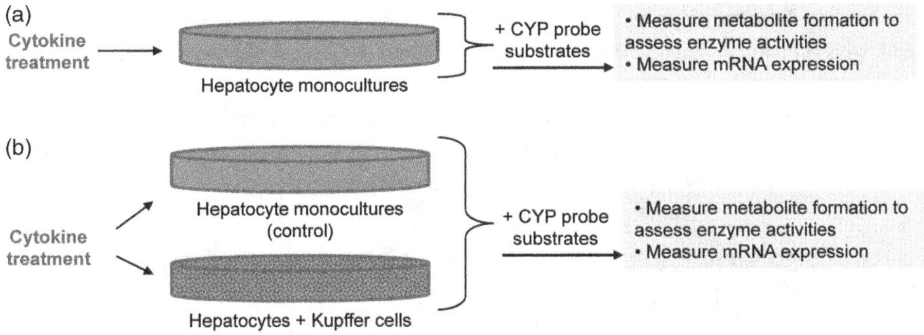

FIGURE 5-4 General experimental design of *in vitro assays*. Hepatocyte monocultures (a) and hepatocyte-Kupffer cell co-cultures (b) are exposed to cytokines of interest for 24 to 96 h. The cultures are then incubated with CYP probe substrates, followed by measurement of metabolite formation and mRNA levels to assess enzyme activities and gene expression. (See the color version of this figure in Color Plates section.)

enzymes and transporters is limited. Current practices predominantly involve the use of primary hepatocyte cultures because they are the most prevalent model employed owing to their relatively easy access and commercial availability from various preclinical species as well as from humans. In addition, plateable cryopreserved hepatocytes are commercially available, which have the advantage of reduced interassay variability. This allows an easier comparison of the effect of different test molecules than in hepatocytes freshly isolated from different donors. A general experimental design for assessing the effects of cytokines on CYP enzymes in a hepatocyte monoculture *in vitro* system is illustrated in Figure 5-4a. Hepatocyte monocultures are cultivated and then incubated with cytokines of interest for the desired period (~24–96 h). The monocultures are then exposed to CYP probe substrates, followed by measurement of metabolite formation and mRNA levels to assess enzyme activities and gene expression.

A limitation of the use of hepatocyte monocultures in studying the effect of cytokines is that typically one cytokine is studied at a time, whereas *in vivo* the liver is exposed to a wide range of immunomodulatory molecules. The question therefore is warranted whether data generated in hepatocytes are representative of *in vivo* conditions or, more important, of disease states in which the concentration of individual cytokines can be grossly altered. As a strategy to overcome this limitation, co-cultures of hepatocytes with either purified Kupffer cells[13] or nonparenchymal cell (NPC) mixtures containing immature hepatocytes, Kupffer cells, biliary cells, stellate cells, and hepatic stem cells are currently being explored.[22] These co-culture systems may offer an advantage over primary hepatocytes by providing a physiological environment that is more closely representative of an *in vivo* scenario.

A general experimental design for assessing the effects of cytokines on CYP enzymes in a hepatocyte–Kupffer cell co-culture *in vitro* system is illustrated in Figure 5-4b. Hepatocytes are co-cultured with Kupffer cells (or NPC) and then incubated with cytokines of interest for the desired period (~24–96 h).

The co-cultures are then exposed to CYP probe substrates followed by measurement of metabolite formation and mRNA levels to assess enzyme activities and gene expression.

The limited availability of *in vitro* tools for studying the impact of cytokines on liver enzymes and transporters highlights a need for more sophisticated assay systems and advocates for exploring new technologies.

5.4 EFFECTS OF CYTOKINES ON METABOLIZING ENZYMES AND TRANSPORTERS

5.4.1 IL-1 Investigations

5.4.1.1 Effects of IL-1 on Cytochrome P450 Enzymes As summarized in Table 5-1, the interaction of IL-1β on CYP enzymes has been widely observed. Barker and coworkers showed in isolated rat hepatocytes that IL-1β (1–4 U/mL, 24 h) substantially reduced 2,3,7,8-tetrachlorodibenzo-*p*-dioxin-induced CYP1A1 and CYP1A2 mRNA expression by approximately 65% and 95%, respectively.[9] The inhibitory effects were dose dependent, with IC_{50} at 2 and 0.5 U/mL for CYP1A1 and CYP1A2, respectively.

Similar observations of CYP enzyme suppression were made in human primary hepatocytes by numerous investigators. Abdel-Razzak and coworkers reported a reduction in activity of CYP1A2 (12–89%) by IL-1β (100 U/mL, 72 h).[10] The interaction was believed to be at the transcriptional level owing to a corresponding decrease in CYP1A2 mRNA (60–93%). IL-1β (100 U/mL, 24 h) also showed suppression of both basal levels and phenobarbital- and bilirubin-mediated induction of CYP2B6 mRNA.[11] Aitken and Morgan were unable to detect similar suppression of CYP2B6 mRNA in their experimental studies with IL-1β (5 ng/mL, 24 and 40 h) but did observe reduced CYP2B6 protein levels and a corresponding increased NO production.[12] Collectively, these results suggest that the IL-1β interaction with CYP2B6 may be mediated by various mechanisms, including alteration of gene transcription, inhibition of protein translation, and/or increased enzyme degradation associated with elevated NO production.

In terms of the CYP2C family, CYP2C8 and CYP2C9 mRNA expression was shown to be downregulated by IL-1β, and a significant reduction in CYP2C9 protein levels was also observed.[11,12] Similar to CYP2B6, a corresponding increase in NO production was detected in parallel with a reduction in CYP2C9 protein after IL-1β exposure.[12] This was not seen for CYP2C8 mRNA inhibition, however. These results suggest that IL-1β may interact with CYP2C8 at the level of gene transcription, whereas it may regulate CYP2C9 through both gene transcription and protein expression/degradation. In contrast to CYP2C8 and CYP2C9, CYP2C18 and CYP2C19 mRNA levels were unaffected by IL-1β.[12]

Analogous to the other CYP enzyme families, IL-1β also negatively affected CYP3A4 mRNA expression, protein levels, and enzyme activity. Abdel-Razzak's group reported that IL-1β (100 U/mL, 72 h) produced significant reduction in gene expression (60–93%) and CYP3A4 activity (30–90%).[10] Similarly, Assenat and coworkers reported IL-1β suppression of both basal and phenobarbital- and bilirubin-mediated induction of CYP3A4 mRNA expression (100 U/mL, 24 h).[11] Aitken and Morgan also demonstrated significant downregulation of CYP3A4 mRNA expression (>95%) accompanied with a corresponding decrease in CYP3A4 protein after IL-1β exposure (5 ng/mL, 24 h and 40 h).[12] Thus IL-1β regulation of CYP3A4 appears to be most likely at the transcriptional level.

Assenat and coworkers hypothesized that the mechanism by which IL-1β interacts with CYP3A4 and other CYP enzymes at the level of gene transcription may potentially be through the downregulation of CAR expression.[11] This was supported by demonstrating that IL-1β activated NF-κB, which led to the downregulation of glucocorticoid receptor (GR) transactivation of the CAR promoter, resulting in suppressed CAR mRNA expression and decreased CAR-mediated gene transcription. Another potential regulatory mechanism associated with IL-1β may involve the downregulation of RXRα and PXR mRNA, leading to altered gene expression of CYP enzyme.[38]

Post-transcriptionally, NO-induced enzyme degradation may also be involved in the regulation of the CYP3A family.[37] Although mRNA levels were unaffected by IL-1β in primary rat hepatocytes, Lee and coworkers detected a 40% downregulation of CYP3A1 protein.[37] This inhibition in protein expression was reversed by NO synthase inhibitors.

5.4.1.2 Effects of IL-1 on Phase II Metabolizing Enzymes Although not as extensively studied, various phase II metabolic enzymes have also been shown to also be affected by proinflammatory cytokines (Table 5-2). Abdel-Razzak and coworkers reported that in human primary hepatocytes IL-1β was a potent suppressor of various enzymes including epoxide hydrolase.[10] Assenat's group showed that IL-1β (100 U/mL, 24 h) suppressed both basal expressions and phenobarbital- and bilirubin-induced UGT1A1, GSTA1, and GSTA2 mRNA.[11] The authors suggest that similar to CYP3A4, IL-1β regulation of phase II metabolic enzymes may also involve the activation of NF-κB, leading to inhibition of GR-mediated CAR mRNA expression.

In rat primary hepatocytes, IL-1β (20–300 U/mL, 8–48 h) produces a marked decrease in basal mRNA levels of rGSTA2 and rGSTM1, with a corresponding decrease in rGSTM1 protein expression (~44%) and activity (47–78%).[39] In addition, IL-1β also suppressed 3-methylcholanthrene-, phenobarbital-, and/or oltipraz-induced expression of rGSTA2 and rGSTM1 mRNAs and proteins. Associated with these findings a shorter mRNA half-life was observed, suggesting the effects of IL-1β on GST were the result of mRNA destabilization rather than reduced gene transcription. In contrast to rGSTA2 and rGSTM1, IL-1β exposure led to increased rGSTP1 mRNA levels and had no effect on the chemically induced expression of both rGSTP1 mRNA and protein.

5.4.1.3 Effects of IL-1 on Drug Transporters In addition to metabolizing enzymes, drug transporters are also affected by proinflammatory cytokines (Table 5-3). Sukhai and coworkers observed that IL-1β (0–50 ng/mL, 72 h) had no

TABLE 5-2 Effects of Cytokines on Phase II Metabolic Enzymes

	Phase II Enzymes									
Cytokines	GSTA1	GSTA2	UGT1A1	UGT2B7	SULT1A1	SULT2A1	CES2	Epoxide hydrolase	Model	Reference
IL-1β								↓r	Human hepatocytes	10
		↓a,p,r (rGSTA2, rGSTM1)							Rat hepatocytes	39
	↓r	↓r	↓r						Human hepatocytes	11
IL-2	↔a								Rat hepatocytes	39
IL-4								↔r	Human hepatocytes	10
	↔a								Rat hepatocytes	39
	↓r (GSTA1/A2)								Human hepatocytes	40
IL-6								↓r	Human hepatocytes	10
IL-6	↔a								Rat hepatocytes	39
	↔r (GSTA1/A2)								Human hepatocytes	40
TNF-α								↓r	Human hepatocytes	10
IFN-γ								↔r	Human hepatocytes	10
IFN-γ/α	↔a								Rat hepatocytes	39
IFN-α2b				↓p,r	↓r	↔r	↔r		Human hepatocytes-NPC co-culture	22

Downregulation (↓) or no effect (↔) on activity (a), protein levels (p), and/or mRNA (r) expression; CES: carboxylesterase; GST: glutathione *s*-transferase; IL: interleukin; IFN: interferon; NPC: nonparenchymal cells; SULT: sulfotransferase; TNF: tumor necrosis factor; UGT: UDP-glucuronosyltransferase..

effect on Mdr1a or Mdr1b mRNA in primary rat hepatocytes, but P-glycoprotein (P-gp) protein expression was significantly reduced (32–75%), which also corresponded with a reduction in efflux activity.[42] Because no effects were observed on Mdr1 gene expression, the effect of IL-1β on P-gp is potentially mediated through either protein synthesis and/or protein stability in rat hepatocytes. Diao's group detected a significant decrease in levels of MRP2 protein in sandwich-cultured (SC) rat hepatocytes but not mRNA in the presence of IL-1β (1 ng/mL, 24 and 48 h).[44] They also saw downregulation of both Bsep gene expression (Abcb11 mRNA) and protein levels in the presence of IL-1β.

General IL-1β suppression of transporters was also observed in human primary hepatocytes and HepaRG cells by Le Vee and coworkers whereby mRNA expression, protein levels, and transporter activities were reduced for both organic anion transporting polypeptide (OATP) 1B1 and sodium-taurocholate co-transporting polypeptide (NTCP).[43] Analogous observations were also made by Assenat's group: IL-1β (100 U/mL, 24 h) suppressed both basal and phenobarbital- and bilirubin-induced expression of OATP1B1 mRNA expression.[11] Besides OATP1B1 and NTCP, downregulation of gene expression was also detected for other uptake transporters such as OATP1B3 and OATP2B1 and efflux pumps such as multidrug resistance protein (MRP) 2, MRP3, and MRP4 and breast cancer resistance protein (BCRP).[43] The reduction in MRP2 gene expression was similarly confirmed by Diao and coworkers as was reduced protein levels after IL-1β exposure in SC human hepatocytes.[44] It is interesting that IL-1β also diminished BSEP gene expression, but protein levels were significantly induced.[44] This finding was contradictory to the reduction in BSEP protein expression observed in HepaRG cells by Le Vee and colleagues.[43] The discrepancy highlights one of the challenges of the *in vitro* system; interlab variability can limit data reproducibility and limit result interpretation.

5.4.2 IL-2 Investigations

5.4.2.1 Effects of IL-2 on Cytochrome P450 Enzymes Tinel and coworkers reported exposure to IL-2 in rat hepatocytes caused a decrease in mRNA levels of CYP3A2 and CYP2C11 and a corresponding reduction in protein expression of CYP1A1 (35%), CYP2B1/2 (35%), CYP2C11 (36%), CYP2D1 (26%), and CYP3A (56%).[14] The interaction of IL-2 on the various CYP enzymes may primarily be at the transcriptional level.

In the clinic, IL-2 has been observed to suppress both activity and protein levels of CYP3A.[51,52] *In vitro*, using cultured primary human hepatocytes, the inhibitory effect was transient with suppression observed at 48 h but returned to baseline at 72 h.[13] However, when Kupffer cells were co-cultured with hepatocytes at physiologically relevant ratios of 10:1 and 10:4 (hepatocytes to Kupffer cells), representing "normal" and "inflamed" states, respectively, IL-2 showed prolonged suppression of CYP3A activity (50–70%) out to 72 h in the co-cultures. Results suggest that the co-culture system may be more physiological representative and thus can offer an advantage for studying the effects of

TABLE 5-3 Effects of Cytokines on Transporters

	Transporters						
Cytokines	OAT2	OCT1	OATP1B1	OATP1B3	OATP2B1	NTCP	MRP2/ Mrp2
IL-1β			↓r ↓a,p,r	↓r	↓r	↓a,p,r	↓r ↔r↓p ↓p,r
IL-2							
IL-6	↓r	↓a,r	↓a,p,r (OATP)	↓r	↓r	↓a,p,r	↓p,r ↓p,r ↓p,r
Oncostatin	↓r	↓r	↓a,p,r (OATP)	↓r	↓p,r	↓a,p,r	↓r↔p
TNF-α	↓r	↓a,r	↓a,p,r	↓a,r	↓a,r	↓a,p,r	↔p,r ↓p,r ↓p,r
IFN-γ	↔r	↔r	↓r	↓r	↓r	↓r	↓r
IFN-α2b	↓p,r	↔r	↔r	↔r	↔r		↑r

Upregulation (↑), downregulation (↓), or no effect (↔) on activity (a), protein levels (p), and/or mRNA (r) expression; BCRP: breast cancer resistance protein; BSEP: bile salt export pump; IL: interleukin; IFN: interferon; MDR1 P-gp: multi-drug resistant 1 P-glycoprotein; MRP: multidrug resistance protein; NPC: nonparenchymal cells; NTCP: sodium-taurocholate co-transporting polypeptide; OAT: organic anion transporter; OATP: organic anion transporting polypeptide; OCT: organic cation transporter; SC: sandwich cultured; TNF: tumor necrosis factor.

MRP3	MRP4	MDR1 P-gp/ Mdr1a P-gp/ Mdr1b P-gp	BCRP	BSEP/ Bsep	Model	References
		↓r			Human Caco-2	41
		↔r↓a,p			Rat hepatocytes	42
					Human hepatocytes	11
↓r	↓r		↓p	↓p	Human hepatocytes and hepatoma HepaRG cells	43
				↓p,r	SC rat hepatocytes	44
				↓r↑p	SC human hepatocytes	44
		↓a,r			Caco-2 (human)	45
		↓a,p,r			Rat hepatocytes	42
↔r↑p	↓r	↓r↔p	↓p,r	↔r	Human hepatocytes	46
				↓r↑p	SC human hepatocytes	44
				↑r↔p	SC rat hepatocytes	44
↔p,r	↔r	↔p,r	↓p,r	↔r	Human hepatocytes and hepatoma HepaRG	47
		↑a,p,r			Rat hepatocytes	48
		↑r			Rat hepatocytes and hepatoma H-4-II-E cells	34
		↔r			Human Caco-2	41
		↓a,r↔p			Human Caco-2	49
↔r↑p	↔r	↔p,r	↔r↑p	↓r	Human hepatocytes	46
				↓r↔p	SC human hepatocytes	44
				↓p,r	SC rat hepatocytes	44
		↓r			Human Caco-2	41
		↑p,r↔a			Human Caco-2	49
↓r		↓r	↓r	↓r	Human hepatocytes	50
		↔r	↔r		Human hepatocytes-NPC co-culture	22

cytokines, whereas hepatocytes alone are insufficient. The authors proposed that IL-2 effects on CYP3A may be through multiple mechanisms, including: direct binding to the hepatocytes IL-2R to produce a transient suppression of CYP3A; indirectly stimulating Kupffer cells to secret other proinflammatory cytokines, such as IL-1, IL-6, and TNF-α;[52] and indirectly stimulating Kupffer cells to release other proinflammatory mediators, such as prostaglandins, prostacyclins, and eicosanoids, which have been shown to regulate CYP activity. Further work is needed to investigate these hypotheses and the applicability of the hepatocyte-Kupffer cell co-culture system in general.

5.4.2.2 Effects of IL-2 on Drug Transporters In Caco-2 cells, IL-2 significantly reduced P-gp transport of rhodamine-123 (21%) (100–5000 IU/mL, 24–72 h) and strongly suppressed MDR1 mRNA levels (1000 IU/mL, 72 h).[45] Currently, the mechanism by which IL-2 interacts with P-gp is unknown.

5.4.3 IL-4 Investigations

Limited ADME-related work has been reported for cytokine IL-4. Abdel-Razzak and colleagues noted variable effects of IL-4 (150 U/mL, 72 h) on CYP enzymes in human primary hepatocytes, in contrast to the usual suppressive effects observed with other proinflammatory cytokines. In their investigation, IL-4 appeared to have no effect on CYP1A2 and CYP2C gene expression, whereas CYP2E1 mRNA levels were significantly induced by five-fold.[10] Effects of IL-4 on CYP3A mRNA expression and activity were also not observed.

5.4.4 IL-6 Investigations

5.4.4.1 Effects of IL-6 on Cytochrome P450 Enzymes As one of the major proinflammatory cytokines, IL-6 effects on CYP enzymes have been widely reported. Initially, in contrast to results from IL-1β exposure, early work by Barker and coworkers did not show an effect of IL-6 on 2,3,7,8-tetrachlorodibenzo-*p*-dioxin-induced CYP1A mRNA expression in rat hepatocytes.[9] However, later investigations in human primary hepatocytes produced consistent inhibitory effects by IL-6 on various CYP enzymes. The discrepancy highlights inherent interlab variability owing to a lack of standardization of experimental design and/or species evaluated, which is a limiting factor of the *in vitro* system. IL-6 suppression of CYP1A2 activity (18–90%) has been reported by Abdel-Razzak's group (50 U/mL, 72 h),[10] Guillen's group (100 U/ml, 48 h),[15] and Dickmann's group (0.005–50 ng/mL, 24–96 h).[18] In parallel with decreased enzyme activity, similar inhibition of CYP1A2 mRNA levels (25–%)[18] and protein expression were also detected.[15] It is interesting that Dickmann and coworkers demonstrated that the incorporation of an IL-6 neutralizing mAb can completely block IL-6 suppression of CYP1A2 activity.[18]

In regard to the mechanism of interaction, Guillen and coworkers noted that measured NO levels were unchanged in cytokine-treated and untreated hepatocytes, suggesting that IL-6 interaction with CYP1A2 may not be attributed to enzyme degradation associated with NO biosynthesis.[15] Abdel-Razzak and coworkers suggested that the presence of an IL-6-response element in the 5′ flanking regions of the CYP1A1/1A2 genes may play a role in the transcriptional regulation of CYP genes by IL-6.[10]

In addition to CYP1A2, IL-6 also had significant impact on CYP2B6 enzyme. Guillen and coworkers reported a 48% reduction in CYP2B6 activity by IL-6 (100 U/mL, 48 h),[15] which was consistent with results from Aitken and Morgan, who found IL-6 exposure (10 ng/mL, 24 h and 40 h) in human primary hepatocytes led to substantial suppression of both CYP2B6 mRNA (80%) and protein levels (25–50%).[12] Pascussi and colleagues also confirmed IL-6- mediated (100 U/mL, 24 h) inhibition of both CYP2B6 basal mRNA levels (60%) as well as phenobarbital- and rifampicin-induced mRNA levels (80–90%).[16] Guillen's group proposed that, similar to the interaction with CYP1A2, the effect of IL-6 on CYP2B6 may not be through protein degradation but most likely mediated by transcriptional regulation.[15] This was supported by Pascussi's group, who reported rapid and marked decrease in PXR (~80%) and CAR (~60%) mRNA expression in the presence of IL-6 (100 U/mL, 6 h).[16] These results suggest that the IL-6-mediated suppression of both PXR and CAR could potentially be the regulatory mechanism associated with the effects of IL-6 on rifampicin- and phenobarbital-induced CYP2B6 expression.

Consistent with the other CYP enzymes, IL-6 inhibition of the CYP2C family has also been observed. The mRNA levels of CYP2C8, CYP2C9, and CYP2C19 in general were decreased (25–80%) in the presence of IL-6.[10,12,16,18] The suppression of rifampicin- and phenobarbital-induced CYP2C8 and CYP2C9 mRNA expression (80–90%) was also observed after IL-6 exposure.[16] In addition, similar changes in CYP2C9 protein levels (25–50% suppression) were reported by Aitken and Morgan.[12] In contrast to the general inhibitory effects on the CYP2C family, IL-6 exposure did not affect CYP2C18, perhaps owing to the low basal levels of expression of this enzyme in the liver.[12]

The effects of IL-6 on the CYP3A family have also been extensively studied. Suppression of CYP3A4 activity (20–80%), mRNA expression (25–90%), and protein levels (25–50%) have been reported by various labs.[10,12,13,15,18] Dickmann and coworkers also observed that, similar to CYP1A2, the incubation with an IL-6 neutralizing mAb can partially prevent IL-6 suppression of CYP3A4 activity.[18] In addition to basal CYP3A4 levels, Pascussi's group[16] and Dickmann's group[18] also reported that IL-6 exposure can lead to decreased rifampicin- and phenobarbital-induced CYP3A4 gene expression. However, this was observed only at supraphysiological concentrations. Guillen and coworkers observed no changes in NO levels with IL-6 treatment that could cause enzyme degradation.[15] Pascussi and coworkers proposed that, similar to CYP2B6, IL-6-mediated suppression of both PXR and CAR genes could potentially be the regulatory mechanism by which IL-6 mediates the inhibition of rifampicin- and phenobarbital-induced CYP3A4 expression.[16]

5.4.4.2 Effects of IL-6 on Phase II Metabolizing Enzymes

For phase II metabolic enzymes, IL-6 had inhibitory effects on epoxide hydrolase mRNA expression (>40%)[10] but no interaction with glutathione *s*-transferase (GST) enzymes[40] when studied in human primary hepatocytes.

5.4.4.3 Effects of IL-6 on Drug Transporters

Unlike IL-1β, IL-6 (0–50 ng/mL, 24 h) treatment of rat hepatocytes significantly downregulated Mdr1b mRNA expression, suppressed P-gp protein levels (20–38%), and reduced P-gp-mediated rhodamine-123 transport.[42] It is believed that transcriptional modulation of the Mdr1 gene may be responsible for IL-6 effects on P-gp protein and activity. Mechanisms of interaction were not investigated, but the authors proposed that owing to similarities shared between CYP and Mdr1 gene families in regard to substrates, inducers, and inhibitors there could be parallelism in regulatory mechanisms of IL-6 on Mdr genes.[53]

In addition to P-gp, Le Vee and coworkers observed that IL-6 (0.1–10 ng/mL, 8–48 h) treatment of primary human hepatocytes suppressed mRNA levels of major influx and efflux transporters, including NTCP, OATP1B1, OATP1B3, OATP2B1, OCT1, OAT2, MRP2, MRP4, and BCRP.[46] The effects on these transporters were mostly concentration dependent (NTCP, OATP1B1, OATP1B3, OATP2B1, OCT1, and BCRP) and time dependent (MDR1). On the other hand, both BSEP and MRP3 gene expression were unaffected by IL-6. In regard to protein levels, IL-6 (10 ng/mL, 48 h) did not alter P-gp but did reduce NTCP, OATP1B1, BCRP, and MRP2, and induced MRP3. Consistent with results of mRNA and protein expression, NCTP, OATPs, and OCT1 activities were also significantly reduced in human primary hepatocytes. Although the mechanisms of the effects were unexplored, modulation of gene transcription leading to suppressed protein and activity levels may explain how IL-6 interacts with NCTP, OATP, and OCT1.

Similar to Le Vee and coworkers, Diao's group observed significant IL-6 (100 ng/mL, 24–48 h) downregulation of both mRNA and protein levels of MRP2/Mrp2 in SC human and rat hepatocytes.[44] On the other hand, variable results were reported with BSEP in SC human hepatocytes: IL-6 suppressed mRNA transcription while inducing protein levels. In SC rat hepatocytes, however, *Bsep* gene expression was induced but no effects were seen on protein levels. Collectively, results demonstrate a differential effect of BSEP/Bsep expression between SC human and rat hepatocytes most likely attributed to species-dependent signaling pathways.

The mechanism of IL-6 regulation was uninvestigated, but it is believed to involve both transcriptional mechanisms associated with the phosphorylation of extracellular-reactive kinase (ERK), Jun kinase (JNK), and p38 mitogen-activated protein kinase (MAPK) as well as post-transcriptional mechanisms.[26,43,54]

5.4.5 Oncostatin

5.4.5.1 Effects of Oncostatin on Cytochrome P450 Enzymes

Oncostatin M (OSM) is a multifunctional cytokine that is related to IL-6. Both have overlapping biological effects in many cellular systems owing to similarity in structure, function,[55,56] and signal transduction pathway activation.[57–59] OSM acts on a wide variety of cells: It regulates growth of certain tumor and nontumor-derived cell lines,[60,61] induces cell differentiation,[55,62] and upregulates low-density lipoprotein in liver cells[63] and acute phase protein in hepatocytes (i.e., α1-antichymotrypsin, haptoglobin, and α1-acid glycoprotein).[64]

Guillen and coworkers demonstrated that in human primary hepatocytes exposure to OSM (25–50 U/mL) produced maximal reductions in CYP1A2 (~55%) and CYP3A4 (~30%) activity after 48 h of treatment.[15] Comparable decreases were also observed for CYP2A6 (30%) and CYP2B6 (62%). The inhibitory effects of OSM on CYP enzymes were also detected after induction of both CYP1A2 and CYP3A4 by 3-methylcholanthrene (MC) and dexamethasone (DEX), respectively. MC-induced CYP1A2 showed a sixfold increase in activity compared to an uninduced control. In the presence of OSM (25 U/mL), only 43% of the induced CYP1A2 activity was detected. Similarly, DEX-induced CYP3A4 activity was increased twofold, and in the presence of OSM 47% of induced CYP3A4 activity was observed.

Reported OSM suppression of CYP activities was consistent with observed protein levels. The presence of OSM in increasing concentration (up to 50 U/mL) resulted in decreased protein synthesis of both CYP1A2 and CYP3A4. After MC- and DEX-induction of CYP1A2 and CYP3A4, respectively, protein levels were increased by ~60% (CYP1A2) and ~100% (CYP3A4) of uninduced control. Introducing OSM after induction inhibited the synthesis of both CYP1A2 and CYP3A4 enzymes to levels that were similar to uninduced cells. OSM presence did not alter NO synthesis; therefore, its effect on CYP protein levels was not attributed to NO.

Parallel to activities and protein synthesis, significant reduction of CYP1A2 and CYP3A4 mRNA expression occurs in the presence of OSM (50 U/mL).[15] This suggests that OSM regulation of CYP enzymes may occur at the transcriptional level, perhaps by similar mechanisms as the related cytokine IL-6, whereby PXR and CAR mRNA levels were altered.[16]

5.4.5.2 Effects of Oncostatin on Drug Transporters In addition to CYP enzymes, Le Vee and coworkers also evaluated the effects of OSM on sinusoidal and canalicular drug transporters in human primary hepatocytes.[47] Exposure to OSM (0.01–20 ng/mL, 8–48 h) produced no effects on P-gp, MRP3, MRP4, and BSEP mRNA expression. In contrast, OSM exposure led to the downregulation of mRNA levels of major sinusoidal solute carrier (SLC) influx transporters (NTCP, OATP1B1, OATP1B3, OATP2B1, OCT1, and OAT2) as well as ATP-binding cassette (ABC) efflux transporters (MRP2 and BCRP). The OSM EC_{50} of mRNA expression ranged between 0.12 and 1.16 ng/mL in human primary hepatocytes. Correspondingly, the protein expression of NTCP, OATP1B1, OATP2B1, and BCRP along with transport activities of NTCP and OATP were also inhibited after OSM exposure.

Rank correlation analysis of suppressive effects of OSM toward drug transporters showed that OSM effects were highly correlated with those of IL-6 and, in a more moderate manner, with those of IL-1β. By contrast, they were not correlated with those of TNF-α and IFN-γ. Type II OSM receptor has been implicated as a mediator of OSM transcriptional regulation of these various transporters. Knocked-down expression of the type II OSM receptor gp130 and OSMRβ subunits with siRNA transfection has counteracted the downregulation of NTCP, OATP1B1, and OATP2B1 mRNA expression. Le Vee and coworkers hypothesized that the signaling pathway that is activated by this receptor, especially by its signal-transducing component gp130 subunit, may play a major role in OSM-related changes in transporter expression.[47] This is further supported by the fact that altered transporter expression owing to OSM treatment was highly correlated with inhibitory effects triggered by IL-6 and that the IL-6 receptor also uses the gp130 subunit for signal transduction.[65] OSM and IL-6 are known to activate several distinct signaling pathways through the gp130 subunit, including the Janus kinase/signal transducer and activator of transcription cascade, the MAPK pathway, and the phosphatidylinositol 3-kinase pathway.[66] In agreement with these data, the MAPK ERK was found to be activated by OSM.[47]

5.4.6 TNF-α Investigations

5.4.6.1 Effects of TNF-α on Cytochrome P450 Enzymes Abdel-Razzak and coworkers observed substantial inhibition of CYP1A2 (30–85%) and 3A activities (24–90%) by TNF-α (50 U/mL, 72 h) in human primary hepatocytes.[10] Analogous reduction in mRNA expression of these enzymes (30–80%) was also detected, indicating a probable transcriptional component to TNF-α effects on these enzymes.[10,12]

Similar to CYP1A2 and CYP3A, gene expression of CYP2C and CYP2E1 was also negatively affected by the presence of TNF-α, with 30–80% inhibition observed.[10] On the other hand, Aitken and Morgan observed only minimal interactions of TNF-α (10 ng/mL, 24 h and 40 h) with mRNA expression of CYP2B6, CYP2C9, CYP2C18, and CYP2C19 in human primary hepatocytes.[12] Despite minor effects on gene expression, CYP2B6 and CYP2C9 protein levels were inhibited, suggesting that TNF-α may interact with these enzymes through modification of protein translation and/or enzyme degradation. The differential effects of TNF-α on various CYP enzymes highlights potentially different regulatory mechanisms involved. TNF-α may regulate CYP3A4 at the transcriptional level, possibly via NF-κB interference with the PXR-RXR complex as suggested by Gu and coworkers (Figure 5-3b) and, in contrast, affect the translation and/or stability of either CYP2B6 or CYP2C9 enzymes.[19]

5.4.6.2 Effects of TNF-α on Phase II Metabolizing Enzymes Limited studies have been dedicated to evaluating the effects of TNF-α on the phase II metabolic enzymes. Abdel-Razzak's group has observed, however, a decrease of epoxide hydrolase mRNA levels in human primary hepatocytes.[10]

5.4.6.3 Effects of TNF-α on Drug Transporters Several studies have increased the understanding of the interaction between TNF-α and the various drug transporters. In rat primary hepatocytes, Hirsch-Ernst and coworkers observed that

TNF-α (4000 U/ml, 72 h) exposure led to a twofold induction of Mdr1b mRNA and P-gp protein and a corresponding decrease in intracellular accumulation of the P-gp substrate rhodamine-123.[48] It was postulated that the induction of Mdr1b mRNA may have been due to elevated oxygen species associated with the presence of TNF-α. Reactive oxygen species have been implicated as a mediator of various TNF-α effects, including activation of NF-κB,[67] stimulation of gene transcription,[68] and alteration of protein phosphorylation.[67] After treating rat primary hepatocytes with antioxidants, Hirsch-Ernst and coworkers detected marked suppression of TNF-α-induced Mdr1b Pgp mRNA and protein levels.[48]

Inhibition of Mdr1 mRNA (56%) was measured in human Caco-2 cells after treatment with TNF-α (1–10 ng/mL, 24–96 h) by Belliard's group.[49] In contrast to the results from rat hepatocytes, however, the suppressed gene expression did not translate to a decreased P-gp protein level but did decrease (20%) unidirectional transport of rhodamine-123. The discrepancy in TNF-α induction/suppression of Mdr1 may be attributed to differences in the experimental models (rat hepatocytes vs. human carcinoma cell lines) and/or differences in regulatory mechanisms in rat and human.

Besides P-gp, TNF-α interaction with MRP2 and BSEP were assessed by Diao and colleagues in both SC human and rat hepatocyte models.[44] After TNF-α treatment (100 ng/mL, 24 and 48 h), significant suppression of both mRNA and protein levels were measured for MRP2/Mrp2 in SC human and rat hepatocytes. On the other hand, variable results were detected with BSEP in SC human hepatocytes as TNF-α-suppressed mRNA transcription but showed no effects on protein. In SC rat hepatocytes, however, both Bsep mRNA and protein levels were reduced by TNF-α. Similar to IL-6, results demonstrate a differential effect of BSEP/Bsep expression between SC human and rat hepatocytes that may be attributed to species-dependent signal transduction pathways. Similar to IL-1β and IL-6, regulation by TNF-α may involve transcriptional mechanism associated with the activation of ERK, JNK, and p38 MAPK phosphorylation.[26]

Le Vee's group observed that in human primary hepatocytes incubation with TNF-α (0.01–100 ng/mL, 8–48 h) led to significant reduction of NTCP, OATP2B1, OATP1B1, OATP1B3, OCT1, OAT2, and BSEP mRNA, whereas mRNA levels of MDR1, MRP2, MRP3, MRP4, and BCRP were unchanged.[46] The effects were mostly concentration dependent (0.01–100 ng/mL), with marked inhibition of gene expression at ~1 ng/mL TNF-α. In regard to protein expression, TNF-α incubation did not alter MRP2 and P-gp protein levels (corresponding to unaltered gene expression), but NTCP and OATP1B1 levels were reduced and BCRP and MRP3 were induced. Consistent with mRNA and protein expression results, NCTP, OATP, and OCT1 activities were also significantly decreased in human primary hepatocytes. For these transporters, although the mechanism of interaction was unexplored, experimental outcomes suggest that regulation by TNF-α may be at the transcriptional level.

5.4.7 IFN Investigations

5.4.7.1 Effects of IFN-γ on Cytochrome P450 Enzymes The interaction of IFN with CYP1A2 has been extensively observed in human primary hepatocytes.

In general, downregulation of CYP1A2 basal activity (29–53%) and 3-methylcholanthrene-induced activity (32%) was observed in the presence of IFN-γ (50–300 U/mL, 24–72 h).[10,15,20] The decrease in activity was consistent with lowered mRNA (up to ~40%)[10]and diminished basal and 3-methylcholanthrene-induced CYP1A2 protein expression (72%).[15,20] These results suggest that the interaction of IFN-γ with CYP1A2 is most likely mediated through gene transcription. However, post-transcriptional regulation may also play a contributing factor because IFN-γ has previously been shown to suppress CYP1A2 activity in parallel with markedly elevated NO production.[20] Besides IFN-γ, IFN-α2b has also been shown to produce analogous inhibition of CYP1A2. Similarly, IFN-α2b can also interact with the interferon receptor to influence the upregulation or downregulation of various genes in the liver.[69]

Similar to CYP1A2, suppression of CYP2A6 and CYP2B6 activities were also observed in the presence of IFN-γ (300 U/mL, 24 h).[15,20] This decrease was in parallel to elevated NO production, which was partially reversed by the presence of an inhibitor of NO biosynthesis (N^G-monomethyl-L-arginine). Associated with the activity suppression, Aitken and Morgan also saw significant inhibition of CYP2B6 mRNA (80%) and protein levels (~65%) after exposure to IFN (10 ng/mL, 24 h).[12] It appears that IFN may regulate CYP2A6 and CYP2B6 by both gene transcription (as evident by changes in mRNA levels) and enzyme stability (influenced by increased NO biosynthesis).

Like other CYP enzymes, IFN downregulated CYP3A activity (21–55%) but had no effect on gene expression after incubation with IFN-γ (50 U/mL, 72 h).[10] On the other hand, IFN-γ at 300 U/mL for 24–48 h elicited suppression of both basal and DEX-induced CYP3A4 activity (20–34%), along with reduced mRNA ($\geq$ 80%) and protein levels (~ 50%).[12,15,20] Discrepancy in gene expression results may be attributed to differences in donors and/or experimental conditions. The noted changes in mRNA expression suggest a transcriptional component in IFN regulation of CYP3A4. In addition, Donato and coworkers detected elevated NO biosynthesis in parallel with CYP3A4 inhibition and a partial reversal of the inhibition after incorporation of N^G-monomethyl-L-arginine to culture media, similar to CYP1A2.[20] These findings suggest that enzyme stability may have a minor role in IFN regulation of the CYP3A family.

In contrast to the other CYP enzymes, IFN in general does not appear to have much interaction with the CYP2C family. With the exception of CYP2C8, for which significant mRNA reduction was detected, no effects were observed on the gene expression of CYP2C9, CYP2C18, and CYP2C19 after IFN-γ (10 ng/mL, 24 and 40 h) incubation in primary human hepatocytes.[10,12]

Chen's group assessed the impact of IFN-α2b on the expression of various drug-metabolizing enzymes in freshly prepared human primary hepatocytes co-cultured with nonparenchymal cells (which included a mixture of immature hepatocytes, Kupffer cells, biliary cells, stellate cells, and hepatic stem cells).[22] At therapeutically relevant concentrations (1000–3000 U/mL), IFN-α2b was shown to downregulate mRNA expression of several metabolizing enzymes, including CYP1A2,

CYP2B6, and CYP2E1. The observed downregulation of mRNA was as high as 50%, with a corresponding decrease in protein expression. Conversely, significant upregulation of CYP3A4 mRNA and protein expression was detected. The observed induction of CYP3A4 in this experimental system was unexpected and in contrast to CYP3A4 suppression detected by Guillen's group[15] with IFN and Aitken and Morgan[12] with IFN-γ. Possible mechanisms mediating this effect are currently unknown and require further elucidation. In contrast to CYP3A4, IFN-α2b suppression of CYP1A2, CYP2B6, and CYP2E1 mRNA and protein expression could potentially be attributed to regulation at the transcriptional level owing to substantial inhibition of CYP gene expression.

5.4.7.2 Effects of IFN-γ on Phase II Metabolizing Enzymes In human primary hepatocytes co-cultured with nonparenchymal cells, IFN-α2b (at therapeutically relevant concentrations of 1000–3000 IU/mL) was shown to have an impact on the expression of various phase II metabolizing enzymes, including the downregulation of UDP-glucuronosyltransferase (UGT2B7) and sulfotransferase (SULT1A1) mRNA expression.[22] Similar to gene expression, UGT2B7 protein levels were also inhibited. Although the regulatory mechanism remains to be elucidated, transcriptional modulation leading to altered protein expression is most likely a method by which IFN-α2b interacts with the various phase II metabolic enzymes. In contrast to UGT2B7 and SULT1A1, IFN-α2b showed no significant effect on the mRNA expression of SULT2A1, carboxylesterase 2 (CES2)[22] or epoxide hydrolase.[10]

5.4.7.3 Effects of IFN-γ on Drug Transporters Belliard and coworkers observed induction of both MDR1 mRNA (2.5-fold) and P-gp protein levels when Caco-2 cells were incubated with IFN-γ (10 ng/mL, 24–72 h).[49] The increased protein expression, however, did not alter P-gp efflux activity. Confocal laser scanning microscopy showed altered distribution of P-gp with more co-localization with lateral F-actin, which may explain the absence in correlation between the upregulation of P-gp expression by IFN-γ and the absence of a change in P-gp activity.

In human primary hepatocytes, while gene expression of both OAT2 and OCT1 remained unchanged, a general downregulation of mRNA levels of both sinusoidal influx transporters (NTCP, OATP1B1, OATP1B3, and OATP2B1) and efflux transporters (MDR1, MRP2, MRP3, BCRP, and BSEP) was detected after treatment with IFN-γ (10 ng/mL, 24 h).[50] The opposite was observed when human hepatocytes were co-cultured with nonparenchymal cells and exposed to IFN-α2b.[22] In general, there was no significant effect on the mRNA expression of OATP1B1, OATP1B3, OATP2B1, OCT1, P-gp, and BCRP at therapeutically relevant IFN-α2b concentrations (1000–3000 IU/mL). The only exception was a significant decrease in both mRNA and protein expression of OAT2. The mechanism of interaction is most likely at the level of gene transcription, leading to altered protein expression.

5.5 SUMMARY AND CONCLUSION

In general, with further characterization and validation, the primary hepatocyte monoculture system does add value as an *in vitro* tool for studying the effects of proinflammatory cytokines on metabolizing enzymes and drug transporters. As an isolated system, hepatocyte monocultures are useful for exploring the mechanisms of interaction between cytokines and drug-metabolizing enzymes and transporters. Furthermore, the commercial availability of hepatocytes from numerous preclinical species and from humans makes them an easily accessible experimental tool to facilitate better understanding of the effects of proinflammatory cytokines in potential toxicological species as well as potential future clinical extrapolation. In addition, the commercial availability of plateable cryopreserved hepatocytes also offers the advantage of reducing interassay variability.

On the other hand, despite some advantages current utility of the hepatocyte monoculture system has limitations (Figure 5-5), including incomplete mechanistic understanding of CYP downregulation and lack of agreement on positive controls representing various pathways. It should also be recognized that mimicking the complex effects of disease state on ADME genes *in vitro* may not be feasible. Furthermore, currently only qualitative assessment of the effects of proinflammatory cytokines on CYP enzymes and transporters can be ascertained in this experimental model. The literature reviewed here illustrates that the overall downregulation of CYP enzymes and transporters can be detected when assessed with the major proinflammatory cytokines (IL-1, IL-6, TNF-α, and IFN-γ), however, the limitation of this *in vitro* model is more apparent when extrapolation of these data to clinical

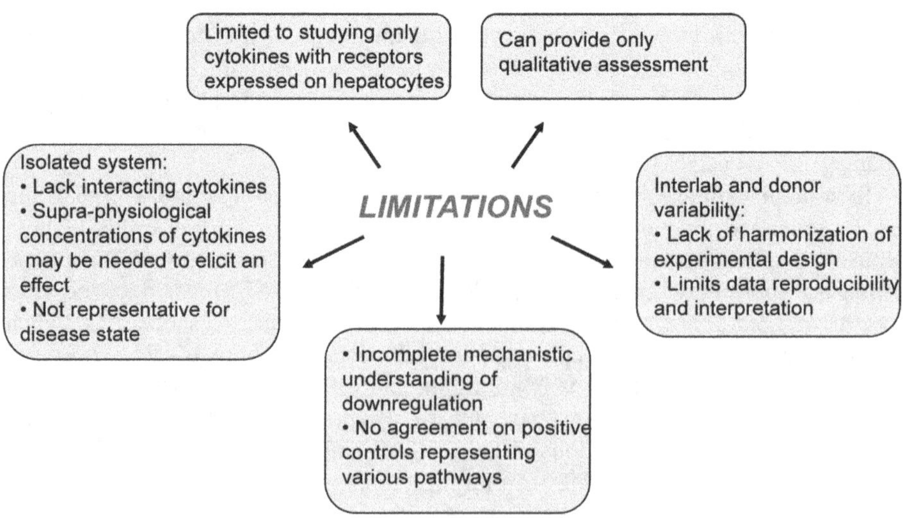

FIGURE 5-5 Limitations of hepatocyte monocultures.

significance is needed. In addition, the lack of standardization of experimental design and study conditions across different labs also limits study reproducibility and data interpretation. Likely, the hepatocyte monoculture system can facilitate assessment only with cytokines that have receptors expressed on the hepatocytes. This limits the utility of the experimental system for cytokines that do not bind directly to receptors expressed on hepatocytes but can still render an effect on CYP enzymes and transporters indirectly through stimulation of nonparenchymal cells (i.e., Kupffer cells). These cells could release proinflammatory cytokines that can in turn act on the hepatocytes. Last, because the monoculture system is an isolated system with no interacting cytokines, supra-physiological concentrations of studied cytokines may be needed to elicit an effect. As a result of all the drawbacks from the hepatocyte monoculture experimental system, interest is gaining for a more inclusive and physiologically representative *in vitro* model for studying the effects of proinflammatory cytokines on the metabolic disposition of SM drugs. To this goal, work by Sunman and coworkers and Chen and coworkers attempted to implement nonparenchymal cells (including Kupffer cells) with hepatocytes in a co-culture system.[13,22] The paucity of literature on this co-culture system indicates that experience (or success) with this experimental model is limited across industry and academia. An additional complication associated with this model is the difficulty of obtaining relatively pure cellular fractions. Thus further characterization and validation of this *in vitro* co-culture model is necessary before its value can be determined.

It is well recognized across academia and industry that the limited amount of *in vitro* data that have been generated thus far limits the use of the hepatocyte monoculture model as a predictive tool for clinical DDIs. In general, these limitations are related to the lack of acceptable and uniformly available models that have been demonstrated to provide not only reasonable *in vitro–in vivo* correlations but also mechanistic rationale defining the limitations of such models. Hence,it is now widely recognized that there is a need to generate more *in vitro* data, especially for biologics that have been clinically shown to cause DDIs.

As discussed in this chapter, initial focus on changes in net CYP activities is a reasonable approach because these enzymes are involved in the metabolism of most small molecule therapeutics and can be well characterized by using established probe substrates. This exercise, however, should not preclude generation of gene expression data in parallel because this technique allows assessing gene expression of any gene of interest (particularly phase II enzymes and transporters). A primary challenge for *in vitro–in vivo* correlations will be to extend data generated *in vitro* to predict whether novel biologics will cause a net change in drug metabolism related to enzymes and transporters in patients. A particular challenge would be the difficulty in predicting DDIs that may be affected or may be variable when the DDI is evaluated relative to the stage of disease.

In conclusion, assessing changes of enzyme and transporter activity and gene expression in novel *in vitro* systems that could predict potential SM and biologic

DDIs is a valid and important effort, particularly because the clinical assessment of such DDIs is long, difficult, and expensive.

REFERENCES

1. Morgan ET. Impact of infectious and inflammatory disease on cytochrome P450-mediated drug metabolism and pharmacokinetics. *Clin Pharmacol Ther* **85**(4), 434–38 (2009).
2. Petrovic V, Teng S, Piquette-Miller M. Regulation of drug transporters during infection and inflammation. *Mol Interv* **7**, 99–111 (2007).
3. Williams, SJ, Baird-Lambert JA, Farrell JC. Inhibition of theophylline metabolism by interferon. *Lancet* **330**, 939–41 (1987).
4. Schmitt C, Kuhn B, Zhang X, et al. Disease-drug-drug interaction involving tocilizumab and simvastatin in patients with rheumatoid arthritis. *Clin Pharmacol Ther* **89**(5), 735–40 (2011).
5. Mahmood I, Green MD. Drug interaction studies of therapeutic proteins or monoclonal antibodies. *J Clin Pharmacol* **47**, 1540–54 (2007).
6. Seitz K, Zhou H. Pharmacokinetic drug-drug interaction potentials for therapeutic monoclonal antibodies: reality check. *J Clin Pharmacol* **47**, 1104–18 (2007).
7. Lee JI, Zhang L, Men AY, et al. CYP-mediated therapeutic protein-drug interactions clinical findings, proposed mechanisms and regulatory implications. *Clin Pharmacokinet* **49**(5), 295–310 (2010).
8. Huang SM, Zhao H, Lee JI, et al. Therapeutic protein–drug interactions and implications for drug development. *Clin Pharmacol Ther* **87**(4), 497–503 (2010).
9. Barker CW, Fagan JB, Pasco DS. Interleukin-1β suppresses the induction of P4501A1 and P4501A2 mRNAs in isolated hepatocytes. *J Biol Chem* **267**(12), 8050–55 (1992).
10. Abdel-Razzak Z, Loyer P, Fautrel A. et al. Cytokines down-regulate expression of major cytochrome P-450 enzymes in adult human hepatocytes in primary culture. *Mol Pharmacol* **44**, 707–15 (1993).
11. Assenat E, Gerbal-Chaloin S, Larrey D, et al. Interleukin-1β inhibits CAR-induced expression of hepatic genes involved in drug and bilirubin clearance. *Hepatology* **40**, 951–60 (2004).
12. Aitken AE, Morgan ET. Gene-specific effects of inflammatory cytokines on cytochrome P450 2C, 2B6 and 3A4 mRNA levels in human hepatocytes. *Drug Metab Dispos* **35**, 1687–93 (2007).
13. Sunman JA, Hawke RL, LeCluyse EL, Kashuba ADM. Kupffer cell-mediated IL-2 suppression of CYP3A activity in hepatocytes. *Drug Metab Dispos* **32**, 359–63 (2004).
14. Tinel M, Robin MA, Doostzadeh J, et al. The interleukin-2 receptor down-regulates the expression of cytochrome P450 in cultured rat hepatocytes. *Gastroenterology* **109**, 1589–99 (1995).
15. Guillen MI, Donato MT, Jover R, et al. Oncostatin M down-regulates basal and induced cytochromes P450 in human hepatocytes. *J Pharmacol Exper Ther* **285**, 127–34 (1998).
16. Pascussi JM, Gerbal-Chaloin S, Pichard-Garcia L, et al. Interleukin-6 negatively regulates the expression of pregnane X receptor and constitutively activated receptor in primary human hepatocytes. *Biochem Biophys Res Commun* **274**, 707–13 (2000).

17. Gubbins PO, Melchert RB, McConnell SA, et al. Effect of interleukin-6 on the hepatic metabolism of itraconazole and its metabolite hydroxyitraconazole using primary human hepatocytes. *Pharmacology* **67**, 195–201 (2003).
18. Dickmann LJ, Patel SK, Rock DA, et al. Effects of interleukin-6 (IL-6) and an anti-IL-6 monoclonal antibody on drug-metabolizing enzymes in human hepatocyte culture. *Drug Metab Dispos* **39**, 1415–22 (2011).
19. Gu X, Ke S, Liu D, et al. Role of NF-κB in regulation of PXR-mediated gene expression: a mechanism for the suppression of cytochrome P450 3A4 by proinflammatory agents. *J Biol Chem* **281**(26), 17882–89 (2006).
20. Donato MT, Guillen MI, Jover R, et al. Nitric oxide-mediated inhibition of cytochrome P450 by interferon-γ in human hepatocytes. *J Pharmacol Exper Ther* **281**, 484–90 (1997).
21. Flaman AS, Gravel C, Hashem AM, et al. The effect of interferon-α on the expression of cytochrome P450 3A4 in human hepatoma cells. *Toxicol Appl Pharmacol* **253**, 130–36 (2011).
22. Chen C, Han YH, Yang Z, Rodrigues AD. Effect of interferon-α2b on the expression of various drug metabolizing enzymes and transporters in cocultures of freshly prepared human primary hepatocytes. *Xenobiotica* **41**(6), 476–85 (2011).
23. Aitken AE, Lee CM, Morgan ET. Roles of nitric oxide in inflammatory down-regulation of human cytochrome P450. *Free Radical Biol Med* **44**(6), 1161–68 (2008).
24. Morgan ET. Regulation of cytochrome P450 by inflammatory mediators: why and how? *Drug Metab Disp* **29**, 207–12 (2001).
25. Aitken AE, Richardson TA, Morgan ET. Regulation of drug-metabolizing enzymes and transporters in Inflammation. *Ann Rev Pharmacol Toxicol* **46**, 123–49 (2006).
26. Fardel O, Le Vee M. Regulation of human hepatic drug transporter expression by pro-inflammatory cytokines. *Expert Opin Drug Metab Toxicol* **5**, 1469–481 (2009).
27. Liptrott NJ, Owen A. The role of cytokines in the regulation of drug disposition: extended functional pleiotropism? *Expert Opin Drug Metab Toxicol* **7**(3), 341–52 (2011).
28. Na SY, Kang BY, Chung SW, et al. Retinoids inhibit interleukin-12 production in macrophages through physical associations of retinoid X receptor and NF-κB. *J Biol Chem* **274**, 7674–80 (1999).
29. Xie W, Barwick JL, Downes M, et al. Humanized xenobiotic response in mice expressing nuclear receptor SXR. *Nature* **406**, 435–39 (2000).
30. Urquhart BL, Tirona RG, Kim RB. Nuclear receptors and the regulation of drug-metabolizing enzymes and drug transporters: Implications for interindividual variability in response to drugs. *J Clin Pharmacol* **47**, 566–78 (2007).
31. Ke S, Rabson AB, Germino JF, et al. Mechanism of suppression of cytochrome P-450 1A1 expression by tumor necrosis factor-alpha and lipopolysaccharide. *J Biol Chem* **276**, 39638–644 (2001).
32. Ghezzi P, Bianchi M, Gianera L, et al. Role of reactive oxygen intermediates in the interferon-mediated depression of hepatic drug metabolism and protective effect of *N*-acetylcysteine in mice. *Cancer Res* **45**, 3444–47 (1985).
33. Cribb AE, Renton KW. Dissociation of xanthine oxidase induction and cytochrome P450 depression during interferon induction in the rat. *Biochem Pharmacol* **46**(11), 2114–17 (1993).

34. Ros JE, Schuetz JD, Geuken M, et al. Induction of *Mdr1b* expression by tumor necrosis factor-α in rat liver cells is independent of p53 but requires NF-κB signaling. *Hepatology* **33**, 1425–31 (2001).

35. Zhou G, Kuo MT. NF-kappaB-mediated induction of *Mdr1b* expression by insulin in rat hepatoma cells. *J Biol Chem* **272**, 15174–83 (1997).

36. Zhou G, Kuo MT. Wild-type p53-mediated induction of rat *Mdr1b* expression by the anticancer drug daunorubicin. *J Biol Chem* **273**, 15387–94 (1998).

37. Lee CM, Pohl J, Morgan ET. Dual mechanisms of CYP3A protein regulation by proinflammatory cytokine stimulation in primary hepatocyte cultures. *Drug Metab Dispos* **37**, 865–72 (2009).

38. Assenat E, Gerbal-Chaloin S, Maurel P, et al. Is nuclear factor kappa-B the missing link between inflammation, cancer and alteration in hepatic drug metabolism in patients with cancer? *Eur J Cancer* **42**, 785–92 (2006).

39. Maheo, K., Antras-Ferry, J., Morel, F., et al. Modulation of glutathione s-transferase subunits A2, M1, and P1 expression by interleukin-1beta in rat hepatocytes in primary culture. *J Biol Chem* **272**(26), 16125–32 (1997).

40. Langouet S, Corcos L, Abdel-Razzak Z, et al. Up-regulation of glutathione *s*-transferases alpha by interleukin 4 in human hepatocytes in primary culture. *Biochem Biophys Res Commun* **216**(3), 793–800 (1995).

41. Bertilsson PM, Olsson P, Magnusson KE. Cytokines influence mRNA expression of cytochrome P450 3A4 and MDR1 in intestinal cells. *J Pharm Sci* **90**(5), 638–46 (2001).

42. Sukhai M, Yong A, Pak A, Piquette-Miller M. Decreased expression of P-glycoprotein in interleukin-1β and interleukin-6 treated rat hepatocytes. *Inflamm Res* **50**, 362–70 (2001).

43. Le Vee M, Gripon P, Stieger B, Fardel O. Down-regulation of organic anion transporter expression in human hepatocytes exposed to the proinflammatory cytokine interleukin-1β. *Drug Metab Dispos* **36**, 217–22 (2008).

44. Diao L, Li N, Brayman TG, et al. Regulation of MRP2/ABCC2 and BSEP/ABCB11 expression in sandwich cultured human and rat hepatocytes exposed to inflammatory cytokines TNF-α, IL-6, and IL-1β. *J Biol Chem* **285**(41), 31185–92 (2010).

45. Belliard AM, Tardivel S, Farinotti R, et al. Effect of hr-IL-2 treatment on intestinal P-glycoprotein expression and activity in Caco-2 cells. *J Pharm Pharmacol* **54**(8), 1103–09 (2002).

46. Le Vee M, Lecureur V, Stieger B, Fardel O. Regulation of drug transporter expression in human hepatocytes exposed to the proinflammatory cytokines tumor necrosis factor-α or interleukin-6. *Drug Metab Dispos* **37**, 685–93 (2009).

47. Le Vee M, Jouan E, Stieger B, et al. Regulation of drug transporter expression by oncostatin M in human hepatocytes. *Biochem Pharmacol* **82**, 304–11 (2011).

48. Hirsch-Ernst KI, Ziemann C, Foth H, et al. Induction of mdr1b mRNA and P-glycoprotein expression by tumor necrosis factor alpha in primary rat hepatocyte cultures. *Journal of Cellular Physiology*, **176**, 506–515, (1998).

49. Belliard AM, Lacour B, Farinotti R, Leroy C. Effect of tumor necrosis factor-alpha and interferon-gamma on intestinal P-glycoprotein expression, activity, and localization in Caco-2 cells. *J Pharm Sci* **93**, 1524–36 (2004).

50. Le Vee M, Jouan E, Moreau A, Fardel O. Regulation of drug transporter mRNA expression by interferon-c in primary human hepatocytes. *Fundam Clin Pharmacol* **25**, 99–103 (2011).

51. Elkahwaji J, Robin MA, Berson A, et al. Decrease in hepatic cytochrome P450 after interleukin-2 immunotherapy. *Biochem Pharmacol* **57**, 951–54 (1999).

52. Piscitelli SC, Forrest A, Vogel S, et al. Pharmacokinetic modeling of recombinant interleukin-2 in patients with human immunodeficiency virus infection. *Clin Pharmacol Ther* **64**, 492–98 (1998).

53. Wacher VJ, Wu CY, Benet LZ. Overlapping substrate specificities and tissue distribution of CYP 3A and PGP: Implications for drug delivery and activity in cancer chemotherapy. *Mol Carcinog* **13**, 129–34 (1995).

54. Ogata S, Kubota Y, Yamashiro T, et al. Signaling pathways regulating IL-1alpha-induced COX-2 expression. *J Dental Res* **86**, 186–91 (2007).

55. Rose TM, Bruce AG. Oncostatin M is a member of a cytokine family that includes leukemia-inhibitory factor, granulocyte colony-stimulating factor, and interleukin 6. *Proc Natl Acad Sci USA* **88**, 8641–45 (1991).

56. Hibi M, Nakajima K, Hirano T. IL-6 cytokine family and signal transduction: a model of the cytokine system. *J Mol Med* **74**, 1–12 (1996).

57. Gearing DP, Comeau MR, Friend DJ, et al. The IL-6 signal transducer, pg130: an oncostatin M receptor and affinity converter for the LIF receptor. *Science* **255**, 1434–37 (1992).

58. Liu J, Modrell B, Aruffo A, et al. Interleukin-6 signal transducer gp130 mediates oncostatin M signaling. *J Biol Chem* **267**, 6763–66 (1992).

59. Kishimoto T, Taga T, Akira S. Cytokine signal transduction. *Cell* **76**, 253–62 (1994).

60. Zarling JM, Shoyab M, Marquardt H, et al. Oncostatin M: a growth regulator produced by differentiated histiocytic lymphoma cells. *Proc Natl Acad Sci U S A* **83**, 9739–43 (1986).

61. Horn D, Fitzpatrick M, Gompper PT, et al. Regulation of cell growth by recombinant oncostatin-M. *Growth Factors* **2**, 157–65 (1990).

62. Brown TJ, Rowe JM, Liu JW, Shoyab M. Regulation of IL-6 expression by oncostatin M. *J Immunol* **174**, 2175–80 (1991).

63. Grove RI, Mazzuco CE Radka SF, et al. Oncostatin M up-regulates low density lipoprotein receptors in HepG2 cells by a novel mechanism. *J Biol Chem* **266**, 194–99 (1991).

64. Richards CD, Brown TJ, Shoyab M, et al. Recombinant oncostatin M stimulates the production of acute phase proteins in hepG2 cells and rat primary hepatocytes in vitro. *J Immunol* **148**,1731–36 (1992)

65. Silver JS, Hunter CA. gp130 at the nexus of inflammation, autoimmunity, and cancer. *J Leukoc Biol* **88**(6), 1145–56 (2010).

66. Nakashima K, Taga T. gp130 and the IL-6 family of cytokines: signaling mechanisms and thrombopoietic activities. *Semin Hematol* **35**(3), 210–21 (1998).

67. Remacle J, Reas M, Toussaint O, et al. Low levels of reactive oxygen species as modulators of cell function. *Mutat Res* **316**, 103–22 (1995).

68. Lo YY, Cruz TT. Involvement of reactive oxygen species in cytokine and growth factor induction of c-*fos* expression in chondrocytes. *J Biol Chem* **270**, 11727–30 (1995).

69. Stark GR, Kerr IM, Williams BR, et al. How cells respond to interferons. *Annu Rev Biochem* **67**, 227–64 (1998).

CHAPTER 6

USE OF ANIMAL MODELS FOR PROJECTION OF CLINICAL DRUG–DRUG INTERACTIONS FOR THERAPEUTIC PROTEINS

EUGENIA KRAYNOV and MARTIN E. DOWTY

6.1 INTRODUCTION

Animal models play an important role in assessing various aspects of a drug candidate during its discovery and clinical development, including anticipated efficacy, pharmacokinetics, and safety. At the same time, there are limitations with animal models in that they do not always reproduce human pathophysiology or disease drivers, pharmacokinetic pathways, or safety mechanisms relevant to humans. Hence the application of animal models must be underpinned by an understanding of the translatability of relevant mechanisms to humans. The most valuable insight into drug–drug interaction (DDI) potential is predicated on a thorough comprehension of the pharmacokinetic (PK) and pharmacodynamic (PD) properties of both interacting drugs and an understanding of their potential to be a DDI victim as well as a perpetrator.

The use of *in vitro* human systems to predict clinical DDIs of small molecules mediated through cytochrome P450 (CYP) metabolism has seen significant value.[1] However, DDI mediated through non-CYP mechanisms requires greater effort to understand *in vitro–in vivo* correlations (IVIVCs). Although the preclinical DDI assessment for therapeutic proteins can theoretically be addressed by using *in vitro* human systems alone, the relevance of the *in vitro* data in the context of a more complex *in vivo* system requires greater validation. *In vivo* animal models can potentially help bridge this gap. However, additional work is needed to address the potential for animal models to provide insight into DDI predictions of therapeutic proteins.

Drug–Drug Interactions for Therapeutic Biologics, First Edition.
Edited by Honghui Zhou and Bernd Meibohm.

6.2 SELECTION OF THE ANIMAL MODEL

In general, a valid animal model should at least have qualitative similarity to humans in drug-metabolizing enzymes and transporters, substrate specificity, response to the inhibitor or inducer (small molecule or therapeutic protein), and disposition mechanisms of both small molecule and therapeutic proteins. These considerations depend on the breadth of questions being asked.

When it comes to evaluation of therapeutic protein–small molecule DDIs, animal models may be used to qualitatively characterize an alteration in absorption, tissue distribution, excretion, or inhibition and induction of metabolism of the small molecule (SM) drug.[2] These processes reflect changes in uptake or efflux transporters and drug-metabolizing enzymes. However, all animal species differ from humans in amino acid sequence identity and substrate specificity of drug-metabolizing enzymes and transporters. The rat is a particularly common species used in preclinical development, although rats have a greater variability in expression and extent of activity in phase I and II drug-metabolizing enzymes and typically metabolize SM drugs considerably faster than do humans.[3–5]

More similarities are found in species more closely related to humans in an evolutionary sense, in particular monkeys.[2] Rhesus and cynomolgus monkeys are used routinely for preclinical pharmacokinetics, pharmacodynamics, and safety evaluations and more recently for the study of *in vivo* drug interactions. For example, cynomolgus monkey CYP3A8 shows 93% homology to human 3A4 (a significant metabolizing enzyme for small molecules), and rhesus monkey CYP3A64 has been shown to be 93% homologous to human CYP3A4 and 83% homologous to CYP3A5.[6]

Genetically modified mouse models in which a specific gene is removed or replaced are powerful tools for various mechanistic studies.[7,8] However, there are important considerations in their successful implementation. For example, modification of a given gene does not always result in the anticipated phenotype. Therefore, interpretation of the significance of the findings from studies using genetically modified mouse models is not always as straightforward as one would expect, especially when it is desired to extrapolate the findings to humans. Nevertheless, knockout and transgenic animal models can play an important role in understanding mechanisms and may become more useful in the future.

For monoclonal antibody (mAb) based therapeutic proteins, species selection depends on mAb cross-reactivity because target-mediated clearance often plays a role in their elimination. A general limitation of rodent models is that therapeutic proteins intended for human use may not cross-react with the rodent target. A rodent-specific surrogate therapeutic protein may provide value in this regard but requires additional effort to develop and assess. In many cases, the pharmacokinetics of therapeutic proteins and its extrapolation to humans is best understood in monkeys because the therapeutic proteins are typically cross-reactive to monkey targets.

Currently no preclinical *in vivo* DDI evaluation is typically being performed for therapeutic proteins. However, given the aforementioned, in most cases the monkey is probably the most appropriate species for preclinical *in vivo* characterization of

DDIs involving therapeutic proteins if such evaluation is attempted. Monkeys generally need to be naive—that is, not previously treated with biotherapeutics (BioTx) of a similar class, to avoid formation of antidrug antibodies (ADAs) confounding the interpretation of these studies.

6.3 STUDY DESIGN

When a therapeutic protein is designed to be coadministered with a narrow therapeutic index (TI) drug, clinical DDI studies may need to be conducted even when there is no known mechanism of interaction. A preliminary evaluation of this interaction in the appropriate animal DDI model can aid in the design of the clinical study. Preclinical *in vivo* evaluation can be conducted for any therapeutic protein intended to be used in combination (with small molecule or another protein therapeutic drug). Combination therapy is now a standard approach in oncology and frequently used in a number of inflammatory and autoimmune diseases.

For a therapeutic protein intended to be used as a single agent and for which no known DDI mechanisms exist, it would be difficult to design an appropriate preclinical DDI study, and in most cases such evaluation is not needed.

Multiple aspects of therapeutic protein PK need to be taken into account when designing an animal DDI study with a small molecule drug. In the case of mAbs, for example, the differences in half-life—typically days for mAbs vs. only hours for small molecules—can affect the time of assessment. An interacting small molecule drug could also alter the expression of the mAb target over time or affect the disposition of the mAb. For mAbs undergoing significant target-mediated elimination characterized by nonlinear pharmacokinetic profiles, the relevance of the animal species and the study design would need to account for this interaction.[9]

From a bioanalytical perspective, the potential interference of each concomitant drug on the quantification of the other should be investigated. Quantification of small molecule drugs typically uses liquid chromatography coupled with tandem mass spectrometry (LC-MS/MS) assays, whereas analytical methods for detection of therapeutic proteins are mostly based on ligand-binding assays. The specificity and selectivity of ligand-binding assays largely rely on the reagents used in the assays and the matrices in which the samples are collected. Because there is no extraction step included in this procedure, coadministered drugs could potentially interfere with the quantification of the therapeutic protein of interest.

6.4 DISEASE MODELS

The only preclinical DDI evaluation being currently performed for therapeutic proteins is for cytokine-mediated DDI using human hepatocytes.[10] However, even this approach is still being optimized, and its translation to the clinical situation is not well understood.

Cytokine-mediated changes in drug-metabolizing enzymes and resulting DDIs with therapeutic proteins can potentially be evaluated in the disease models of inflammation. The adjuvant-induced inflammatory rat model shares common features with human rheumatoid arthritis with respect to development of systemic inflammation and structural changes in bone and cartilage.[11] Adjuvant treatment causes cytokine activation, resulting in a number of hepatic events, such as defenestration of sinusoids, irregular-shaped lysosomes, increased level of α1-acid glycoprotein, and lower CYP levels compared to normal rats.[12–14]

Inflammation alters expression of hepatobiliary transporters: Expression of *Oatp1a1, Oatp1a4*, and *Bsep* is downregulated in adjuvant-treated rats, while the expression of *Mrp3* is upregulated nearly by twofold.[15] Inflammatory cytokines—including tumor necrosis factor α (TNFα), interleukin 1 (IL-1), and IL-6—secreted by macrophages, and Kupffer cells have been characterized as mediators that reduce bile flow and organic anion excretion.[15]

These effects are similar to cytokine effects observed with human hepatic enzymes and transporters *in vitro* [16,17] and may at least partly reproduce the systemic pathophysiology of human rheumatoid arthritis.

In human systemic inflammatory disease, the clearance of drugs that are extensively metabolized in the liver is decreased.[18,19] Anti-inflammatory drugs should restore or normalize the amount and activity of microsomal enzymes. This was nicely demonstrated in the preadjuvant arthritis rat model after treatment with anti-TNFα mAb, infliximab.[20] The same authors earlier reported that preadjuvant arthritis is associated with elevated serum nitrite, reduced hepatic CYP content, and reduced verapamil clearance.[18] The clearance of verapamil is also reduced in patients with rheumatoid arthritis.[18] Treatment with infliximab resulted in a significant increase in total CYP, CYP1A, and CYP3A protein content in preadjuvant arthritic rats. Interestingly verapamil clearance was not significantly affected, which the authors attributed to the lack of effect on plasma protein binding.

Cytokine-mediated DDIs are typically studied *in vitro* in human hepatocytes,[16,17] and clinical DDI studies are usually conducted for high-risk therapeutic proteins. However, the IVIVC in humans is not well understood. Conceptually, this IVIVC link could be initially established using rat models of inflammation. However, the interspecies relationship may be difficult to characterize, and this may therefore reduce the predictive power of animal models. Specifically, rat and human cytokines have sufficiently low homology (e.g., mature human IL-6 shares 41% amino acid sequence identity with mouse and rat IL-6),[21] and therefore antihuman cytokine therapeutic proteins may not recognize the homologous target in a rat model. In addition, human therapeutic are likely to be immunogenic in rodents.

At present it is fair to say that while inflammatory rodent models appear to be rather straightforward, further characterization is needed.

In cases when DDI occurs via more than one pathway, animal models may be better able to reflect a net outcome of the interplay between those human-relevant pathways that otherwise would be difficult to predict from *in vitro* assays by themselves. Even if quantitative prediction from animal models is not understood, it may still be a better qualitative predictor of potential DDI than *in vitro* studies alone. Similar to small molecule DDI, linking *in vitro* data in human and animal models to *in vivo* animals systems, may allow for more predictive DDI characterization.

6.5 EMERGING CHALLENGES

The structural diversity of therapeutic proteins is constantly expanding. Drug modalities representing hybrids of small molecule and proteins pose unique challenges from the DDI perspective. Included in this class are antibody-drug conjugates (ADCs) and various nanoparticles. In the case of ADCs, a small molecule is linked to a carrier antibody that provides targeting specificity through recognition of a specific target on the cell surface. As the antibody–receptor complex gets internalized, the small molecule is delivered to the target tissue where it is released. Nanoparticles can have a small molecule drug encapsulated inside while on the outside they can have specific peptides or proteins, such as mAbs that specifically deliver the small molecule to the target tissue.

These newer therapeutic protein modalities containing small molecule drugs may have a higher probability of being involved in DDI with conventional small molecule drugs. This type of DDI is more challenging to characterize *in vitro* owing to the complex and targeted nature of these drug products. One option would be to evaluate the small molecule component's DDI potential separately from the carrier or targeting moiety. However, this would not truly represent the *in vivo* situation and therefore might not adequately predict clinical behavior. Preclinical animal models may be helpful in the initial characterization of these modalities.

Biodistribution of these compounds should probably be evaluated as part of their DDI evaluation. Factors that influence tissue distribution of therapeutic proteins include size, charge, target and off-target binding properties, and route of administration.[22] For proteins that are above the glomerular filtration cutoff (such as immunoglobulins), the major elimination organs are liver and other highly perfused organs. These proteins cannot easily pass through biological membranes and distribute primarily into the vascular space and extracellular fluids.

6.6 CONCLUSIONS

It is important to appreciate that findings from animal DDI models should initially be characterized for qualitative information only. An appropriately selected and used animal model can be a valuable tool for DDI assessments during the drug discovery and development processes. Although clinical DDI assessment remains a necessary part of the development of therapeutic proteins, additional insight and characterization of preclinical tools, including animal models, may provide greater value in the future.

REFERENCES

1. Obach RS. Predicting drug-drug interactions from in vitro drug metabolism data: challenges and recent advances. *Curr Opin Drug Discov Devel* **12**, 81–89 (2009).
2. Tang C, Prueksaritanont T. Use of *in vivo* animal models to assess pharmacokinetic drug-drug interactions. *Pharm Res* **27**(9), 1772–87 (2010).
3. Nedelcheva V, Gut I. P450 in the rat and man: methods of investigation, substrate specificities and relevance to cancer. *Xenobiotica* **24**, 1151–75 (1994).
4. Smith DA. Species differences in metabolism and pharmacokinetics: are we close to an understanding? *Drug Metab Rev* **23**, 355–73 (1991).
5. Soucek P, Gut I. Cytochromes P450 in rats: structure, functions, properties and relevant human forms. *Xenobiotica* **22**, 83–103 (1992).
6. Carr B, Norcross R, Fang Y, et al. Characterization of the rhesus monkey CYP3A64 enzyme: species comparisons of CYP3A substrate specificity and kinetics using baculovirus-expressed recombinant enzymes. *Drug Metab Dispos* **34**,1703–12 (2006).
7. Ashino T, Arima T, Shioda S, et al. Effect of interleukin-6 neutralization on CYP3A11 and metallothionein-1/2 expressions in arthritic mouse liver. *Eur J Pharmacol* **558**, 199–207 (2007).
8. Lin JH. Applications and limitations of genetically modified mouse models in drug discovery and development. *Curr Drug Metab* **9**, 419–38 (2008).
9. Mager DE. Target-mediated drug disposition and dynamics. *Bioch Pharmacol* **72**, 1–10 (2006).
10. Lee J-I, Zhang L, Men AY, et al. CYP-mediated therapeutic protein-drug interactions: clinical findings, proposed mechanisms and regulatory implications. *Clin Pharmacol* **49**(5), 295–310 (2010).
11. Van Bilsen JH, Wagenaar-Hilbers JP, Boot EP, et al. Searching for the cartilage-associated mimicry epitope in adjuvant arthritis. *Autoimmunity* **35**(3), 201–10 (2002).
12. Monshouwer M, Hoebe KH. Hepatic (dys-)function during inflammation. *Toxicol In Vitro* **17**, 681–86 (2003).
13. Whitehouse MW, Beck FJ. Impaired drug metabolism in rats with adjuvant-induced arthritis: a brief review. *Drug Metab Dispos* **1**, 251–55 (1973).
14. Hung, DY, Siebert GA, Chang P, et al. Hepatic pharmacokinetics of propranolol in rats with adjuvant-induced systemic inflammation. *Am J Physiol Gastrointest Liver Physiol* **290**, G343–51 (2006).
15. Roberts MS, Liu X, Zou Y, et al. Effect of adjuvant-induced systemic inflammation in rats on hepatic disposition kinetics of taurocholate. *Am J Physiol Gastroint Liver Physiol* **300**, G130–36 (2011).
16. Zhang X, Schmitt C, Grange S, et al. Disease-drug interaction studies of tocilizumab with cytochrome p450 substrates *in vitro* and *in vivo*. *Clin Pharmacol Ther* **85**, S59 (2009).
17. Huang S-M, Zhao H, Lee J-I, et al. Therapeutic protein-drug interactions and implications for drug development. *Clin Pharmacol Ther* **87**(4), 497–503 (2010).
18. Mayo PR, Skeith K, Russell AS, Jamali F. Decreased dromotropic response to verapamil despite pronounced increased drug concentration in rheumatoid arthritis. *Br J Clin Pharmacol* **50**, 605–13 (2000).

19. Ling S, Jamali F. Effect of early phase adjuvant arthritis on hepatic P450 enzymes and pharmacokinetics of verapamil: an alternative approach to the use of an animal model of inflammation for pharmacokinetic studies. *Drug Metab Dispos* **33**, 579–86 (2005).
20. Ling S, Jamali F. The effect of infliximab on hepatic cytochrome P450 and pharmacokinetics of verapamil in rats with pre-adjuvant arthritis: a drug–disease and drug–drug interaction. *Basic Clin Pharmacol Toxicol* **105**, 24–29 (2009).
21. Chiu C-P, Moulds C, Coffman L, et al. Multiple biological activities are expressed by a mouse interleukin 6 cDNA clone isolated from bone marrow stromal cells. *Proc Natl Acad Sci USA* **85**, 7099–103 (1988).
22. Vugmeyster Y, Defranco D, Sckut P, et al. Biodistribution of-labeled therapeutic proteins: application in protein drug development beyond oncology. *J Pharm Sci* **99**(2), 1028–45 (2010).

CHAPTER 7

THE COCKTAIL APPROACH AND ITS UTILITY IN DRUG–DRUG INTERACTION ASSESSMENTS FOR THERAPEUTIC PROTEINS

ALEXANDER JETTER and UWE FUHR

7.1 ASSESSMENT OF ENZYME ACTIVITIES USING THE COCKTAIL APPROACH

The cocktail approach is used to assess the effect of drugs on the activities of the main cytochrome P450 (CYP) enyzmes *in vivo*. As for all other drugs, any effect of therapeutic proteins on the metabolism of other (victim) drugs needs to be tested and quantified. The main mechanisms of interaction for small molecules on CYPs are enzyme inhibition by direct interaction with the enzyme, increased synthesis of enzymes via aryl hydrocarbon (Ah)- or retinoid X receptor (RXR) related processes, or decreased turnover of the enzyme (in the case of the cytochrome P450 enzyme CYP2E1). These mechanisms of interaction are related to the structural similarity of victim substrates and other small molecule drugs. Appropriate *in vitro* tests have been established for enzyme inhibition and, with limitations in quantification of the effect, also for the described mechanisms of enzyme induction. In contrast, therapeutic proteins would rarely have a direct action on the enzyme or would trigger classical enzyme induction, which developed as specific defense mechanisms to protect an organism against exposure to toxic nutritional components. Thus *in vitro* or animal studies with therapeutic proteins have limited power to predict clinical interactions because the molecular mechanisms are only partially understood and/or often cannot be mimicked in nonclinical test systems.[1] Because not all potential combinations of a therapeutic protein with other drugs can be tested in a clinical study, the effect of such drugs on drug metabolizing enzymes, mainly

Drug–Drug Interactions for Therapeutic Biologics, First Edition.
Edited by Honghui Zhou and Bernd Meibohm.

CYP enzymes, would usually be assessed using the phenotyping approach. Phenotyping for an enzyme—that is, quantification of its actual *in vivo* activity in an individual—is performed by administration of a selective substrate for the enzyme and subsequent determination of appropriate pharmacokinetic metrics closely reflecting enzyme activity.[2] Pharmacokinetic metrics reflecting individual CYP activity best, depending on the individual phenotyping substrate used, may be metabolic ratios of metabolite over parent drug in plasma or urine, clearance of the parent drug, or partial clearance via a specific pathway. It is desirable to use metrics based on a limited number of blood samples. The standard design of respective phenotyping interaction studies is a crossover design with administration of phenotyping agents in both periods, which is combined with coadministration of the therapeutic protein to be tested at its highest chronic therapeutic dose in one period.[2,3]

Concomitant administration of phenotyping substances in a cocktail to test the effects of drugs including therapeutic proteins on several CYPs at a single occasion has been reported extensively.[2,4,5] This approach offers the advantage that the burden for both investigators and volunteers is considerably decreased. In the case of therapeutic proteins, this is especially important because such studies often need to be conducted in patients (discussed later). Beyond the selection of appropriate phenotyping metrics within a phenotyping cocktail, it should be safeguarded that there is no relevant interaction between individual cocktail components.[6] If an interaction between cocktail components cannot be excluded (e.g., chlorzoxazone, a CYP2E1 substrate, is an inhibitor for CYP3A4),[7] it is also possible to split the cocktail, administering the interacting substance on a different day of the study. Tolerability of such drug cocktails is usually excellent because it is possible to select very low doses.[2] Thus the cocktail phenotyping approach is the method of choice for screening for the potential of therapeutic proteins to change CYP activities *in vivo*.

7.2 GUIDELINES APPLICABLE FOR COCKTAIL DRUG–DRUG INTERACTION STUDIES

Current regulatory guidelines provide recommendations on individual phenotyping agents.[8,9] Recommended CYP phenotyping agents include CYP1A2, theophylline, caffeine; CYP2B6, efavirenz, bupropion; CYP2C8, amodiaquine, cerivastatin, repaglinide, rosiglitazone; CYP2C9, warfarin, tolbutamide; CYP2C19, omeprazole, esomeprazole, lansoprazole, pantoprazole; CYP2D6, metoprolol, desipramine, dextromethorphan, atomoxetine; CYP2E1, chlorzoxazone; and CYP3A4, midazolam, buspirone, felodipine, lovastatin, eletriptan, sildenafil, simvastatin, triazolam.[8,10] The European Medicines Agency (EMA) list is shorter and more specific. The FDA published an additional list in 2011,[11] presenting substrates considered as sensitive and substrates with a narrow therapeutic range (according to the 1999 document "Guidance for Industry: In Vivo Drug Metabolism/Drug Interaction Studies—Study Design, Data Analysis, and Recommendations for Dosing and Labeling"), which is also incorporated in the 2012 draft guideline.[8,10]

In most cases, several different pharmacokinetic metrics may be used to assess activity of a given CYP. However, the 2010 EMA (draft) guideline on the investigation of drug interactions (CPMP/EWP/560/95/Rev. 1—Corr.[10]) and the updated (draft) FDA guidance[8] request the determination of complete area under the curve (AUC) values of phenotyping substrates. The reason for the reluctance to accept other metrics is that validation is in most cases limited to the quantification of baseline activity in healthy individuals and in the absence of factors with major impact on pharmacokinetics. However, any effects of therapeutic proteins on pharmacokinetic processes may not be limited to the activity of the CYP to be investigated. Additional mechanisms could then be identified by a change in the entire concentration vs. time profile of the phenotyping drug.[3]

Cocktail studies are addressed specifically by the EMA guidance; the FDA standpoint is also available.[8,12,13] The FDA clearly accepts data from cocktail studies. For both agencies, the appropriate composition of a cocktail is a prerequisite.[2] The FDA considers data from cocktail studies to supplement other *in vitro* or *in vivo* information, including physiology-based pharmacokinetic modeling and simulation data. Cocktail studies may also be used to determine the need for further *in vivo* studies. In its most recent (draft) guideline,[8] a further implementation of cocktail studies is supported, and it is mentioned that negative results from cocktail studies may eliminate the need to conduct additional studies. For an investigational therapeutic protein that is a cytokine or cytokine modulator, it is requested that studies should be conducted to determine the effects on cytochrome P450 enzymes (or transporters), and the use of the cocktail approach is mentioned to this end.[8]

7.3 COCKTAIL INTERACTION STUDIES WITH THERAPEUTIC PROTEINS: SPECIAL FEATURES

The metabolism and elimination of therapeutic proteins is distinct from that of small molecule drugs, which share metabolism and elimination pathways with most toxic compounds and components of food and environment. Because dietary proteins are predominantly digested in the gut and taken up as amino acids or dipeptides, no particular enzymes or transporters exist for therapeutic proteins. It is therefore unlikely that a therapeutic protein is a substrate of a drug-metabolizing enzyme of phases I and II or of a drug transporter, because the molecules are too large and their structure is not suitable for metabolism or transport by those enzymes and transporters usually encountered in pharmacokinetics. Thus, in general, therapeutic proteins have a lower risk of causing specific metabolic drug interactions mediated by changes in activities of individual enzymes. However, therapeutic proteins may act as modulators of function of these drug-metabolizing and -transporting proteins. Several cytokines, including interleukin 1 (IL-1), IL-6, IL-10, tumor necrosis factor α (TNF-α), and interferon (IFN), have been shown repeatedly to change (usually to downregulate) the expression of various cytochrome P450 enzymes in *vitro*[14–16] and *in vivo*.[17] Accordingly, therapeutic proteins that block the action of such cytokines in pathophysiological processes will also block their effects on drug-metabolizing enzymes. These cytokine-mediated changes in cytochrome P450 enzyme activities take some time to develop because protein expression is affected. This is in contrast to most enzyme inhibitions by small molecule drugs, which are present almost immediately after the inhibitory drug has been administered, because the underlying mechanism is a direct competitive or a noncompetitive inhibitory interaction with the enzyme itself. Translated to the timing of cocktail dosing this would mean that the administration of the probe substrate cocktail should not take place immediately after the first dose, because the expected effect may require several days and/or repeated dosing until it is fully developed.

Another factor that has to be taken into account is the pharmacokinetics of the therapeutic protein. Because therapeutic proteins are often administered subcutaneously or are chemically modified for increased metabolic stability (e.g., by pegylation) their residence time in the body is often much longer than that of small molecules. Because the therapeutic protein is continuously present in the body for days or even weeks, the effect on drug-metabolizing enzymes and drug transporters may be fully present only after several days or weeks. It is therefore crucial to take the biologic half-life of the compound under investigation into account when the timing of the cocktail-phenotyping days is planned. In addition, it may be wise to consider more than one cocktail-phenotyping day during one treatment course with a therapeutic protein.

Because some therapeutic proteins may cause severe adverse effects, including infections and cancer, their use in healthy subjects is not always justifiable. Thus cocktail interaction studies may need to be conducted in patients. However, such studies in patients may pose major challenges in regard to recruitment of participants, further interactions by concomitant medication, and the variability caused by disease processes. This may also include the amelioration in organ function caused by the therapeutic protein. To give an example, interferons used in the treatment of hepatitis, or therapeutic proteins directed against hepatically metastasizing tumors, may lead to improvements in hepatic function during treatment. This may then also translate into increased activities of CYP enzymes, other drug-metabolizing enzymes, and transporters. Another example is the association between the stage of congestive heart failure, TNF-α and IL-6 concentrations, and their inverse relationship to the activity of the CYP enzyme CYP2C19.[17]

In conclusion, therapeutic proteins may lead to changes in the activities of CYP enzymes by mechanisms different from small molecule drugs. These interactions may take longer to develop (and presumably also longer to subside, but this has not yet been studied), and may also necessitate other designs in cocktail drug–drug interaction studies, particularly because of the long half-life of most therapeutic proteins.

7.4 PUBLISHED COCKTAIL INTERACTION STUDIES WITH THERAPEUTIC PROTEINS

In the literature, there are only few drug–drug interaction studies on therapeutic proteins that use the cocktail phenotyping approach, although such an approach has been endorsed by authors of some review articles in recent years[1,18,19] and by regulatory guidelines on drug–drug interaction studies for drug approval.[8]

A very early example of a cocktail interaction study was carried out with IFN-α in 1993.[20] In that study, a drug cocktail consisting of 150 mg theophylline, 250 mg antipyrine, and 250 mg hexobarbitone was administered to seven cancer patients before, with the first dose, and after 2 weeks of IFN-α dosing. Because these drugs are either no longer in use at all, and/or are not established probe drugs for specific enzymes, the findings are difficult to interpret in terms of generalizability to other drug substrates. The 33% decrease in oral theophylline clearance in this study may be interpreted as a decrease in CYP1A2 activity, the enzyme mainly involved in theophylline clearance; but CYP2A6 may also contribute. Antipyrine is a nonspecific CYP substrate, and the nonsignificant 20–21% decrease in antipyrine clearance both after the first dose and after 14 days of treatment cannot be attributed to a change in a single enzyme activity. Hexobarbitone clearance was unchanged, butbecause hexobarbitone is probably stereoselectively metabolized by various CYP enzymes, the significance of this result remains unclear.

Another, more recent, study investigated the effects of multiple doses of pegylated IFN-α2b on the activities of CYP1A2, CYP2C8/2C9, CYP2D6, CYP3A4/3A5, and *N*-acetyltransferase 2 (NAT2) in 26 patients with chronic hepatitis C infection.[21] The cocktail used in the study consisted of 200 mg caffeine for CYP1A2 phenotyping, 500 mg tolbutamide for CYP2C8/2C9 phenotyping, 45 mg dextromethorphan for CYP2D6 activity assessment, and 100 mg dapsone for the assessment of NAT2 activity, all administered orally and simultaneously, while 4.0 mg midazolam for overall CYP3A phenotyping was administered as a syrup 24 h later. The phenotyping procedures were carried out twice, once before and once after 4 weekly doses of pegylated IFN-α2b adjusted to the body weight of the participants. This cocktail is, according to the authors of the paper, validated but unpublished as "data on file" of the pharmaceutical company that initiated the study. Its composition most resembles the Indianapolis cocktail, with the addition of dapsone as a NAT2 probe substrate, a higher dose of dextromethorphan, and the omission of intravenous midazolam, which is used as a probe for hepatic CYP3A activity.[2]

The authors used appropriate bioequivalence testing methods to calculate point estimates of the geometric means ratios for the selected phenotyping parameters and corresponding 90% confidence intervals. They concluded that multiple doses of pegylated IFN-α2b moderately increased CYP2D6 activity by 67%, but variability was huge (25–123% increase in activity), although genetically poor metabolizers of CYP2D6 had been excluded from study participation. In addition, the authors observed a minor increase in CYP2C8/2C9 activity by 28% (90% confidence interval 15–42%), which was statistically significant but of presumably limited clinical importance, because one border of the 90% confidence interval was within the bioequivalence zone. They did not observe statistically significant changes in overall CYP3A activity and NAT2 activity, but a minimal, but statistically significant decrease in CYP1A2 activity to 92% was noted (90% confidence interval 87–98%).

Since the 90% confidence interval is still entirely within the 80–125% bioequivalence zone, this finding is certainly not clinically relevant.

It is interesting that these findings do not entirely match with the information from studies with pegylated IFN-α2a or IFN-α2b. Namely, CYP2D6 activity is not influenced by pegylated IFN-α2a, while a larger effect on CYP1A2 activity (in the order of a 25% reduction in activity) is mentioned. Possible explanations include the differences in study populations (the present cocktail study was carried out in hepatitis C patients, whereas the studies with IFN-α2a were done with healthy subjects) and an indirect effect mediated by a general amelioration in hepatic function in the patients with chronic hepatitis C by IFN-α2b administration. This study is cited in the U.S. prescribing information, and it is the basis of the precautions for the use of CYP2C9 and CYP2D6 substrates together with pegylated IFN-α2b.

Although this example nicely shows that a cocktail phenotyping study is suitable for the clinical assessment of the relevance of interactions between a therapeutic protein and drug-metabolizing enzymes, there is no other published *in vivo* cocktail interaction study investigating the effect of a therapeutic protein on enzymes and transporters. A search of the U.S. National Institutes of Health clinical study register (www.clinicaltrials.gov) revealed one other cocktail study on the interaction potential of a therapeutic protein. However, that study, which aimed at investigating the effect of briakinumab (ABT-874), a IgG-based monoclonal antibody targeting IL-12/23, was stopped before enrollment, perhaps in the course of the withdrawal of filing of briakinumab at the U.S. and the European medicines agencies because of safety issues. The study was planned to be carried out in patients with plaque psoriasis, the target population for briakinumab, and the scheduled cocktail included caffeine, tolbutamide, omeprazole, metoprolol, and midazolam, for the assessment of the activities of CYP1A2, CYP2C9, CYP2C19, CYP2D6, and CYP3A4. The cocktail was scheduled to be administered twice, once before, and once 14 days after a single dose of briakinumab.[22]

7.5 CONCLUSIONS

In summary, it has been shown that the cocktail approach is suitable for the study of interactions between therapeutic proteins and drugs in humans, namely for the study of the effect of therapeutic proteins on CYP enzyme activities. This is the area in which the FDA's current draft guideline on drug interaction studies recommends the use of cocktail phenotyping studies as an alternative to individual therapeutic protein–drug interaction studies.[8]

REFERENCES

1. Huang SM, Zhao H, Lee JI, et al. Therapeutic protein-drug interactions and implications for drug development. *Clin Pharmacol Ther* **87**, 497–503 (2010).

2. Fuhr U, Jetter A, Kirchheiner J. Appropriate phenotyping procedures for drug metabolizing enzymes and transporters in humans and their simultaneous use in the "cocktail" approach. *Clin Pharmacol Ther* **81**, 270–83 (2007).
3. Zadoyan G, Rokitta D, Klement S, et al. Effect of Ginkgo biloba special extract EGb 761(R) on human cytochrome P450 activity: a cocktail interaction study in healthy volunteers. *Eur J Clin Pharmacol* **68**, 553–60 (2012).
4. Tanaka E, Kurata N, Yasuhara H. How useful is the "cocktail approach" for evaluating human hepatic drug metabolizing capacity using cytochrome P450 phenotyping probes in vivo? *J Clin Pharm Ther* **28**, 157–65 (2003).
5. Zhou H, Tong Z, McLeod JF. "Cocktail" approaches and strategies in drug development: valuable tool or flawed science? *J Clin Pharmacol* **44**, 120–34 (2004).
6. Turpault S, Brian W, Van Horn R, et al. Pharmacokinetic assessment of a five-probe cocktail for CYPs 1A2, 2C9, 2C19, 2D6 and 3A. *Br J Clin Pharmacol* **68**, 928–35 (2009).
7. Palmer JL, Scott RJ, Gibson A, et al. An interaction between the cytochrome P450 probe substrates chlorzoxazone (CYP2E1) and midazolam (CYP3A). *Br J Clin Pharmacol* **52**, 555–61 (2001).
8. U.S. Department of Health and Human Services, U.S. Food and Drug administration, Center for Drug Evaluation and Research, (CDER). Draft guidance for industry: drug interaction studies—study design, data analysis, implications for dosing, and labeling recommendations. Silver Spring, MD, 2012. Available at www.fda.gov/downloads/Drugs/GuidanceComplianceRegulatoryInformation/Guidances/UCM292362.pdf.
9. European Medicines Agency. Note for guidance on the investigation of drug interactions. London, 1997. Available at www.ema.europa.eu/docs/en_GB/document_library/Scientific_guideline/2009/09/WC500002966.pdf.
10. European Medicines Agency. Draft Guideline on the investigation of drug interactions. London, 2010. Available at www.ema.europa.eu/docs/en_GB/document_library/Scientific_guideline/2010/05/WC500090112.pdf.
11. U.S. Department of Health and Human Services, U.S. Food and Drug administration. Drug development and drug interactions: table of substrates, inhibitors and inducers. Silver Spring, MD, 2011. Available at www.fda.gov/Drugs/DevelopmentApprovalProcess/DevelopmentResources/DrugInteractionsLabeling/ucm093664.htm.
12. Zhang L. A regulatory perspective on the utilization of cocktail approach in assessing drug interactions. Paper presented at the FIP Pharmaceutical Sciences World Congress—American Association of Pharmaceutical Scientists Annual Meeting, New Orleans, LA, November 16, 2010. Available at www.fda.gov/downloads/Drugs/DevelopmentApprovalProcess/DevelopmentResources/DrugInteractionsLabeling/UCM237160.htm.
13. Zhang L, Reynolds KS, Zhao P, Huang SM. Drug interactions evaluation: an integrated part of risk assessment of therapeutics. *Toxicol Appl Pharmacol* **243**, 134–45 (2010).
14. Morgan ET. Regulation of cytochromes P450 during inflammation and infection. *Drug Metab Rev* **29**, 1129–88 (1997).
15. Dickmann LJ, Patel SK, Rock DA, et al. Effects of interleukin-6 (IL-6) and an anti-IL-6 monoclonal antibody on drug-metabolizing enzymes in human hepatocyte culture. *Drug Metab Dispos* **39**, 1415–22 (2011).

16. Morgan ET. Impact of infectious and inflammatory disease on cytochrome P450-mediated drug metabolism and pharmacokinetics. *Clin Pharmacol Ther* **85**, 434–38 (2009).

17. Frye RF, Schneider VM, Frye CS, Feldman AM. Plasma levels of TNF-alpha and IL-6 are inversely related to cytochrome P450-dependent drug metabolism in patients with congestive heart failure. *J Card Fail* **8**, 315–19 (2002).

18. Girish S, Martin SW, Peterson MC, et al. AAPS workshop report: strategies to address therapeutic protein-drug interactions during clinical development. *AAPS J* **13**, 405–16 (2011).

19. Lee JI, Zhang L, Men AY, et al. CYP-mediated therapeutic protein-drug interactions: clinical findings, proposed mechanisms and regulatory implications. *Clin Pharmacokinet* **49**, 295–310 (2010).

20. Israel BC, Blouin RA, McIntyre W, Shedlofsky SI. Effects of interferon-alpha monotherapy on hepatic drug metabolism in cancer patients. *Br J Clin Pharmacol* **36**, 229–35 (1993).

21. Gupta SK, Kolz K, Cutler DL. Effects of multiple-dose pegylated interferon alfa-2b on the activity of drug-metabolizing enzymes in persons with chronic hepatitis C. *Eur J Clin Pharmacol* **67**, 591–99 (2011).

22. Abbott. Open label, single dose, non-randomized study to assess the drug to drug interaction of briakinumab on CYP substrates. 2010. Available at http://clinicaltrials.gov/ct2/show/NCT01260844.

CHAPTER 8

LOGISTIC CONSIDERATIONS IN STUDY DESIGN FOR BIOLOGIC DRUG–DRUG INTERACTION ASSESSMENTS

MIN ZHU and YU-NIEN (TOM) SUN

8.1 INTRODUCTION

Evaluation of the drug interaction potential of biologics or therapeutic proteins (TPs) with other TPs or small molecule drugs is an area of emerging research. Different from the pharmacokinetics (PK) drug–drug interaction (DDI) among small molecule drugs in which there are clear mechanisms associated with metabolic pathways mediated by cytochrome P450 (CYP450) enzymes or transporters, there is limited understanding of the mechanistic basis for TP–drug interactions to date. Therefore the design of such interaction studies is challenging in many cases. As described in the literature,[1–3] the common TP elimination mechanisms may include filtration (e.g., into urine), secretion (e.g., into the bile), biotransformation (e.g., catabolism), and target-mediated disposition. Renal elimination, which is one of the primary pathways of small-molecule drugs, is relatively unimportant for large molecules, such as immunoglobin G (IgGs) with a molecular weight of ~150 kDa that prevents efficient filtration through the glomerulus. Secretion into the bile is an important pathway of elimination of IgA antibodies, but this route is not a significant contributor to the elimination of IgG antibodies. The majority of IgG elimination occurs via intracellular catabolism, following fluid-phase or receptor-mediated endocytosis. TP elimination mechanisms distinguish themselves from the hepatic metabolism that is the dominant mechanism for small molecules. In addition, as TPs may be viewed as foreign substances by the body, immune responses that lead to the generation of endogenous antibodies against the TPs may be triggered, which can increase the clearance of TPs.

Drug–Drug Interactions for Therapeutic Biologics, First Edition.
Edited by Honghui Zhou and Bernd Meibohm.

According to the BIO Survey in 2010, potential TP-DDIs were most evaluated in the therapeutic areas of immunology and oncology.[4] More specifically, TPs as anti-inflammatory agents were often evaluated as potential victims for DDIs. The activities for selected cytochrome P450 enzymes may be altered in the presence of TPs that are cytokines, cytokine modulators, or hormones.[5] Drug- and disease-mediated TP-DDIs have been observed for tocilizumab, which modulates CYP enzymes in patients with rheumatoid arthritis via interleukin 6 (IL-6) specific blockage.[6] It has been demonstrated that methotrexate, an immunosuppressant, will decrease the formation of antidrug antibodies against infliximab and therefore increases its overall exposure.[7] Anticancer TPs were mostly evaluated as potential perpetrators to concomitant medications, particularly for small molecules that are intended to be used in combination therapy and have a narrow therapeutic window (e.g., chemotherapeutics with cytotoxic properties). For example, a DDI evaluation was performed for irinotecan plus bevacizumab in patients with solid tumors, and no significant effect of bevacizumab on the pharmacokinetics of irinotecan or its active metabolite SN38 was observed.[8]

Due to the unique characteristics of biologics in pharmacokinetics and immunogenicity, some design parameters of TP–drug interaction studies are different from those for small molecules. A risk-based drug–drug interaction assessment strategy for TPs should be implemented according to its risk factors and appropriate study designs should be selected.[5,9] The challenges and logistic considerations in study designs for TP–drug interactions are discussed in this chapter.

8.2 CHALLENGES IN THE CONDUCT OF A TP–DRUG INTERACTION STUDY

Therapeutic proteins can be roughly divided into two groups according to indications–namely, oncology TPs (for cancer treatment) and nononcology TPs (or rheumatoid arthritis, psoriasis, asthma, prevention of cardiac ischemic complication or other cardiovascular disease). The challenges in the assessment of TP–drug interactions for the two groups of TPs are discussed separately.

In oncology, drug combination (e.g., TP with chemotherapy) is commonly used for the improvement of anticancer activity. With respect to TP–drug interaction assessments, *in vitro* screening–guided drug interaction studies would have limited value because there is often no known underlying mechanism. Furthermore, the TP–drug interaction study is usually not a standalone but rather a substudy of Phase I–III clinical trials conducted in cancer patients in conjunction with efficacy and safety assessments. As a result, conventional well-controlled DDI study conditions (e.g., a controlled diet, comedications, disease status, dose, dosing time, and schedule) are rarely preserved. TP pharmacokinetics data collection is usually less intensive than that for small molecule drug interaction studies. The number of subjects may not be well balanced across treatment cycles or arms, and the number of qualifying subjects is also affected by deviations of scheduled dose and sampling times due to dose interruptions and reductions or patient discontinuations.[10] Challenges exist in performing noncompartmental pharmacokinetics data analysis because there are often a significant number of data exclusions. Pharmacokinetics assessments are further complicated due to a number of oncology TPs or small molecule drugs exhibiting nonlinear pharmacokinetic properties. Moreover, because a TP-alone arm is uncommon in a drug combination oncology trial, historical data are usually referenced in the TP–drug interaction evaluation either by cross-study comparison or by population pharmacokinetic modeling. Thus a scientifically sound and technically feasible integrated TP–drug interaction evaluation strategy needs to be developed to address these challenges.[11]

In contrast, for nononcology therapeutics such as those in the area of immunology and anti-inflammation, TPs can be used as a monotherapy agent. The concomitant drugs can be randomly prescribed medications for symptom relief or for treatment of other comorbidity conditions according to individual patient's needs. The most documented TP–drug interaction involves an indirect mechanism of cytokine-mediated changes in drug metabolizing enzymes that is secondary to changes in inflammation status. Studies have demonstrated the effect of individual cytokines and their modulators on CYP enzymes and transporters.[6,12–14] In this regard, the TP plays the role of a perpetrator and the small molecule drugs are usually victims. This type of interaction is not solely drug related but also disease related. To date, inflammation- or cytokine-mediated inhibition of CYP activities has become an evolving research area for understanding the mechanism of disease–drug interactions and the associated effect on TP–drug interactions.[15] The challenge in the design of such TP–drug interaction studies is that both the inflammation stage and the concomitant drugs can be moving targets. *In vitro* experiments by using different cytokines and selected animal models at the early stage of drug development may provide some insight into the pharmacokinetic interaction potential for coadministered drugs that mainly undergo the CYP-mediated drug disposition, but the magnitude and clinical relevance of the interaction may still need to be evaluated in the target patient population. A strategy of early establishment of a therapeutic exposure range under TP monotherapy may be important for anti-inflammatory agents to assess extent and clinical significance of changes in drug exposures.

Additional logistic considerations of TP–drug interaction assessments are provided in Table 8-1. Clearly, because the assessment is related to the impact on drug efficacy and safety, early strategy planning is essential to reduce logistic burdens and meet regulatory requirements. The recently released "Draft Guidance for Industry: Drug Interaction Studies–Study Design, Data Analysis, Implications for Dosing, and Labeling" from the FDA, clearly indicates that (1) if an investigational TP is a cytokine or cytokine modulator, studies should be conducted to determine the TP's effect on CYP enzymes or transporters; (2) for TPs that will be used in combination with other drugs as a combination therapy, studies should evaluate the effect of each product on the other; and (3) when there are known mechanisms or prior experience with certain pharmacokinetic or pharmacodynamic interactions, appropriate *in vitro* and *in vivo* assessments for possible interactions should be conducted.[16]

TABLE 8-1 Logistic Considerations of TP–Drug Interaction Assessment Approach

Approach of Assessing TP–Drug Interaction	Pros	Cons
No evaluation of TP–drug interaction potential	No burden on clinical trial operation and costs	Lack of data to assess potential TP–drug interaction risk and impact on safety and efficacy Lack of data to fulfill regulatory requirements
Test of TP–drug interaction potential in healthy subjects	Traditional drug interaction designs and conditions can be applied Low operational burden	Drug combination may not be safe to be tested in healthy subjects (e.g., anticancer drugs) Results may not be applicable to target patient population owing to disease–drug interaction (e.g., anti-inflammatory agents)
Estimation of TP–drug interaction in early drug development stage (e.g., Phases I and II)	Data may be used to address potential DDI issue and assist in better design of later stage clinical trials May be executable with limited number of patients and at low cost May identify large signal with an estimation study	Adds some operational burden to clinical trials All tested drugs, including marketed drugs, will be treated as IMPs in Europe and governed by associated regulations Estimation study may not have enough power to identify a mild interaction signal
Estimation of TP–drug interaction in late drug development stage (e.g., Phase III)	"No DDI findings for the tested drugs" may be included on a product label	Adds significant operational burden and cost to clinical trials All tested drugs will be treated as IMPs in Europe Study conditions may not be controlled consistently across countries in a global trial Study noise may be increased in a large global trial for detecting a mild interaction signal
Dedicated TP–drug interaction study as postmarketing	Study can be powered to address the TP–drug interaction question definitively	Needs an independent study High operational burden and cost to clinical trials

TABLE 8-1 (*Continued*)

Approach of Assessing TP–Drug Interaction	Pros	Cons
commitment or during drug development	A claim of "lack of TP–drug interaction" may be acceptable on a product label	Patient enrollment may be low owing to intensive PK sample collection and other logistic issues Data evaluability may not be high because of patient dropout and change in dose and regimen owing to side effects (e.g., oncology drugs)

DDI: drug–drug interaction; IMP: investigational medicinal product; PK: pharmacokinetics; TP: therapeutic protein.

8.3 TP–DRUG INTERACTION STUDY DESIGN

Unlike a small molecule drug–drug interaction study in which a two-way drug interaction (i.e., A → B *and* B → A) can be evaluated in the same study, a TP–drug interaction assessment may need to be evaluated as a one-way interaction (i.e., A → B *or* B → A), especially in oncology. Therefore, an important consideration in the design of a clinical TP–drug interaction study is to clearly define the aim of the study and then design the study components accordingly. As mentioned in the previous section, most of the TP–drug interaction studies for oncology and anti-inflammatory therapeutics need to be conducted in patient populations and strictly controlling the potential influential factors (e.g., other comedications, dose and schedule, and disease status) may be challenging. Thus conventional drug interaction study settings generally applicable to most small molecule drugs may not be practical for the TP–drug interaction assessment.

In the oncology research area, because there is no clear mechanism-based TP–drug interaction identified yet, the value of *in vitro* screening and preclinical assessments for guiding a clinical TP–drug interaction study planning has not been well established. To date, inclusion of clinical TP–drug interaction components in Phase I–III studies in patient populations for fixed drug combinations has become a common approach for screening purposes and for identifying if exposures of any drug in a fixed combination are considerably affected, especially for drugs with narrow therapeutic windows. The study is usually performed on a small scale for providing initial data for assessing the risk level of a PK interaction. If a high risk is identified, a confirmative study or formal TP–drug interaction study may be triggered, depending on the clinical relevance of the interaction. This approach appears to be accepted by the regulatory agencies.

In the nononcology research areas, *in vitro* screening and *in vivo* cocktail studies may be applicable for risk assessments at a case-by-case basis. With the *in vitro* evaluation, the impact of cytokines and cytokine modulators on CYP enzymes, mRNA, protein, and activities can be assessed in a concentration-dependent manner. Nevertheless, there are many uncontrolled factors in the experimental conditions, and the *in vitro* system could be donor dependent. Uncertainty of the translatability of nonclinical results to clinical impact may be high.[17] With respect to the *in vivo* cocktail approach, a number of probe drugs for enzymes and transporters can be administered together for simultaneous assessment of interaction between small molecule drugs and TPs.[15] The approach may be efficient for screening TP–drug interaction potential and may be used when no *in vitro* assays can be properly established or for verifying *in vitro* screening results or for planning further clinical assessments. The challenges of this approach may be related to the lack of ideal probe substrates, the validity of cocktail studies, the composition of the cocktails, the extrapolation of results to other drugs, and the differentiation of PK and pharmacodynamic (PD) interactions. Further TP–drug interaction studies may be needed for labeling purposes to provide the magnitude of change in exposure for a particular drug of interest in the cocktail.

With respect to clinical TP–drug interaction assessments, there are at least three types of studies commonly employed–namely, screening studies, estimation studies, and dedicated studies–depending on the purpose of the assessment, phase of clinical development, disease areas, TP–drug interaction risks, and sample size. The pros and cons of these types of study are discussed in the following sections.

8.3.1 Screening Study for TP–Drug Interactions

A screening study is usually conducted in patients with a relatively sparse PK sampling scheme to determine if any TP–drug interaction signals are apparent. The study is commonly designed for TPs for which the mechanism of interaction is unknown or an interaction is not expected based on the clearance mechanism of the drugs involved.

The screening study for oncology therapeutics aims at testing the effect of the coadministered drug in a fixed combination with the TP (D $\rightarrow$ TP). Sparse TP PK samples are usually collected from all subjects in a trial for evaluating the safety or efficacy while the PK samples of concomitant drugs in the fixed combination are not collected. The TP concentration time data are analyzed preferably via population PK modeling with pooled historical data of monotherapy and the data from the combination study to assess covariate effects of coadministered drugs on the TP clearance. The assessment is feasible in most of clinical trials for screening signals and to check if the TP clearance was considerably affected. If any concerns are raised based on an observed change in drug exposure, additional data analysis may be required, such as including data from additional trials to examine confounding factors because the change in exposure may be the result of not only a TP–drug interaction but also of other factors (e.g., cancer type or disease stage). Besides population PK analysis, summary statistics for comparison of observed TP concentrations in a drug combination study with those in monotherapy studies are also commonly performed to identify a TP–drug interaction signal. The screening study is usually conducted during Phases I and II trials. Further TP–drug interaction studies may be needed if clinically meaningful changes in TP exposure was to be identified. For nononcology TPs, cocktail studies can be used for screening in order to test the effect of the TP on small molecule drugs (TP → D) because the small molecule drugs are commonly the victims.

8.3.2 Estimation Study for TP–Drug Interaction

An estimation study is commonly used for a TP–drug interaction assessment when the mechanism of PK interaction is unknown or the interaction is not expected based on the clearance mechanisms of the drug involved. Unlike a screening study, an estimation study collects PK profiles over a specified interval for concomitantly administrated drugs used in combinations, either for small or large molecule drugs, and relative sparse PK samples (e.g., before and after infusion) for the tested TP. The aim of the study is to evaluate the TP–drug interaction potential with a predefined sample size, which is commonly used for the evaluation of the effect of a TP on coadministered drugs (TP → D). Nevertheless, the sample size for an estimation study is usually 6–15 evaluable subjects per arm, depending on the intersubject variability in drug exposure. This is not a powered study for hypothesis testing. The advantage of this kind of study is to test TP–drug interactions in a small group of patients so the operational challenge is relatively low while it still provides evidence for risk assessment. If remarkable changes in drug exposure are identified, the study data can be used for hypothesis generation and provide valuable information for the design of a formal study. The limitation of this type of study is that the effect of coadministered drugs on TPs (D → TP) may need to be assessed by cross-study comparison or population PK modeling if there is no TP alone arm or testing period in the trials.

8.3.3 Dedicated Study for TP–Drug Interaction

The dedicated study for the TP–drug interaction assessment is not frequently conducted for oncology TPs but may be conducted for nononcology TPs. In this type of study, intensive PK samples for both TP and comediations are collected. There is not much difference between the dedicated TP–drug interaction study and the traditional drug–drug interaction study for small molecule drugs except the TP–drug interaction studies are mostly conducted in patient populations to test the effect of drug–disease interactions (e.g., inflammation) or fixed drug combinations (e.g., oncology). The study aims are hypothesis testing with sufficient detection power calculated according to results obtained from a screening study or estimation study. Although this type of TP–drug interaction study can provide confirmative results, the logistic issues are apparent as it usually requires much larger sample size, depending on intersubject variability of TPs, the predefined no-effect boundary, and the rates of evaluable individuals after accounting for patient dropout, dose reduction, and dose interruptions.

8.4 TIMING OF TP–DRUG INTERACTION STUDY

The question when to conduct the TP–drug interaction study may be associated with therapeutic areas and the risk level of individual therapeutic proteins. For instance, in the oncology area, it is preferable to include TP–drug interaction assessments in the early Phases I and II trials because this can be the first time that a drug combination is tested, trials are relatively small in size, and operational issues can be addressed easier. It is important that obtaining the interaction data early can help establishing a development strategy and guide later phase study designs. If clinically meaningful interaction signals are identified in early trials, expanded assessments may be further considered. The limitation of early assessments is that the optimal drug combination may be modified in later stages of development or the tested drug combination may not proceed to later phase due to ending of the development program in certain indications. Although it is possible to test TP–drug interactions in late phase studies for product registration, it significantly increases the complexity of pivotal Phase III trials and is thus likely to have more confounding factors and causes difficulty in data interpretation.[10]

In contrast, for TPs for the treatment of immunology- and inflammation-related diseases, most clinical trials in the early phase of clinical development are monotherapy trials. The risk in a change in drug exposure may be from an inflammation-mediated disease–drug interaction rather than a direct TP-mediated PK interaction with randomly prescribed medications. However, because the treatment of inflammation with TPs may affect the magnitude of disease–drug interactions, the treatment-associated TP–drug interaction effect may still need to be monitored if patients are taking narrow therapeutic drugs along with the TPs. Owing to the complexity of the interaction, this type of TP–drug interaction study is not easy to design becuase the extent of interaction could change with treatment and disease status. Standalone trials can be conducted for nononcology TPs by using probe substrates (e.g., midazolam for CYP3A4) to examine the risk level. This type of study is commonly planned in the middle or late phase of development or the postmarketing stage based on clinical significance of effect.

8.5 STRATEGIC PLANNING OF TP–DRUG INTERACTION STUDIES

Owing to uncertainties in drug development, strategic planning of TP–drug interaction assessments at the nonclinical, clinical Phases I–III, and postmarketing stages is important for an increase of research effectiveness. A stepwise strategy may be needed for planning purposes, which should be focused on the therapeutic area and specific drug, according to the characteristics of the molecules. To elaborate this, examples for general TP–drug interaction assessment strategies in the oncology and anti-inflammation areas are further discussed.

8.5.1 Oncology

The TPs for cancer treatment are frequently used in fixed drug combinations (e.g., with chemotherapy) for enhancement of treatment effect and survival benefit. Because there is no known mechanism-based TP–drug interaction identified to date, either for TPs as victim or perpetrators, the study planning heavily relies on clinical screening with data from both monotherapy and fixed drug combination trials for initial risk assessment. There is no one size fits all strategy in such assessments. Although, there is no clear interaction mechanism, evidence of lack of interaction is still required by regulatory agencies.

The strategy of interaction assessment should be clearly described in a clinical pharmacology plan of an investigational TP at the early phase of drug development. The plan may include the purpose of assessment (e.g., evaluate the TP as a victim or a perpetrator, or both), method of assessment (e.g., population analysis approach, noncompartmental analysis, or a combined approach), criteria of analysis (e.g., what is the clinical significance with a defined magnitude change in exposure?), study planning (alignment in study conditions, sampling strategy, and study population across studies), data integration plan (e.g., identification of studies from which data will be used for the TP–drug interaction evaluation), covariate data collection plan (guarantee that common covariates are collected consistently across studies), communication strategy (alignment with clinical teams), and labeling strategy. It is preferable to collect the TP–drug interaction data in Phases I and II; avoiding Phase II, however, reduces operational challenges. Strategic and systemic planning of TP–drug interaction assessment is important in the oncology area because a cross-study comparison is hardly avoidable, especially for TPs with long half-lives.

For small molecule drugs (e.g., chemotherapy agents) that are commonly used in combination with TPs for cancer treatment, it would be useful to establish an integrated internal database across different oncology programs for effectively using all available data.[18] For instance, it is common that selected chemotherapy backbone treatments (e.g., paclitaxel, cisplatin) are tested repeatedly for the same or similar indications (e.g., lung and colon cancers) for different investigational TPs. However, the sample size of each trial might be small (e.g., $n = 5$–7 evaluable subjects) owing to operational difficulties, and missing samples are common because patients are unable to go back to clinics to have blood drawn and often have dose delays or reductions. Therefore, it would be useful to collect PK data of a specific chemotherapy agent with a standardized sample collection scheme for all oncology programs and the data from the control arms (e.g., paclitaxel backbone cohort) of each study in the same patient population may be combined for data analysis to increase detection power and establish a no-effect boundary for a test agent.[19,20]. The established exposure range can be used for all other relevant studies in which the agent was included in drug combinations.

For TPs, PK data will be collected from most subjects, thus sample size will not be an issue. As a result of rare standalone drug interaction trials, TP monotherapy data, usually collected in the first-in-human study, need to be included in the data analysis. It is important to predefine a no-effect boundary as a reference exposure range for each drug in a drug combination based on pharmacokinetics, safety, and efficacy information. If the exposures in the testing cohort were within the ranges, lack of TP–drug interaction effect may be claimed. Otherwise, a detailed covariate analysis will be performed to identify if the change in exposure is owing to a TP–drug interaction or nondrug interaction factors (e.g., cancer type, disease stage, biomarker related subpopulation). If there is clear evidence that the change in exposure is due to nondrug interaction factors, lack of TP–drug interaction may be claimed. If the change in exposure could be the result of a TP–drug interaction and the magnitude of change may have a clinical consequence, further evaluation may be considered. All evaluation results of different TP–drug combinations in a TP development program can be summarized to provide an overall picture of interaction potential for an investigational TP. The results of the TP–drug interaction assessment via planned analysis should be provided on the drug product label with a detailed description of how data were collected and analyzed. This approach may be applicable not only to oncology TPs but also to nononcology TPs that have no clear mechanism of interaction.

8.5.2 Immunology and Inflammation

In contrast, in nononcology areas, such as immunology and anti-inflammation, the TPs are frequently used as monotherapy. Although *in vitro* screening and *in vivo* cocktail approaches may be useful for initial risk assessment, clinical studies under actual treatment conditions in target populations will provide more convincing evidence. Because drugs mainly metabolized by CYP enzymes may be affected by the disease–drug interaction, adequate recording of concomitantly used drugs over a sufficiently long period (e.g., >6 months) may provide some evidence for assessing risk of TP–drug interactions. Covariate analysis via population PK modeling is commonly used for identifying the interaction signals with the individually prescribed and long-term used comedications.[15] If a significant effect is identified and change in exposure may have a clinical consequence, further evaluation of a drug interaction may be considered.

8.6 CONSIDERATIONS IN STUDY DESIGN

8.6.1 Study Population

In drug–drug interaction studies for small molecules, it is acceptable to conduct the study in healthy subjects because the purpose of the study is to evaluate the direct effect of one drug on the disposition of another concomitantly administered drug for which drug metabolism and elimination pathways may be affected. However, the assessment of oncology TP–drug interaction is rarely conducted in healthy subjects owing to safety considerations. In the immunology therapeutic area, because inflammation-mediated disease–drug interaction governs the mechanism of TP–drug interaction, the target patient population has to be evaluated to assess the magnitude of disease–drug interaction over time. As a result, the conduct of TP–drug interaction studies in target patient populations is appropriate in most cases.

8.6.2 Selection of Interacting Drugs

The selection of interacting drugs for investigation is mainly based on the potential of concomitant usage, PK, and PD properties and the therapeutic window of the selected drugs. For instance, if an investigational TP will be used in a combination of two chemotherapy agents and one supportive hormonal agent, and the hormone drug has a wide safety window and is prescribed at a low dose, the selected testing drugs for the interaction study can be the TP and the two chemotherapy agents. In the inflammation area, because each patient may be taking his or her own long-term medications along with the TP treatment, the drugs with narrow therapeutic windows will be the choice of probe drugs for the assessment. The primary consideration in drug selection should be based on the drug elimination mechanism.

8.6.3 Type of Study Design

The parallel design and the one-sequence crossover are the commonly selected study designs in TP–drug interaction evaluations (Figure 8-1). The design should be chosen based on the PK properties of the TP and coadministered drugs as well as the potential interaction mechanism.

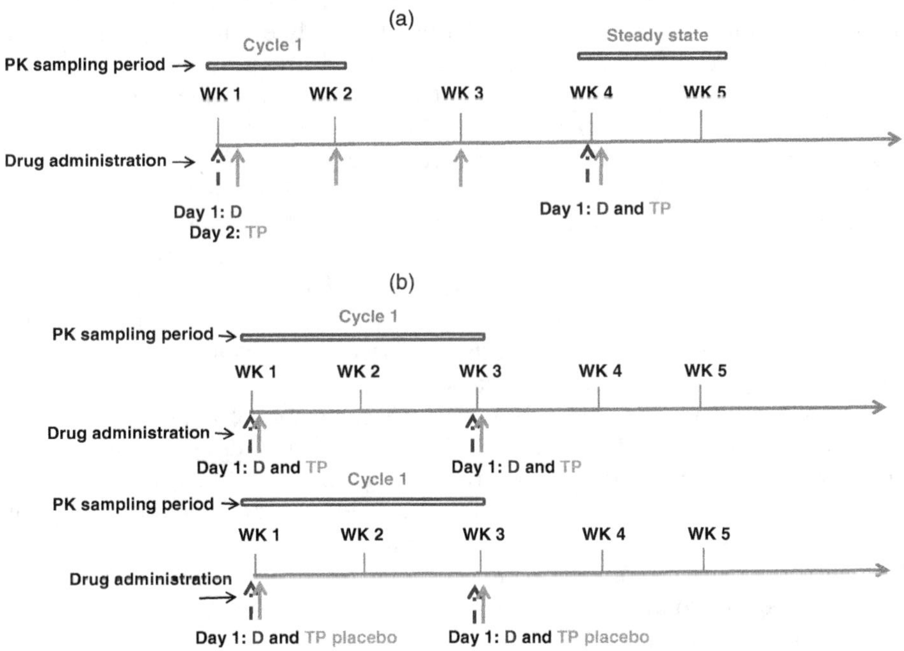

FIGURE 8-1 Examples of (a) one-sequence crossover and (b) parallel designs. D: small molecule drug; PK: pharmacokinetics; TP: therapeutic protein; WK: week. (See the color version of this figure in Color Plates section.)

The one-sequence crossover (sequential) design is common for testing the effect of a TP on a small molecule drug (TP → D), such as administration of a chemotherapy agent (D) on day 1 and TP on day 2 in one cycle followed by administration of both D and TP on day 1 in a cycle at steady state (Figure 8-1a). Each subject serves as his or her own control to minimize variability and allow for the achievement of sufficient statistical power with a reasonable sample size.

Examples exist for the evaluation of two-way interactions of both TP and small molecule drugs with one sequence crossover design in a single study. In oncology, a study was conducted for the evaluation of the interaction between trastuzumab and paclitaxel in which paclitaxel was administered alone in cycle 1 to collect paclitaxel concentrations, followed by paclitaxel coadministration with trastuzumab with assessment of PK profiles of both drugs in cycle 4, and then trastuzumab monotherapy in cycle 10 or beyond for collection of trastuzumab-alone concentration data.[21] In the inflammation area, the interaction between digoxin and etanercept was assessed with a one-way, crossover, three-period design (digoxin → digoxin + etanercept → etanercept).[22]

The sequential one-way crossover study is ideal for short-half life ($t_{½}$) drugs (e.g., hours) and TPs (e.g., a few days). The advantage of this design is that the study requires fewer subjects when compared to that with parallel design. A limitation of this type of study is that it requires longer study duration so that it may be suitable only for patient populations that have long life expectancy. However, if patients drop out from the study for any reason, imbalanced patient numbers over the study period may occur, which will affect the study evaluation and power of assessment. Moreover, this type of study cannot avoid a sequence effect, especially when dose interruption occurs, and it cannot be used to evaluate the effect of D → TP if there is no TP-alone arm or testing period in the study. In this case, a cross-study comparison for TP exposure needs to be made.

Many TP products have long half-lives (weeks) as well as nonlinear PK owing to target-mediated clearance, or they form antiproduct antibodies that may alter their own clearance. These unique TP characteristics produce unavoidable sequence effects that preclude the use of the crossover designs for assessing TP–drug interaction effects. More complicated studies can be designed to assess the sequence effect. As an alternative for avoiding the sequence effect, parallel designs can be considered for the TP–drug interaction assessments.

A parallel design includes two arms (Figure 8-1b), one for examining drug exposures when administered alone and the other for examining drug exposures in the presence of TP. When a TP has a relative long half-life (10–20 days) and patients have relative short life expectancy (< 6 months, with end stage cancer), this type of design may become the only choice for the drug interaction assessment. Intensive PK samples are usually collected in the first cycle of treatment rather than at steady-state owing to the short life expectancy of the patients. Sparse PK samples can be collected continuously in the following cycles for examining drug accumulation and steady-state concentrations.[18] The study is more appropriate for Phase II oncology studies in which chemobackbone is included in a positive control arm. PK samples of chemotherapy agent from both arms can be collected to compare exposures with and without the TP. The assessment of effect of chemotherapy on TP may need a cross-study comparison by either noncompartmental analysis (NCA) or population PK modeling owing to lack of a TP alone arm in the same study.

When the TP elimination half-life is relatively short, it is possible to evaluate TP $\rightarrow$ D and D $\rightarrow$ TP effects with a parallel design in a single study. This was demonstrated in an interaction study between cetuximab ($t_{1/2} = 5$ days) and irinotecan ($t_{1/2} < 1$ day). In this study, two test groups were designed in parallel. Group A was to test the effect of cetuximab on the PK of irinotecan, in which irinotecan was administered alone followed by the drug combination. Group B was to test the effect of irinotecan on the PK of cetuximab in which cetuximab was administered alone followed by the drug combination.[23]

The advantage of the parallel design is the study duration is relatively short when compared to the crossover design, and it is relatively easy to randomize patients in the parallel arms to minimize bias in patient selection owing to baseline characteristics. There is no sequence effect for this type of design, and the design is commonly used to examine TP–drug interaction effects for D + TP and TP + TP combinations in either open-label or double-blind studies. The disadvantage of the design is that much more patients are needed, especially for highly variable TPs, to achieve the same detection power as with the crossover design.

A hybrid approach with a one-way sequential design for a Phase Ib study (e.g., chemo followed by chemo + TP, open label) staggered with a double-blind parallel design in the Phase II portion of the study for testing the effect of TP on chemotherapy is not uncommon in oncology trials.[10] The study is more appropriate for Phase Ib/II trials in which insufficient data on chemotherapy are collected in the Phase Ib portion and the data will be continuously collected in the Phase II portion. The assessment of the effect of chemotherapy on TP may still need a cross-study comparison.

8.6.4 Regimen Selection

In oncology, because TPs are commonly used as add-on therapy, the regimen of the backbone treatment (e.g., chemotherapy or target therapy) becomes an important determining factor for the choice of regimen. For instance, the TP regimen will be once every 3 weeks (Q3W) for the treatment of lung cancer if coadministered paclitaxel is given Q3W. On the other hand, if a small molecule drug is given daily, then the regimen of TP can be flexible (e.g., QW, Q2W, Q3W), according to the half-life of the TP and the safety and efficacy profiles. In immunology, in which TPs are frequently used as monotherapy, the choice of regimen is mainly based on the PK, efficacy, and safety profiles of the TP.

8.6.5 Sampling Scheme

Depending on the PK characteristics of the tested drugs, intensive PK samples that enable assessment of the AUC of the tested drugs are usually collected for drugs with short half-lives, and relatively sparse PK samples are collected for TPs owing to their much longer half-lives. Combined sampling schemes–for example, collection of intensive sampling in cycle 1 and sparse sampling at steady state–are common for TPs. Population PK modeling and simulation can be used to optimize sampling schemes. To evaluate the clinical relevance of change in drug exposure parameters, samples for safety, pharmacodynamics, and efficacy measures may be collected along with the concentration samples. If an interaction was observed, clinical relevance of the interaction can be evaluated in conjunction with safety and efficacy data for supporting data interpretation.

8.6.6 Endpoints for TP–Drug Interaction Evaluation

The peak (C_{max}) and trough (C_{min}) concentration are common PK endpoints for TPs in TP–drug interaction assessment when only sparse concentration data are collected. With intensive concentration sample collection and model-based data analysis, the AUC and clearance can be the estimated exposure and PK parameters. Because intensive PK samples can be collected for small molecules, C_{max}, C_{min}, and AUC are the commonly estimated exposure endpoints. The most relevant exposure parameters should be the ones associated with safety and efficacy events. If a TP–drug interaction signal is observed, an assessment based on a joint evaluation of PK, safety, and efficacy data should be the primary consideration in data interpretation.

8.6.7 Sample Size

For a screening TP–drug interaction study, sample size can be < 10 subjects per arm, with intensive sampling for small molecule drugs and several hundred subjects with sparse sampling for TPs. All subjects with relevant data can be included in the data analysis. For an estimation study, the TP–drug interaction evaluation is usually performed with 10–20 patients per arm in a PK subgroup assessment of clinical trials. The sample size of fully powered TP–drug interaction studies for hypothesis testing is much larger, depending on power of detection, intersubject variability in TP exposure (e.g., CV = 30–50%), and the preset criteria of no-effect boundary.

8.6.8 Criteria for TP–Drug Interaction Evaluation

One of the important aspects in strategic planning and perspective study design is to predefine acceptable criteria of no-effect boundary. Prior knowledge and literature information from molecules with similar structures and elimination mechanisms will be valuable for developing the acceptable clinical criteria. In fact, experience in development of such criteria is very limited for TPs so far. It requires collective knowledge of the disease, comorbidities, cytokine levels under physiological and disease conditions, immunogenicity, biological target, pharmacology of the molecules, PK and PD of involved drugs, and the various mechanisms involved in the

TP clearance. Accumulation of such knowledge on a continuous basis will help properly establish the criteria.[11]

For small molecule drugs, the nominal 80–125% criterion is commonly accepted for no-effect boundary of parameter estimates. However, if the same criterion is applied to TPs, a larger sample size is needed not only because of the large intersubject variability in TP exposures but also because of the studies needed to be conducted in patients in the outpatient setting. In oncology, for instance, patients participate in PK evaluation voluntarily. They may not come to clinics for a blood draw according to protocol-specified time as a reslut of individual difficulties. Data are commonly missed; changes of dosing schedule and reduction of dose levels may be needed for management of adverse events. These factors may lead to an increase in the number of unevaluable subjects so that an extra subjects are need to make up for unpredictable events.

In the TP–drug interaction assessment, the no-effect boundary may be considered to be slightly larger for TPs (e.g., 70–143%) than that for small molecule drugs if it is clinically justifiable. Eventually the determining factors for the acceptable criteria should be associated with the therapeutic window established for the drugs of interest.

8.7 DATA ANALYSIS

A data analysis approach is closely related to the study design. For dedicated TP-DI studies, intensive PK samples are collected in a sufficient number of subjects. Noncompartmental analysis (NCA) is the primary data analysis method, which is straightforward, makes few assumptions, and is accepted by regulatory agencies. For screening and estimation TP-DI studies, depending on the sampling scheme and number of subjects with evaluable samples, both NCA and population PK modeling approaches as well as a combination of the two approaches may be applied. For the prospectively designed Phase IIb or Phase III studies with DDI components, the primary analysis approach is population PK modeling. Each analysis approach has advantages and limitations, which are discussed in the following sections.

8.7.1 Noncompartment Analysis

As mentioned, the NCA method is simple, descriptive, less resource constrained, and widely accepted by regulatory agencies. Therefore it should be the first choice when intensive PK samples are collected. This is especially true when an interaction study is conducted in healthy subjects. However, when a TP–drug interaction study is conducted in patients, whether this method is still applicable depends on the number of evaluable samples collected in a dosing interval and whether the original sampling scheme is closely followed. If the preset dosing conditions have been greatly modified, it would be difficult to provide summary statistics and make reasonable comparisons to assess the interaction effect. In such cases, the population PK modeling approach will be more appropriate for further analyzing the data.

8.7.2 Population Pharmacokinetic Modeling

The great advantages of the population PK modeling approach may include the following:

- The method is less affected by sampling conditions and dose modifications and interruption.
- Data from different trials can be combined to increase detection power and accuracy of parameter estimation and provide a comprehensive evaluation of interaction potential for drugs of interest.
- Both sparse and intensive PK data can be used, and multiple covariates can be evaluated at the same time.
- The magnitude of multiple covariate effects on drug exposure or clearance can be ranked, thus the effect contributed by coadministered drugs can be quantified.
- The effect on drug metabolite(s), if any, in the TP–drug interaction can be further evaluated.

The effect of one drug on the other is usually assessed as a covariate of drug clearance. The method is able to differentiate the effect of a coadministered drug from other factors (e.g., baseline characteristics, demographics, lab values, biomarkers, disease status, etc.) if sufficient information is available. With combination of efficacy and safety data, a PK-PD model can be further established for the evaluation of the clinical relevance of such an effect. The model is used not only to describe observed data but also to simulate data under untested study conditions and for an improved design of the next interaction studies (e.g., to select an optimal sampling scheme and to overcome operational difficulties). The simulation may help evaluate logic of dosing schedules for coadministered drugs to minimize the interaction. Model-based simulations with a large number of subjects may increase the confidence of the analysis.

Concerns of using population PK analysis may include the following:

- The modeling approach makes assumptions and may not be accepted by all regulatory agencies around the world.
- It requires much longer time or more resources for data analysis when compared to NCA.
- When there are too few data and parameter estimates have a large uncertainty, the modeling result may be inconclusive.
- The interaction assessment may be affected if dosing records for tested TP and coadministered drugs are incomplete.
- The interaction effect may be underestimated or overestimated when using sparse data pooled from multiple studies and when certain study factors are imbalanced across studies.

- Studies generating PK data were not designed for the drug interaction assessment, thus the evaluation results may be considered as exploratory.

The lack of a signal can occur if the studies are underpowered to detect the interaction signal. Therefore, the modeling approach will be best used when the study is designed for an interaction evaluation, and potential confounding factors are controlled as much as possible.

8.8 PROSPECTIVELY DESIGN OF TP–DRUG INTERACTION STUDY

The screening and estimation types of TP–drug interaction studies, as mentioned in the early part of this chapter, are exploratory in nature for hypothesis generation. The findings may need to be confirmed, especially when changes in drug exposure have potential clinical impact. It is possible to prospectively design the interaction study so that the evaluation results can be considered as confirmatory from the statistical perspective.

The analysis approach requires (1) establishing a model structure for the drug of interest with historical monotherapy data with incorporation of prior knowledge, (2) simulation of concentration-time profiles based on conditions set in the TP interaction study protocol, (3) analysis of the simulated data with the established model and making adjustments in the model structure if needed (the sample size of the study needs to be reasonable, which will be estimated based on the number of subjects in each arm, number of samples per subject, and variability in drug exposure), (4) the selected model structure and no-effect boundary of drug exposure need to be prespecified in the data analysis section of the protocol or in the related statistical analysis plan, and (5) the analysis of the data from the prospectively designed interaction with the prespecified model and examination if the observed data fall within the expected concentration range. Additional sensitivity analysis and ad-hoc analysis may be performed based on the outcomes of analysis. If the observed data are within the specified concentration range, lack of interaction can be confirmed. If significant interaction is observed, the clinical relevance of the effect will be further evaluated in conjunction with assessments of efficacy and safety data. A publication on how to perform confirmatory analysis with population PK modeling is recommended for further reading.[24]

8.9 SUMMARY

Assessment of TP–drug interactions is an emerging research area of clinical pharmacology. The logistic considerations in conducting a TP–drug interaction study are being realistic and being practical. The FDA is currently revising its NDA and Biologics License Application (BLA) recommendations to industry sponsors to recognize the accumulation of TP–drug interaction experiences.[5,16] The current guidance does acknowledge the challenges associated with the use of *in vitro* and *in vivo* data for assessing the risk of TP–drug interactions as well as the design and conduct of such studies. Scientific investigation in this emerging area is needed to improve study design and analysis methods. Prospectively designing TP–drug interaction studies with integrated data analysis may improve the power of detection and reduce certain operational challenges. A further understanding of the mechanisms of TP–drug interaction is needed to better design studies and to ensure patient safety.

REFERENCES

1. Lobo ED, Hansen RJ, Balthasar JP. Antibody pharmacokinetics and pharmacodynamics. *J Pharm Sci* **93**, 2645–68 (2004).
2. Wang W, Wang EQ, Balthasar JP. Monoclonal antibody pharmacokinetics and pharmacodynamics. *Clin Pharmacol Ther* **84**, 548–58 (2008).
3. Kraynov E, Martin SW, Hurst S, et al. How current understanding of clearance mechanisms and pharmacodynamics of therapeutic proteins can be applied for evaluation of their drug-drug interaction potential. *Drug Metab Dispos* **39**, 1779–83 (2010).
4. Zhou H. A recent BIO survey of therapeutic protein-drug interaction evaluation. Paper presented at the AAPS Workshop on Strategies to Address TPDI During Clinical Development, San Francisco, May 15–16, 2010.
5. Huang SM, Zhao H, Lee JI, et al. Therapeutic protein-drug interactions and implications for drug development. *Clin Pharmacol Ther* **87**, 497–503 (2010).
6. Morgan ET. Impact of infectious and inflammatory disease on cytochrome P450-mediated drug metabolism and pharmacokinetics. *Clin Pharmacol Therap* **85**, 434–38 (2009).
7. Aikawa NE, de Carvalho JF, Silva CAA, Bonfá E. Immunogenicity of anti-TNF-α agents in autoimmune diseases. *Clin Rev Allergy Immunol* **38**, 82–89 (2010).
8. Denlinger CS, Blanchard R, Xu L, et al. Pharmacokinetic analysis of irinotecan plus bevacizumab in patients with advanced solid tumors. *Cancer Chemother Pharmacol* **65**, 97–105 (2009).
9. Sun Y-N. Risk-based drug-drug interaction assessment strategy for therapeutic proteins. Paper presented at the Symposium for Therapeutic Proteins: From Bench to Bedside; 39th American College of Clinical Pharmacology Annual Meeting, Baltimore, MD, September 12–14, 2010.
10. Zhu M, Sun Y-N. Drug-drug interaction of oncology biologics—study design and logistics in oncology setting. Paper presented at the AAPS National Biotech Conference, Seattle, June 24–27, 2009.
11. Zhu M. An integrated approach to assess therapeutic protein drug interaction in oncology setting—current strategy. Paper presented at the TPDI Workshop, AAPS National Biotechnology Conference, San Francisco, May 17–21, 2010.
12. Strehlau J, Pape L, Offner G, et al. Interleukin-2 receptor antibody induced alterations of ciclosporin dose requirements in paediatric transplant recipients. *Lancet* **356**, 1327–28 (2000).
13. Prandota J. Important role of proinflammatory cytokines/other endogenous substances in drug-induced hepatotoxicity: depression of drug metabolism during infections/

inflammation states, and genetic polymorphisms of drug-metabolizing enzymes/cytokines may markedly contribute to this pathology. *Am J Ther* **12**, 254–61 (2005).

14. Morgan ET, Goralski KB, Piquette-Miller M, Renton, et al. Regulation of drug-metabolizing enzymes and transporters in infection, inflammation, and cancer. *Drug Metab Dispos* **36**, 205–16 (2008).
15. Zhou H. Clinical strategies to assess therapeutic protein-drug interaction (TP-DIs) for therapeutic proteins in immunology. Paper presented at the at AAPS National Biotechnology Conference, San Franciso, May 16–18, 2011.
16. US Food and Drug Administration, US Department of Health and Human Services, Center for Drug Evaluation and Research, Center for Biologics Evaluation and Research (CBER). Draft guidance for industry: drug interaction studies—study design, data analysis, implications for dosing, and labeling. Silver Spring, MD, 2012. Available at www.fda.gov/downloads/Drugs/GuidanceComplianceRegulatoryInformation/Guidances/UCM292362.pdf.
17. Kenny J. Cytokine mediated regulation of human cytochrome P450s—can we translate in vitro evaluation to clinical relevance? Paper presented at the AAPS National Biotechnology Conference, San Francisco, May 16–18, 2011.
18. Zhu M. Clinical strategies to assess TP-DDI in oncology: study design and data analysis. Paper presented at the AAPS National Biotechnology Conference, San Francisco, May 16–18, 2011.
19. Girish S, Martin SW, Peterson MC, et al. AAPS workshop report: strategies to address therapeutic protein–drug interactions during clinical development. *AAPS J* **13**, 405–16 (2011).
20. Zhu M, Jonsson EN, Hsu C-P, Sun Y-N. Establish a model system for evaluation of drug interaction in oncology: paclitaxel example. Paper presented at the annual American Conference of Pharmacometrics Meeting, San Diego, CA, April 3–6, 2011.
21. Leland-Jones B, Gelmon K, Ayoub JP, et al. Pharmacokinetics, safety and efficacy of trastuzumab administered every three weeks in combination with paclitaxel. *J Clin Oncol* **21**, 3965–71 (2003).
22. Zhou H, Parks V, Patat A, et al. Absence of a clinical relevant interaction between etanercept and digoxin. *J Clin Pharmcol* **44**, 1244–51 (2004).
23. Delbaldo C, Pierga JY, Dieras V, et al. Pharmacokinetic profile of cetuximab (Erbitux^{TM}) alone and in combination with irinotecan in patients with advanced EGFR-positive adenocarcinoma. *Eur J Can* **41**1739–45 (2005).
24. Hu C, Zhang, J, Zhou, H. Confirmatory analysis for phase III population pharmacokinetics. *Pharm Stat* **10**, 14–26 (2011).

CHAPTER 9

STATISTICAL CONSIDERATIONS IN ASSESSING DRUG–DRUG INTERACTIONS FOR THERAPEUTIC BIOLOGICS

CHUANPU HU and HONGHUI ZHOU

9.1 INTRODUCTION

Therapeutic biologics have different mechanisms of action from small molecules, usually making drug–drug interactions (DDIs) less likely, although this may depend on the specific circumstances. That is, the *a priori* expectation of specific DDIs is different with therapeutic biologics. In addition, therapeutic biologics usually have long half-lives, preventing the effective use of the usual crossover trials to assess drug–drug interactions. These issues influence the design and analysis of DDI assessments. Furthermore, population pharmacokinetics modeling has been used increasingly in drug development and has promising potential in DDI assessment. This chapter discusses the associated statistical issues, particularly to ensure appropriate power and type I error.

DDIs can adversely affect the safety or efficacy of a treatment. Therefore a new drug's DDI potential needs to be fully assessed. Namely, the objective is to obtain estimates of AUC and C_{max} ratios of interest and their corresponding 90% confidence intervals (CIs). The 90% CIs falling within a predetermined interval–for example, (0.80, 1.25)–indicates the absence of a pharmacokinetic DDI. Dedicated DDI study designs may be the conventional 2 × 2 crossover, single sequence crossover, or parallel designs. A specifically selected sequence of *in vitro* and *in vivo* studies can generally define a new experimental drug's clinically relevant DDI potential. The FDA has published a draft guidance on the design and analysis of DDI studies.[1] While AUC assessments are the mainstay of DDI studies, C_{max} is not assessed in all situations.

Drug–Drug Interactions for Therapeutic Biologics, First Edition.
Edited by Honghui Zhou and Bernd Meibohm.

Therapeutic biologics often have long elimination half-lives of up to 3 weeks, making conventional crossover studies with adequate washout periods difficult to conduct.[2] A single sequence crossover study with the investigated therapeutic biologic added to the second period is still possible, but a potential confounding time effect cannot be excluded. Some other unique challenges of designing and conducting DDI studies of therapeutic biologics were acknowledged by the FDA draft guidance. For example, classical biotransformation studies are not required routinely, and *in vitro* metabolism-based interactions studies are not generally predictive. On the other hand, there is a concern for the appropriate assessment of metabolic DDIs involving certain biologics, such as type I interferons that inhibit CYP1A2.[3]

There is not yet a uniformly accepted approach regarding DDI assessments for therapeutic biologics. Ideally, an integrated risk assessment should be made in consideration of existing data and knowledge, and clinical relevance. Important factors include therapeutic target, therapeutic window, and mechanisms of drug action.[4]

Population pharmacokinetics (PK) modeling is a powerful tool to synthesize multiple sources of information, including occasional sparse data, to assess PK along with its between-subject variability and to determine the influence of covariates. Because concomitant medication is a type of covariate, population PK can also be used for DDI assessments. It has special appeal in light of the difficulties of conducting conventional DDI studies for therapeutic biologics and has been used in multiple drug development programs.[5–16]

The methodology of population PK is still evolving, along with its applications. The FDA guidance document acknowledged its benefits in assessing DDI, and emphasized the importance of the associated study design and analysis planning. Most applications adopt the exploratory approach where multiple forms of structural, between-subject and within-subject variability models are attempted and one final, best-fitting model is chosen. Conclusions are then based on the final model. In principle, the exploratory approach is hypothesis generating, and the findings need to be confirmed later. While p-values may be generated, they are for the purpose of selecting the final model and do not have the appropriate original statistical interpretation of type I error (i.e., false positive rates). On the other hand, it is often desirable for DDI assessment to be confirmatory, especially in regulatory settings. That is, evidence of DDI, or the lack thereof, needs to be quantifiable with controllable type I error. Using population PK modeling in this setting requires special considerations.[17] In the following sections, the necessary theory is introduced along with an example for its application.

9.2 METHODOLOGY FOR DRUG–DRUG INTERACTION ASSESSMENTS

9.2.1 Conventional DDI Assessment

The statistical methodology for assessing DDI is based on that for bioequivalence. The most common purpose of bioequivalence assessments is to evaluate the comparability of bioavailability of several drug formulations. Standard bioequivalence studies are 2 × 2 crossover studies, and subjects are densely sampled so that individual PK exposure parameters, AUC and $C_{\max}$, can be determined with precision. These individual PK parameters are then log-transformed to be used with standard models, adjusting for sequence and period effects, to obtain a 90% confidence interval of log(test/reference ratio). The confidence interval is then back-transformed to the original scale to obtain the confidence interval for the test/reference ratio of average PK parameters. Confidence intervals falling within (0.80, 1.25) indicate bioequivalence. For the statistical theory, see Ref. 18 for details.

DDI studies assess the influence on pharmacokinetics by other drugs instead of formulation effects. Such studies are usually single-sequence crossover or parallel studies, but the assessment method remains the same–that is, whether the confidence intervals of AUC and $C_{\max}$ ratios fall within (0.80, 1.25). Of resemblance are PK similarity assessments between subject populations (e.g., healthy volunteers vs. patients). The single-sequence design is more convenient than the 2 × 2 crossover, although it faces the theoretical risk that a potential period effect may confound the DDI effect. The parallel design is the most convenient, but in addition to the risk of confounding period effect, it also requires larger sample sizes due to uncontrolled between-subject variability.

Testing whether the DDI effect is significant is not appropriate. The FDA draft guidance states that "because small . . . differences can be statistically significant ($p < 0.05$), but not clinically relevant." There is also a further reason to this–namely that inadequate sample sizes can result in statistically insignificance even though the real effect is clinically relevant.

Preliminary testing—that is, testing the statistical significance of some terms such as period or sequence effect to decide whether to include them in the analysis model—is generally not conducted. This would result in poor statistical behavior of the overall test owing to correlation between the preliminary test and the final test.[19]

9.2.2 Population Pharmacokinetics Based DDI Assessments

Population PK has been used for DDI assessment, typically by assessing a drug interaction effect on clearance (CL) or apparent CL.[6–15] Statistical significance of, for example, a DDI effect on CL, has been used on occasion as the deciding criterion; however, this is not appropriate for the same reason as with traditional DDI assessments. The current FDA draft guidance also discourages the use of this simplistic statistical significance criterion by stating: "Population analyses should focus on excluding a specific clinically meaningful PK change." In some other instances, empirical Bayesian individual parameter estimates (e.g., for CL) have been used; however, this is also inappropriate because the individual estimates suffer from shrinkage and are not independent.[20,21]

The appropriate population PK based DDI assessment should be performed as follows: Under the PK model, the effect of the interacting drug on any model parameter can be determined along with 90% CI. Specifically, let θ be the parameter, CL, and θ_i be the parameter value of the ith subject, and $X_i = 0$ or 1 be the indicator of whether subject i has the concomitant medication, and β represents the DDI effect. The parameter estimate of β, along with its 90% CI, can be obtained. The DDI effect is most commonly modeled as

$$\theta_i = \beta^{X_i}\theta \tag{9-1}$$

with constraint $\beta > 0$, or

$$\theta_i = (1 + \beta X_i)\theta \tag{9-2}$$

with constraint $\beta > -1$.

Usually, a DDI effect on AUC can be considered the same as that on CL or apparent CL owing to the relationship AUC = Dose/CL.

A DDI effect on $C_{\max}$, if needed, can be determined from the DDI effect on model parameters, at least in theory. In practice, this may be most easily done by reparameterizing the model and using $C_{\max}$ as a parameter. For example, in an intravenous (IV) one-compartment model with a single dose, the DDI effect on $C_{\max}$ can be considered the same as that on volume V owing to the relationship $C_{\max}$ = Dose/V. Then, the model can be reparameterized using (K_a, AUC, $C_{\max}$) instead of (K_a, V, CL). If reparameterization cannot be easily done (e.g., for two- or three-compartment models with first-order absorption), the assessment can be done using the bootstrap analysis outlined later in the chapter.

It is possible to obtain an approximate sense of the sample size needed.[22] Simulation is a general option that can be used to assess sample size and design characteristics.[23]

9.2.2.1 *Similarity and Difference between Traditional and Population PK Based Assessments* In principle, the traditional method assesses the drug interaction effect on the geometric means of AUC and $C_{\max}$, whereas the population PK based model assesses the drug interaction effect on the population average effect. The relationship between the geometric means and the population average AUC and $C_{\max}$ therefore deserves elaboration.

Commonly used population PK models assume between-individual variability of CL as log-normally distributed, which implies that the geometric mean AUC in the population equals the AUC of the average individual.[17] Likewise, under the commonly used assumption of the central volume V being lognormally distributed, the geometric mean $C_{\max}$ in the population (approximately) equals the $C_{\max}$ of the average individual for IV dosing regimens. However, no such general conclusions for $C_{\max}$ holds under oral or subcutaneous (SC) dosing regimens, regardless of distributional assumptions for the rest of the PK parameters.[20] In such situations, the computation of $C_{\max}$ depends on all other model parameters. Unlike for CL, in most situations no single model parameter would amount to the drug-interaction effect on $C_{\max}$.

Obtaining the average C_{max} from population PK models with absorption terms is difficult, because the distribution of C_{max} usually cannot be directly calculated from the between-subject variability distribution of model parameters. Despite the difference, there is no *a priori* reason to prefer the geometric means, and assessing the drug interaction effect on the C_{max} of the average individual, as computed from the estimated population PK model parameters, is reasonable.[20]

9.2.2.2 Confidence Interval Construction under Population Pharmacokinetic Modeling Given a population PK model and data, estimates of DDI effects and the corresponding 90% CIs can be computed.

A simple way to obtain the 90% CI of a DDI effect β is to use its estimate β′ and standard error SE(β′), as

$$(\beta' - \Phi(0.95) * \text{SE}(\beta'),\ \beta' + \Phi(0.95) * \text{SE}(\beta'))$$

where Φ is the cumulative standard normal distribution, and $\Phi(0.95) \approx 1.645$. This involves a normal approximation that, in theory, requires that β is parameterized on the scale $(-\infty, +\infty)$. Therefore, using the following parameterization of the DDI effect β

$$\theta_i = \exp(\beta)^{X_i}\theta \tag{9-3}$$

instead of Eq. 9-1 or 9-2 can make the normal approximation more tenable. If estimation is precise–that is, SE(β′) is small relative to β—the difference between using Eq. 9-3 vs. Eq. 9-1 or 9-2 will be small.

The standard errors of parameter estimates from nonlinear mixed-effect models are approximate, not exact. Bootstrap is an alternative method for generating standard errors and CIs of parameter estimates.[24] It is often believed to be more accurate, although some evidence suggests that it behaves similarly as the standard errors of parameter estimates.[25,26] In the case that the DDI effect on C_{max} needs to be assessed but cannot be formulated as a model parameter, assessment based on standard error of parameter estimate is not possible. However bootstrap can still be used, because in each bootstrap run, the predicted C_{max} with and without the interacting drug, and thus their ratio, can be calculated from the population model parameters.

Bootstrap Confidence Interval Construction Bootstrap can be used to generate 90% CIs by the percentile bootstrap method, which uses the 5% and 95% percentiles of the bootstrap runs. Specifically, in each bootstrap run, a bootstrap data set can be generated in which the subjects were resampled with replacement. Parameter estimates can be obtained for the bootstrap data set, and thus a ratio of AUC and C_{max} can be determined. Results of all bootstrap runs can be assembled and the 5% and 95% percentiles used to construct the 90% bootstrap confidence intervals. In general, bootstrap results are also approximate, and the precision depends on whether the numbers of subjects in the resampled categories (i.e., studies) are large enough to properly represent the entire population. In addition, the following issues are important for its implementation:

Stratifications. More stratification variables usually lead to more similar bootstrap datasets and thus narrower CIs. Therefore the proper level of stratification needs to be maintained to ensure correct statistical interpretations. The principle of bootstrap relies on the assumption that the resampling distribution correctly represents the true population distribution. This implies that, in general, design factors, and only design factors should be chosen as stratification factors. For example, *study* should always be a stratification factor–that is, the respective numbers of subjects in each study from a multistudy dataset should be kept unchanged. A more complex and subtle factor is the ratio of subjects with and without the interacting drug in each study. Often it is not a design variable and therefore should not be stratified. However if the study had been included in the DDI analysis because it has a particular ratio of subjects with and without concomitant medication, then the ratio should be a stratification variable. It may happen when all design factors are stratified, as they should be, the number of subjects within some strata becomes too small to accurately represent the population distribution. This is a potential limitation to bootstrap use.

Number of bootstrap runs to conduct. This may depend on the desired precision and the type of data, and thus is difficult to say in general. It has been suggested that at least 2000 samples may be needed.[24] In the authors' experience, 2000–5000 will generally suffice.

Which bootstrap runs to include. It may be argued that the runs that "failed" in certain ways (e.g., unable to converge or provided standard errors, parameter estimates near boundary) should be excluded. On the other hand, such failed runs correspond to data that are more likely to be atypical, such as at the tail ends of the distribution. In estimating the 90% confidence interval, these types of samples are of particular interest, and runs need to be restarted until convergence is achieved. A counterargument to restarting the failed runs is that the successful runs were not paid similar attention, thus the bias, if any, caused by ignoring failed runs may not be so severe. It may also be argued that when the software converges successfully but fails to produce standard error of parameter estimates, it presents only an algebraic problem unrelated to the quality of convergence and thus, such runs should be included. The problem can be alleviated, to a certain extent, by using multiple starting estimates for each run to improve the convergence rate. A conservative solution may be to include all runs. Future investigations are necessary to determine how to best deal with failed runs.

9.2.2.3 Influence of Model Searching on DDI Effect Assessment The use of population PK requires the selection of a population PK model, which consists of three components: a structural model, a variability model (for both between-subject and within-subject variability), and a covariate model. Influences of these components on DDI assessments are elaborated in the subsections below. Most applications of population PK were conducted using the exploratory approach, which selects the model that best fits the data among the many possible alternative models, either explicitly or implicitly. While this approach has intuitive appeal, the inferences drawn from the final selected model do not possess statistical validity.[20] Even though at times some hypothesis tests may be conducted, they are generally used as a tool for model selection. Because the model and hypothesis tests are not prespecified, the associated p-values and CIs are not statistically interpretable. The source of the problem is similar to that of the earlier mentioned preliminary testing in bioequivalence (BE) analyses. Further reasons that model exploration may adversely affect DDI analysis include overfitting and selection bias.[27–30]

In theory, inferences could appropriately be made by including the model search into the considerations. For example, bootstrap could be done by including the entire model search process instead of only the final selected model. This has not been described yet, but this argument shows that model search usually leads to increased uncertainty in the results.

Influence of Structural Model on DDI Effect Assessment The AUC (and thus the ratio of with to without interacting drug) estimate should be relatively robust because models usually represent the average concentration reasonably well. For C_{max}, however, the opposite is true, because fitting simpler structural models to the same data generally leads to smaller C_{max} estimates. Therefore, a simpler model is more likely to bias the DDI assessment of C_{max} toward lack of DDI. This argument could be extended to say that any population PK model will likely underestimate C_{max} because the reality is always more complex. This may be true, but these findings should be placed into perspective. The traditional use of measured C_{max} also underestimates the true C_{max}, because T_{max} may not be among the sampling times. A practical prerequisite of using population PK is that sufficient data are available to describe the absorption process.

Influence of Random Effect Model on DDI Effect Assessment The within subject random effect model, also called residual error model, has the effect of weighting concentration measurements differently during the model fitting. Commonly used models are the additive-plus-proportional, proportional, and lognormal model,[29] which take the form as following, respectively:

$$y = p + p * \varepsilon_1 + \varepsilon_2 \tag{9-4}$$

$$y = p + p * \varepsilon_1 \tag{9-5}$$

$$\log(y) = \log(p) + \varepsilon_1 \tag{9-6}$$

where y is the observed concentration, p is model prediction, and ε_1, ε_2 are independently normally distributed. These, in increasing order, weigh smaller concentrations more. Therefore model (9-4) may result in a C_{max} estimation closer to observed data than models (9-5) and (9-6). Choice among these models, especially between models (9-6) and (9-5), is still of much debate, and currently there is no consensus on which one to use. It may be prudent to base the choice on the type of assay and prior experience with the class of compounds.

Other models have occasionally appeared in literature, at times selected by goodness-of-fit criteria, but caution is necessary. For example, the log-likelihoods of model (9-5) and (9-6) have been compared, which is not statistically meaningful. Another example lies in the attempted approach to combine models (9-4) and (9-5), which takes the form of

$$\log(y) = \log(p) + \varepsilon_1 + \log(p) * \varepsilon_2.$$

This form has the variance of the total residual error on the log scale at the minimum when the prediction $p = 1$, which is difficult to justify. Yet another example has a type of hybrid residual error model as

$$\log(y) = \log(p) + \sqrt{\varepsilon^2 + \varepsilon^2/p^2}$$

This model has the property that, as the prediction approaches zero, the variance approaches infinity. This is also difficult to justify, especially when predose concentration measurements suggest limited variability. Inappropriate residual error models may lead to biased estimation of $C_{\max}$ and AUC.

Many between-subject random effect model parameters, such as correlations among random effects, are of limited interest in DDI assessment. In theory, there is no value to formally test them because there is no *a priori* reason that they should not exist, and any statistical insignificance would only be due to the lack of power to detect the effects. However, it seems plausible that including such effects could cost degrees of freedom and may dilute the power for assessing DDI effects. We suggest that such terms are not included, especially in situations in which power is lacking, pending further investigation.

Influence of Covariate Models on DDI Effect Assessment The influence of including additional covariate effects (e.g., gender) in a population PK model is discussed here. In traditional DDI assessments of standard crossover studies, adjusting for gender has no effect because each subject serves as his or her own comparison. Including gender in a population PK assessment for DDI is analogous to adjusting for gender in BE assessments of parallel studies, which adds one term to the linear model and costs only one degree of freedom but can potentially improve the error estimate and thus improve the power of the DDI assessment. However, in population PK modeling it is less clear where to add the gender effect, unless prior information is available. Adding gender influence to all model parameters not only reduces more degrees of freedom but also increases the complexity of parameter estimation. Bootstrap analyses could be particularly sensitive to model instability, because a large number of model runs are needed and convergence could become problematic. Furthermore, casual correlations between gender and interacting drug usage, if present, can also confound DDI assessments.

In general, the statistical behavior of an analysis becomes worse the more exploration is allowed.[19,28] Intuitively, more model exploration would adversely affect the power of concluding the similarity of PK, because some degrees of freedom must be sacrificed for exploration. The issue would be much more apparent if an exploratory analysis had arrived at some exotic model, such as one that had ethnic origin influence absorption rate that is both weak from its pharmacological mechanism and numerically unstable.

9.2.2.4 Confirmatory Population Pharmacokinetic Modeling for DDI Assessments A confirmatory population PK approach has been recently proposed that achieves the appropriate inferences by prespecifying the analysis model based on available information and the analysis objective.[30,31] The quality of the confirmatory analysis will clearly depend on the suitableness of the prespecified model for the DDI analysis. At most stages of the DDI analysis, data from prior Phase I and II studies are available and provide good information about the PK structural and variability models and even information for some relevant covariates, such as weight. Depending on the analysis objective, all components of the model do not have to be "correct" to yield appropriate results.[29] For example, assessing DDI effects on CL is often robust to model misspecification of between-subject random effect components on the absorption rate constant in a one-compartment model with first-order absorption.[30,32]

Typical steps of DDI assessments using confirmatory population PK in Phase III studies involve the following:

1. *Determine which DDI effects need to be assessed based on mechanistic rationale and whether the sample size is sufficient.* The sample size needed for a reasonably precise determination of the effect size depends on the number of subjects with and without the concomitant medication, along with the number of PK concentration measurements and PK between- and within-subject variability.
2. *Prespecify the population PK structural and variability models based on available information and current Phase III study design.* This can frequently be done by using a previously built population PK model using data from Phases I and II to simulate datasets under the current Phase III study design. Model exploration may be used at this time on the simulated data to see what choice of the structural model, variability model, or even limited covariate models, such as weight can be best supported from the current Phase III study design. The model most stable under the simulated data can be used as the population PK analysis model for the DDI assessment. Most often, a one-compartment model will be selected. As the observed PK concentration data of the current Phase III study have not been used to select the model, they are prespecified and thus the analysis is confirmatory.

3. *Prespecify the covariate model.* Most often, the concomitant medication effects are on clearance. Mechanistically meaningful covariates based on prior information should be retained, such as weight on clearance and volume when existing in the previously built population PK model.
4. *Prespecify sensitivity analyses.* Alternative competing assessments, if needed. Such analyses, if conducted at all, should be limited, given the confirmatory objective.
5. *Determine the magnitude of a relevant DDI effect.* For example, try 70–143%. Determine the power of the DDI assessment accordingly, likely through simulations, and assess only those DDI effects that have sufficient power. If a clear criterion cannot easily be determined, the rule of thumb of at least 20 subjects per effect estimated might suffice, and the resulting 90% CIs may be compared with the nominal 80–125% criterion.

Some of the above steps, especially 1–3, are interrelated and should perhaps be better viewed as one joint step with multiple components. Step 5, determining the criterion of DDI, affects the power assessment but not the assessment method, thus it may be less crucial in practical implementations For further details, see Ref. 29.

In rare situations unexpected complexities may arise, including substantial noncompliance or model nonconvergence. Deviations from the prespecified model would be necessary in these situations to deal with such irregularities but should be maintained at a minimal level and should stay true to the principle of the prespecified analysis.

Because of the many assumptions, a population PK model, whether prespecified or obtained by exploratory search, is unlikely to be "true." However, the relevant question for most applications of modeling is never whether the final model is right or wrong. As has been repeatedly quoted, "All models are wrong, but some are useful." The focus should be on the DDI assessment results instead of on the model–that is, the purpose of the analysis is to obtain not the best description of the data but the most credible estimates along with the associated standard errors for DDI effects.

9.2.3 Comparison between Traditional and Population PK Based DDI Assessments

The traditional and the population PK approaches have different advantages. The traditional assessment employs fewer assumptions, is robust to model misspecification, better protects against type I error, and is easy to interpret. The population PK approach can be used in situations not originally designed for DDI assessments, such as, Phase III studies, and to incorporate information from multiple sources by using meta-analyses. A population PK assessment based on large Phase III studies also may better reflect the DDI effect in actual clinical practice. However, population PK is more complex in methodology and thus requires more expertise in design, analysis, and evaluation stages. Data/design quality (e.g., variability in actual dose/sampling times) may adversely affect study power. However this should not affect type I errors and thus presents only a sponsor risk.

9.3 POPULATION PHARMACOKINETICS FOR DRUG–DRUG INTERACTION ASSESSMENTS: USTEKINUMAB

An example of using confirmatory population PK for DDI assessments in phase III clinical studies is described in the following. Ustekinumab is a human monoclonal antibody that binds with high specificity and affinity to the shared p40 subunit of IL-12 and IL-23 and blocks interaction with the IL-12Rβ1 cell surface receptor. The pharmacokinetic properties of ustekinumab have been well characterized after single-dose SC administration (0.27–2.7 mg/kg) and single-dose intravenous administration (0.1–5.0 mg/kg) in patients with psoriasis. The PK of ustekinumab was linear over the wide dose range studied.

9.3.1.1 Study Design Data from two large Phase III trials of ustekinumab in patients with moderate to severe psoriasis (PHOENIX 1 and PHOENIX 2) were used for the DDI assessment.[33,34] Both were multicenter, randomized, double-blind, placebo-controlled, parallel design studies. The study designs were complex, and consisted of placebo-controlled, placebo crossover, dose optimization, randomized withdrawal, and long-term extension periods. Briefly, in PHOENIX 1, a total of 766 patients were assigned to receive SC injections of ustekinumab 45 mg, ustekinumab 90 mg, or placebo at weeks 0 and 4, followed by active treatment every 12 weeks or placebo crossover to ustekinumab 45 mg or 90 mg starting at week 12 (weeks 12–40), followed by randomized withdrawal (weeks 40–76). In PHOENIX 2, a total of 1230 patients were assigned to the same treatment groups and design until week 28, followed by dose schedule optimization (weeks 28–52). Blood samples for the measurement of serum ustekinumab concentrations were collected at weeks 0, 4, 12, 16, 24, 28, 40, 44, 48, and 52 in PHOENIX 1 and at weeks 0, 4, 12, 16, 20, 24, and 28 in PHOENIX 2. At visits when patients received the study agent, blood samples were collected prior to study agent administration. A validated electrochemiluminescent immunoassay method, with a lower limit of quantification of 0.17 μg/mL at a minimum required 1 : 10 dilution, was used to measure serum ustekinumab concentrations. Multiple covariates were collected in the trials, including demographics and concomitant medications.

9.3.2 Confirmatory Population Pharmacokinetic Analysis Plan

Several key elements were considered in deciding the covariates to be included in the analysis. These include inherent pharmacokinetic behavior, intended indication and patient population, intended product label, and anticipated important/relevant covariates as learned from similar or related molecules. Before beginning the analysis, a list of the covariates to be included was developed based on physiological relevance, pharmacological importance, baseline disease severity and/or status, geographic location, past use of relevant medications, relevant concurrent comorbidities, and most commonly used concurrent medications. The following covariates were included: weight, immune response (positive/nonpositive), age (<65 vs. ≥ 65 years), sex, race (Caucasian, Black, Asian, other), creatinine clearance, geographic region, past use of biologics (yes/no), past use of methotrexate (yes/no), past use of cyclosporine (yes/no), duration of disease, baseline disease scores (PASI), baseline physician global assessment score, baseline dermatology quality index, diabetes status (yes/no), hypertension history (yes/no), lipidemia history (yes/no), alcohol history (yes/no), smoking status (yes/no), and the following concomitant medicines: amlodipine, atenolol, atorvastatin, hydrochlorothiazide, lisinopril, metoprolol, ramipril, metformin, citalopram, and influenza vaccine. The list was developed with the assurance that at least 20 subjects were available for each covariate effect parameter estimated. The exact numbers of subjects who had the concomitant medications were 67, 69, 61, 140, 110, 74, 57, 76, 46, and 106, respectively.

The following analysis plan and rationale were specified before beginning the analysis on observed concentrations.

Base structural, random-effect and covariate model. A previous population PK model based on a combined prior Phase I/II dataset was available, with a structural model of a one-compartment model with extravascular absorption. It had between-subject variance on absorption rate constant (K_a), apparent clearance (CL/F) and apparent volume of distribution (V/F), and an additive-plus-proportional residual error model. In addition, CL/F was proportional to weight raised to the power of 0.75 and V/F proportional to weight. This model was used to simulate data with the actual dosing information and sampling times in the Phase III studies. The best random-effect structure was developed from the simulated data, based on the usual fitting criteria. Base-model exploration on simulated data showed that, as expected, between-subject random-effect on K_a could not be identified. Influence of weight on CL/F and V/F was deemed estimable. Therefore, the prespecified base model had between-subject random effects on CL/F and V/F and also weight effects on them, with an additive-plus-proportional residual error model.

Covariate model and DDI assessments. The full model consisting of the covariates listed was used on CL/F only. That is, no covariate models were used on V/F or K_a.

9.3.3 Covariate and Drug–Drug Interaction Effect Assessment

The prespecified model was implemented with NONMEM (version VI, level 1, ICON Development Solutions, Ellicott City, MD). Model estimation was stable. The parameter estimates and their associated standard errors were used to compute the 90% CIs for all covariate effects on CL/F, including the concomitant medications. The 90% CIs are plotted in Figure 9-1. All concomitant medication effects had 90% CIs within (0.8, 1.25), therefore the lack of DDI was concluded for them.

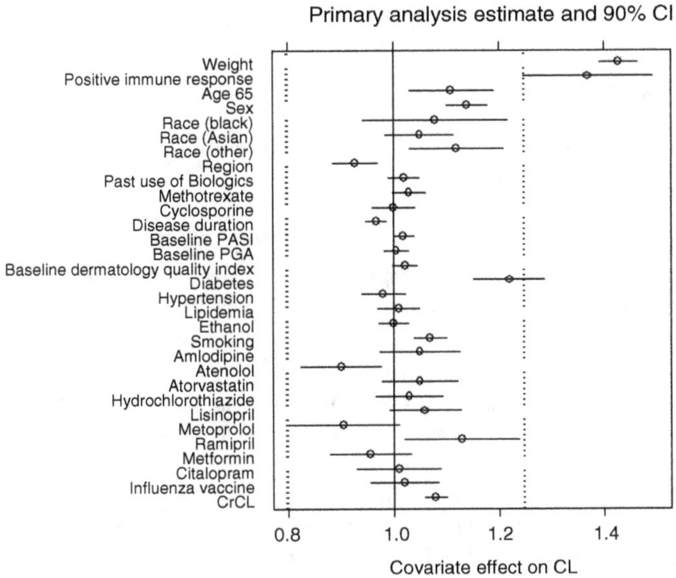

FIGURE 9-1 The 90% confidence intervals (CI) of covariate and concomitant medication effects. CL: apparent clearance; CrCL: creatinine clearance; PASI: Psoriasis Area and Severity Index score; PGA: Physician's Global Assessment score.

The confirmatory population PK approach gave statistically appropriate 90% CIs for all covariate effects on CL/F, including the concomitant medications.

9.4 SUMMARY

DDI assessments for therapeutic biologics have unique challenges. Conventional DDI assessment is still the most robust and least controversial approach. Although it can be inconvenient to conduct, risk-based approaches can be used to guide the assessment. In contrast, the population PK approach has the ability to integrate information from multiple sources, and its potential in DDI assessment has yet to be fully utilized. Adopting the confirmatory approach ensures methodological appropriateness and increases the confidence in its conclusions. The population PK approach may be most useful when AUC is the main focus. The methodology of population PK is still evolving, and special attention needs to be paid to fulfill the objectives of DDI assessments and maintain statistical appropriateness. In principle, population PK assessment should fully parallel the traditional assessments, both in prespecifying the analysis model and ensuring sufficient study design and sample size.

REFERENCES

1. US Food and Drug Administration, US Department of Health and Human Services, Center for Drug Evaluation and Research, Center for Biologics Evaluation and Research

(CBER). Draft guidance for industry: drug interaction studies–study design, data analysis, implications for dosing, and labeling. Silver Spring, MD, 2012. Available at www.fda.gov/downloads/Drugs/GuidanceComplianceRegulatoryInformation/Guidances/UCM292362.pdf.

2. Zhou H, Mascelli MA. Mechanisms of monoclonal antibody-drug interactions. *Annu Rev Pharmacol Toxicol* **51**, 359–72 (2011).
3. Seitz K, Zhou H. Pharmacokinetic drug-drug interaction potentials for therapeutic monoclonal antibodies: reality check. *J Clin Pharm* **47**(9), 1104–18 (2007).
4. Zhou H, Davis HM. Risk-based strategy for the assessment of pharmacokinetic drug-drug interactions for therapeutic monoclonal antibodies. *Drug Discov Today* **14**, 891–98 (2009).
5. Zhou H. Population-based assessments of clinical drug-drug interactions: qualitative indices or quantitative measures? *J Clin Pharmacol* **46**, 1268–89 (2006).
6. Delavenne X, Laporte B, Demasles S, et al. Investigation of PK-PD drug-drug interaction between acenocoumarol and amoxicillin plus clavulanic acid. *Fundam Clin Pharmacol* **23**, 127–35 (2009).
7. de Jonge ME, Huitema ADR, Rodenhuis S, Beijnen JH. Integrated population pharmacokinetic model of both cyclophosphamide and thiotepa suggesting a mutual drug–drug interaction. *J Pharmacokinet and Pharmacodyn* **31**(2), 135–56 (2004).
8. Mirochnick M, Cooper E, Capparelli E, et al. Population pharmacokinetics of dapsone in children with human immunodeficiency virus infection. *Clin Pharmacol Ther* **70**(1), 24–32 (2011).
9. de Maat MMR, Huitema ADR, Mulder JW, et al. Drug interaction of fluvoxamine and fluoxetine with nevirapine in HIV-1-infected individuals. *Clin Drug Invest* **23**(10), 629–37 (2003).
10. Lehr T, Staab A, Trommeshauser D, et al. Semi-mechanistic population pharmacokinetic drug-drug interaction modelling of a long half-life substrate and itraconazole. *Clin Pharmacokinet* **49**(1), 53–66 (2010).
11. Sallas WM, Milosavljev S, D'Souza J, Hossain M. Pharmacokinetic drug interactions in children taking oxcarbazepine. *Clin Pharmacol Ther* **74**(2), 138–49 (2003).
12. Joerger M, Huitema ADR, Boogerd W, et al. Interactions of serum albumin, valproic acid and carbamazepine with the pharmacokinetics of phenytoin in cancer patients. *Basic Clin Pharmacol Toxicol* **99**, 133–40 (2006).
13. Bockbrader HN, Burger P, Knapp L. Pregabalin effect on steady-state pharmacokinetics of carbamazepine, lamotrigine, phenobarbital, phenytoin, topiramate, valproate, and tiagabine. *Epilepsia* **52**(2), 405–09 (2011).
14. Bouillon T, Bruhn J, Radu-Radulescu L, et al. Non–steady state analysis of the pharmacokinetic interaction between propofol and remifentanil. *Anesthesiology* **97**, 1350–62 (2002).
15. Vucicevic K, Miljkovic B, Pokrajac M, et al. The influence of drug–drug interaction and patients' characteristics on valproic acid's clearance in adults with epilepsy using nonlinear mixed effects modeling. *Eur J Pharm Sci* **38**, 512–18 (2009).
16. Duan JZ. Applications of population pharmacokinetics in current drug labelling. *J Clin Pharm Ther* **32**, 57–79 (2007).
17. Hu C, Moore K, Kim Y, Sale ME. Statistical issues in a modeling approach to assessing bioequivalence or PK similarity with presence of sparsely sampled subjects. *J Pharmacokinet Pharmacodyn* **33**(4), 321–39 (2004).

18. Schuirmann DJ. A comparison of the two one-sided tests procedure and the power approach for assessing the equivalence of average bioavailability. *J Pharmacokinet Biopharm* **15**(6), 657–80 (1987).
19. Senn S. *Statistical issues in drug development*. Wiley, Chichester, 1997.
20. Hu C, Dong Y. Estimating the predictive quality of dose-response after model selection. *Statis Med* **26**, 3114–39 (2007).
21. Dubois A, Gsteiger S, Pigeolet E, Mentré F. Bioequivalence tests based on individual estimates using non-compartmental or model-based analyses: evaluation of estimates of sample means and type I error for different designs. *Pharm Res* **27**(1), 92–104 (2010).
22. Kang D, Schwartz JB, Verotta D. A sample size computation method for non-linear mixed effects models with applications to pharmacokinetics models. *Statis Med* **23**(16), 2551–66 (2004).
23. Vong C, Bergstrand M, Nyberg J, Karlsson MO. Rapid sample size calculations for a defined likelihood ratio test-based power in mixed effects models. *AAPS J* **14**(2), 176–86 (2012).
24. Efron B, Tibshirani RJ. *An introduction to the bootstrap*. Chapman & Hall, London, 1993.
25. Hu C, Xu Z, Rahman MU, et al. A latent variable approach for modeling categorical endpoints among patients with rheumatoid arthritis treated with golimumab plus methotrexate. *J Pharmacokinet Pharmacodyn* **37**(4), 309–321 (2010).
26. Hu C, Xu Y, Rahman M, et al. A population approach for exposure-response modeling of golimumab in patients with rheumatoid arthritis. *J Clin Pharmacol* **51**(5), 639–48 (2011).
27. Faraway JJ. On the cost of data analysis. *J Comput Graph Stat* **1**, 213–29 (1992).
28. Harrell F. *Regression modeling strategies: with applications to linear models, logistic regression, and survival analysis*. Springer, New York, 2001.
29. Hu C, Zhang J, Zhou H. Confirmatory analysis for phase III population pharmacokinetics. *Pharm Stat* **10**(1), 14–26 (2011).
30. Ribbing J, Jonsson EN. Power, selection bias and predictive performance of the population pharmacokinetic covariate model. *J Pharmacokin Pharmacodyn* **31**(2), 109–34 (2004).
31. Hu C, Zhou H. An improved approach for confirmatory Phase III population pharmacokinetic analysis. *J Clin Pharmacol* **48**(7), 812–22 (2008).
32. Kowalski KG, Hutmacher MM. Design evaluation for a population pharmacokinetic study using clinical trial simulations: a case study. *Stat Med* **20**, 75–91 (2001).
33. Leonardi CL, Kimball AB, Papp KA, et al. Efficacy and safety of ustekinumab, a human interleukin-12/23 monoclonal antibody, in patients with psoriasis: 76-week results from a randomised, double-blind, placebo-controlled trial (PHOENIX 1). *Lancet* **371**(9625), 1665–74 (2008).
34. Papp KA, Langley RG, Lebwohl M, et al. Efficacy and safety of ustekinumab, a human interleukin-12/23 monoclonal antibody, in patients with psoriasis: 52-week results from a randomised, double-blind, placebo-controlled trial (PHOENIX 2). *Lancet* **371**(9625), 1675–84 (2008).

CHAPTER 10

SCIENTIFIC PERSPECTIVES ON THERAPEUTIC PROTEIN DRUG–DRUG INTERACTION ASSESSMENTS*

HONG ZHAO, LEI ZHANG, KELLIE REYNOLDS, and SHIEW-MEI HUANG

10.1 INTRODUCTION

Many intrinsic and extrinsic factors can affect an individual patient's drug exposure and response.[1] The FDA has published a number of guidance documents that recommend how and when to evaluate these factors during drug development.[2] The most recent FDA draft guidance on drug interactions[3] provides advice for *in vitro* and *in vivo* drug interaction studies, including suggestions for study design, dosing strategies, data analysis and interpretation, and the type of information that should be included in medical product labels; it updates the FDA's recommendations on the evaluation of important cytochrome P450 enzyme- and transporter-based drug interactions, and the use of physiologically based pharmacokinetic modeling approaches for drug interaction assessment; and it expands recommendations on therapeutic protein–drug interaction assessment. The recommendations are illustrated via a series of decision trees. This draft guidance and other FDA publications related to drug–drug interactions (DDIs),[4–8] including the website *Drug Development and Drug Interactions*,[9] provide recommendations on how to evaluate drug interaction potential for both small molecules (drugs or Ds) and therapeutic proteins (TPs).

Two review papers published in 2007 contain examples of drug interactions involving TPs.[10,11] Several recent publications summarize the TP-DDI evaluation included in Biologics License Application (BLA) submissions and current strategies for the evaluation of TP-DDI potential during drug development.[12–17]

*The views expressed in this book chapter are those of the authors and do not necessarily reflect the official views of the US Food and Drug Administration.

Drug–Drug Interactions for Therapeutic Biologics, First Edition.
Edited by Honghui Zhou and Bernd Meibohm.

The purpose of this chapter is to describe our review of the literature on drug interactions involving TPs, the FDA's expanded recommendations on TP-DDIs in the updated draft drug interaction guidance, and FDA-approved product labels that include TP-DDI information. The information presented in this chapter includes the following:

- TPs with drug interaction studies conducted during product development
- Types of study designs used to evaluate TP-DDIs
- Impact of study results on product label

10.2 THERAPEUTIC PROTEIN–DRUG INTERACTION STUDIES

Therapeutic proteins are increasingly being combined with small-molecule drugs and/or with other TPs to gain greater clinical benefit. Our survey of 76 new molecular entity TPs that were approved by the FDA up to February 2010 examined drug interaction information in product labels to help determine the types of TPs for which drug interaction studies are informative (Table 10-1).[14] The survey showed that 44 TPs (58%) included some information on drug interactions in the label (including 9 with dedicated TP-DDI studies). The products most likely to have drug interaction information in the label were therapeutic cytokines (9/11, 81%), followed by therapeutic monoclonal antibodies (mAbs), some of which are cytokine modulators (19/29, 66%). Some cytokines are involved in the pathophysiology of various human diseases and are produced during infection and inflammation. Therapeutic cytokines are used in the treatment of certain diseases such as cancer. Various *in vitro* and limited *in vivo* human studies indicate that cytokines and cytokine modulators can differentially influence the expression or activities of specific cytochrome P450 (CYP) enzymes and/or drug transporters through multiple mechanisms[18–23] and therefore have the potential to alter the systemic exposure and/or clinical response of concomitantly administered drugs that are substrates for the

TABLE 10-1 Summary of Drug Interaction Studies in the FDA-Approved Package Inserts for New Molecular Entity Therapeutic Proteins Approved by the End of February 2010[a]

Category	Dedicated Studies	Some Information	No Information	Total
Cytokines	2	7	2	11 (14%)
Growth factors	0	2	8	10 (13%)
Enzymes	1	7	9	17 (22%)
Monoclonal antibodies	6	13	10	29 (38%)
Miscellaneous	0	6	3	9 (12%)
Total	9 (12%)	35 (46%)	32 (42%)	76 (100%)

[a]Adapted from Ref. 14.

affected CYP enzymes or transporters. For example, therapeutic mAbs that modulate cytokine activities can indirectly influence the expression of specific CYP enzymes and drug transporters by affecting cytokine concentrations.[24,25] Examples of TPs that affect CYP enzymes are listed in Table 10-2. In addition, the exposure to therapeutic mAbs can be altered by concomitantly administered immunosuppressive drugs that affect the formation of antiproduct antibodies.[26,27]

TP-drug interactions (either pharmacokinetics [PK] or pharmacodynamic [PD] based) have been observed, and information about these interactions is included in product labeling (Tables 10-3 and 10-4).[26] The FDA's recommendations on the evaluation of DDIs involving small molecule drugs, including study design, data analysis, and dosing and labeling implications have been described in detail.[4–9] Many of the same principles apply to the evaluation of DDI involving TPs. The current thought process for evaluation of TP-DDIs and the types of studies that have been conducted during drug development to evaluate TP-DDI interactions are illustrated in Figure 10-1.[3] General considerations are discussed below:

- If an investigational TP is a cytokine or cytokine modulator that has known effects on CYP enzymes or transporters, studies should be conducted to determine the magnitude of the TP's effects on drugs that are substrates of the affected CYP enzymes or transporters[13,19]

  In vitro and *in vivo* studies have been conducted to determine the TP's effects on CYP enzymes or transporters. When the TP is a cytokine modulator, the effects of both the specific cytokine(s) being modulated and the cytokine modulator on enzymes and transporters may be assessed using *in vitro* models (such as human hepatocytes with the addition of a physiologically relevant

TABLE 10-2 CYP Enzymes with Decreased Activities (Unless Noted) in the Presence of Specific Cytokines, Cytokine Modulators, and Human Growth Hormone Based on *In Vitro* and/or *In Vivo* Human Studies[a]

CYP Enzyme	Cytokines/Cytokine Modulators
CYP1A2	IFNα, IFNα-2b, IFNβ, IL-2, IL-6, hGH[b]
CYP2C8	IL-1
CYP2C9	IL-2, IL-10
CYP2C19	Tocilizumab,[c] IFNα-2b, IFNβ, IL-2, TNFα, IL-6, hGH
CYP2D6	IFNα-2b
CYP2E1	IL-2, IFNα-2b
CYP3A	Basiliximab, muromonab-CD3, tocilizumab,[c] IL-1, IL-2, IL-6, IL-10

[a]Modified from Ref. 13; data from Refs. 14 and 20. All have data from *in vitro* and *in vivo* studies in humans to support the changes in cytochrome P450 (CYP) enzyme activities and/or expression with the exception of human growth hormone (hGH), interleukin 10 (IL-10), muromonab-CD3, and basiliximab, for which only *in vivo* data are available.

[b]The effect of hGH is increased activity.

[c]The effect of tocilizumab, an IL-6 receptor inhibitor, is increased activity.

IFN: interferon; TNF: tumor necrosis factor.

TABLE 10-3 Examples of Labeling Language for Therapeutic Protein–Drug Interactions (Pharmacokinetic-Based)

TP Name (Brand Name) Pharmacologic Class Year of Labeling	Item	Labeling Language
1. CYP enzyme modulation by cytokines and cytokine modulators		
Interferon α-2a (Roferon-A) INF-α 2006	Precautions: drug interactions	Roferon-A has been reported to reduce the clearance of theophylline. The clinical relevance of this interaction is presently unknown.
		α-Interferons may affect the oxidative metabolic process by reducing the activity of hepatic microsomal cytochrome enzymes in the P450 group. Although the clinical relevance is still unclear, this should be taken into account when prescribing concomitant therapy with drugs metabolized by this route.
Peginterferon α-2a (Pegasys) INF-α 2011	Highlights of prescription information: drug interactions	*Drugs metabolized by CYP1A2*: monitor for increased serum levels of theophylline and adjust dose accordingly. (§7.2)
		Methadone: monitor for signs and symptoms of methadone toxicity. (§7.3)
	4 Contraindications	Pegasys/Copegus combination therapy is additionally contraindicated in combination with didanosine. Reports of fatal hepatic failure and peripheral neuropathy, pancreatitis, and symptomatic hyperlactatemia/lactic acidosis have been reported in clinical trials. (§7.4)
	7 Drug interactions	
	7.1 Drugs metabolized by cytochrome P450	There was no effect on the pharmacokinetics of representative drugs metabolized by CYP 2C9, CYP 2C19, CYP 2D6 or CYP 3A4.
		Treatment with Pegasys once weekly for 4 weeks in healthy subjects was associated with an inhibition of P450 1A2 and a 25% increase in theophylline AUC.
	7.2 Theophylline	Treatment with Pegasys once weekly for 4 weeks in healthy subjects was associated with an inhibition of P450 1A2 and a 25% increase in theophylline AUC. Theophylline serum levels should be monitored and appropriate dose adjustments considered for patients given both theophylline and Pegasys.

	7.3 Methadone	In a PK study of HCV subjects concomitantly receiving methadone, treatment with Pegasys once weekly for 4 weeks was associated with methadone levels that were 10–15% higher than at baseline. The clinical significance of this finding is unknown; however, patients should be monitored for the signs and symptoms of methadone toxicity. The PK of concomitant administration of methadone and Pegasys were evaluated in 24 Pegasys naive cCHC subjects (15 male, 9 female) who received 180 μg Pegasys SC weekly. All subjects were on stable methadone maintenance therapy (median dose 95 mg, range 30 mg to 150 mg) before receiving Pegasys. Mean methadone PK parameters were 10–15% higher after 4 weeks of Pegasys treatment compared to baseline. Methadone did not significantly alter the PK of Pegasys compared to a PK study of 6 cCHC subjects not receiving methadone.
Interferon α-2b (Intron A) INF-α 2011	Precautions: drug Interactions	Interactions between Intron A and other drugs have not been fully evaluated. Caution should be exercised when administering Intron A therapy in combination with other potentially myelosuppressive agents such as zidovudine. Concomitant use of INF-α and theophylline decreases theophylline clearance, resulting in a 100% increase in serum theophylline levels.
Peginterferon α-2b (PEG-INTRON) INF-α 2011	Highlights of prescription information: drug interactions	*Drug metabolized by CYP450*: caution with drugs metabolized by CYP2C8/9 (e.g., warfarin, phenytoin) or CYP2D6 (e.g., flecainide). (§7.1) *Methadone*: monitor for increased narcotic effect. (§7.2)
	7 Drug Interactions 7.1 Drugs metabolized by cytochrome P450	When administering PEG-INTRON with medications metabolized by CYP2C8/9 (e.g., warfarin and phenytoin) or CYP2D6 (e.g., flecainide), the therapeutic effect of these substrates may be decreased. (§12.3).
	7.2 Methadone	PEG-INTRON may increase methadone concentrations. (§12.3). The clinical significance of this finding is unknown; however, patients should be monitored for the signs and symptoms of increased narcotic effect.

(*continued*)

TABLE 10-3 (*Continued*)

TP Name (Brand Name) Pharmacologic Class Year of Labeling	Item	Labeling Language
	12 Clinical Pharmacology	
	12.3 Pharmacokinetics: drug interactions	
	Drugs metabolized by cytochrome P450	The pharmacokinetics of representative drugs metabolized by CYP1A2 (caffeine), CYP2C8/9 (tolbutamide), CYP2D6 (dextromethorphan), CYP3A4 (midazolam), and *N*-acetyltransferase (dapsone) were studied in 22 subjects with CHCwho received PEG-INTRON (1.5 μg/kg) once weekly for 4 weeks. PEG-INTRON treatment resulted in a 28% (mean) increase in a measure of CYP2C8/9 activity. PEG-INTRON treatment also resulted in a 66% (mean) increase in a measure of CYP2D6 activity; however, the effect was variable: 13 subjects had an increase, 5 subjects had a decrease, and 4 subjects had no significant change. (§7.1)
		No significant effect was observed on the pharmacokinetics of representative drugs metabolized by CYP1A2, CYP3A4, or *N*-acetyltransferase. The effects of PEG-INTRON on CYP2C19 activity were not assessed.
	Methadone	The pharmacokinetics of concomitant administration of methadone and PEG-INTRON were evaluated in 18 PEG-INTRON-naive CHC subjects receiving 1.5 μg/kg PEG-INTRON SC weekly. All subjects were on stable methadone maintenance therapy receiving ≥40 mg/day before initiating PEG-INTRON. Mean methadone AUC was approximately 16% higher after 4 weeks of PEG-INTRON treatment compared to baseline. In 2 subjects, methadone AUC was approximately double after 4 weeks of PEG-INTRON treatment compared to baseline. (§7.2).
Tocilizumab (Actemra) Human IL-6 receptor blocker 2011	7 Drug interactions 7.2 Interactions with CYP450 substrates	Cytochrome P450s in the liver are downregulated by infection and inflammation stimuli, including cytokines, such as IL-6. Inhibition of IL-6 signaling in RA patients treated with tocilizumab may restore CYP450 activities to higher levels than those in the absence of tocilizumab, leading to increased metabolism of drugs that are CYP450

	substrates. *In vitro* studies showed that tocilizumab has the potential to affect expression of multiple CYP enzymes, including CYPIA2, CYP2B6, CYP2C9, CYP2C19, CYP2D6 and CYP3A4. Its effect on CYP2C8 or transporters is unknown. *In vivo* studies with omeprazole, metabolized by CYP2C19 and CYP3A4, and simvastatin, metabolized by CYP3A4, showed up to a 28% and 57% decrease in exposure 1 week after a single dose of Actemra, respectively. The effect of tocilizumab on CYP enzymes may be clinically relevant for CYP450 substrates with narrow therapeutic index, where the dose is individually adjusted. On initiation or discontinuation of Actemra, in patients being treated with these types of medicinal products, therapeutic monitoring of effect (e.g., warfarin) or drug concentration (e.g., cyclosporine or theophylline) should be performed and the individual dose of the medicinal product adjusted as needed. Prescribers should exercise caution when Actemra is coadministered with CYP3A4 substrate drugs when a decrease in effectiveness is undesirable (e.g., oral contraceptives, lovastatin, atorvastatin.). The effect of tocilizumab on CYP450 enzyme activity may persist for several weeks after stopping therapy. (§12.3).
12 Clinical pharmacology 12.3 Pharmacokinetics: drug interactions	*In vitro* data suggested that IL-6 reduced mRNA expression for several CYP450 isoenzymes, including CYPIA2, CYP2B6, CYP2C9, CYP2C19, CYP2D6, and CYP3A4, and this reduced expression was reversed by co-incubation with tocilizumab at clinically relevant concentrations. Accordingly, inhibition of IL-6 signaling in RA patients treated with tocilizumab may restore CYP450 activities to higher levels than those in the absence of tocilizumab, leading to increased metabolism of drugs that are CYP450 substrates. Its effect on CYP2C8 or transporters (e.g., P-gp) is unknown. This is clinically relevant for CYP450 substrates with a narrow therapeutic index, where the dose is individually adjusted. On initiation of Actemra, in patients being treated with these types of medicinal products, therapeutic monitoring of the effect (e.g., warfarin) or drug concentration (e.g., cyclosporine or theophylline) should be performed and the individual dose of the medicinal product

(*continued*)

TABLE 10-3 (*Continued*)

TP Name (Brand Name) Pharmacologic Class Year of Labeling	Item	Labeling Language
		adjusted as needed. Caution should be exercised when Actemra is coadministered with drugs where decrease in effectiveness is undesirable, such as oral contraceptives (CYP3A4 substrates). (§7.2).
	Simvastatin	Simvastatin is a CYP3A4 and OATPIBI substrate. In 12 RA patients not treated with Actemra and receiving 40 mg simvastatin, exposures of simvastatin and its metabolite, simvastatin acid, was 4- to 10-fold and 2-fold higher, respectively, than the exposures observed in healthy subjects. One week following administration of a single infusion of Actemra (10 mg per kg), exposure of simvastatin and simvastatin acid decreased by 57% and 39%, respectively, to exposures that were similar or slightly higher than those observed in healthy subjects. Exposures of simvastatin and simvastatin acid increased upon withdrawal of Actemra in RA patients. Selection of a particular dose of simvastatin in RA patients should take into account the potentially lower exposures that may result after initiation of Actemra (due to normalization of CYP3A4) or higher exposures after discontinuation of Actemra.
	Omeprazole	Omeprazole is a CYP2C19 and CYP3A4 substrate. In RA patients receiving 10 mg omeprazole, exposure to omeprazole was approximately twofold higher than that observed in healthy subjects. In RA patients receiving 10 mg omeprazole, before and 1 week after Actemra infusion (8 mg per kg), the omeprazole AUC_{inf} decreased by 12% for poor ($n = 5$) and intermediate metabolizers ($n = 5$) and by 28% for extensive metabolizers ($n = 8$) and were slightly higher than those observed in healthy subjects.

	Dextromethorphan	Dextromethorphan is a CYP2D6 and CYP3A4 substrate. In 13 RA patients receiving 30 mg dextromethorphan, exposure to dextromethorphan was comparable to that in healthy subjects. However, exposure to its metabolite, dextrorphan (a CYP3A4 substrate), was a fraction of that observed in healthy subjects. One week following administration of a single infusion of Actemra (8 mg per kg), dextromethorphan exposure was decreased by approximately 5%. However, a larger decrease (29%) in dextrorphan levels was noted after Actemra infusion.
Rilonacept (Arcalyst) IL-1β blocker 2008	7 Drug interactions 7.2 Cytochrome P450 substrates	The formation of CYP450 enzymes is suppressed by increased levels of cytokines (e.g., IL-1) during chronic inflammation. Thus it is expected that for a molecule that binds to IL-1, such as rilonacept, the formation of CYP450 enzymes could be normalized. This is clinically relevant for CYP450 substrates with a narrow therapeutic index, where the dose is individually adjusted (e.g., warfarin). On initiation of Arcalyst, in patients being treated with these types of medicinal products, therapeutic monitoring of the effect or drug concentration should be performed and the individual dose of the medicinal product may need to be adjusted as needed.
Canakinumab (Ilaris) IL-1β blocker 2012	7 Drug interactions	Interactions between Ilaris and other medicinal products have not been investigated in formal studies.
	7.3 Cytochrome P450 substrates	The formation of CYP450 enzymes is suppressed by increased levels of cytokines (e.g., IL-1) during chronic inflammation. Thus it is expected that for a molecule that binds to IL-1, such as canakinumab, the formation of CYP450 enzymes could be normalized. This is clinically relevant for CYP450 substrates with a narrow therapeutic index, where the dose is individually adjusted (e.g., warfarin). On initiation of canakinumab, in patients being treated with these types of medicinal products, therapeutic monitoring of the effect or drug concentration should be performed and the individual dose of the medicinal product may need to be adjusted as needed.
Golimumab (Simponi) TNF blocker	7 Drug interactions 7.4 Cytochrome P450 substrates	The formation of CYP450 enzymes may be suppressed by increased levels of cytokines (e.g., TNF-α) during chronic inflammation. Therefore, it is expected that for a

(*continued*)

TABLE 10-3 (*Continued*)

TP Name (Brand Name) Pharmacologic Class Year of Labeling	Item	Labeling Language
2011		molecule that antagonizes cytokine activity, such as golimumab, the formation of CYP450 enzymes could be normalized. On initiation or discontinuation of Simponi in patients being treated with CYP450 substrates with a narrow therapeutic index, monitoring of the effect (e.g., warfarin) or drug concentration (e.g., cyclosporine or theophylline) is recommended and the individual dose of the drug product may be adjusted as needed.
Infliximab (Remicade) TNF blocker 2011	7 Drug interactions 7.5 Cytochrome P450 substrates	The formation of CYP450 enzymes may be suppressed by increased levels of cytokines (e.g., TNF-α, IL-1, IL-6, IL-10, IFN) during chronic inflammation. Therefore, it is expected that for a molecule that antagonizes cytokine activity, such as infliximab, the formation of CYP450 enzymes could be normalized. On initiation or discontinuation of Remicade in patients being treated with CYP450 substrates with a narrow therapeutic index, monitoring of the effect (e.g., warfarin) or drug concentration (e.g., cyclosporine or theophylline) is recommended, and the individual dose of the drug product may be adjusted as needed.
Ustekinumab (Stelara) human IL12/IL23 antagonist 2012	7 Drug interactions 7.3 Cytochrome P450 substrates	Drug interaction studies have not been conducted with Stelara. The formation of CYP450 enzymes can be altered by increased levels of certain cytokines (e.g., IL-1, IL-6, IL-10, TNFα, IFN) during chronic inflammation. Thus, STELARA®, an antagonist of IL-12 and IL-23, could normalize the formation of CYP450 enzymes. Upon initiation of STELARA® in patients who are receiving concomitant CYP450 substrates, particularly those with a narrow therapeutic index, monitoring for therapeutic effect (e.g., for warfarin) or drug concentration (e.g., for cyclosporine) should be considered and the individual dose of the drug adjusted as needed [*see Clinical Pharmacology (12.3)*].

	12 Clinical pharmacology 12.3 Pharmacokinetics: drug–drug interactions	The effect of IL-12 or IL-23 on the regulation of CYP450 enzymes were evaluated in an *in vitro* study using human hepatocytes, which showed that IL-12 and /or IL-23 at levels of 10 ng/mL did not alter human CYP450 enzyme activities (CYP1A2, 2B6, 2C9, 2D6, or 3A4). However, the clinical relevance of *in vitro* data has not been established. (§7.3).
Belatacept (Nulojix) Selective T-cell costimulation blocker 2011	7 Drug interactions 7.1 Cytochrome P450 Substrates	No formal drug interaction studies have been conducted with Nulojix. Other biologic therapies that are cytokines or cytokine modulators have been shown to affect the expression and/or functional activities of CYP450 enzymes *in vitro* and/or *in vivo*. *In vitro* studies have shown that Nulojix inhibits the production of certain cytokines during an alloimmune response. No studies in kidney transplant patients have been conducted to assess if Nulojix inhibits cytokine production *in vivo*. The potential for Nulojix to alter the systemic concentrations of drugs that are CYP450 substrates has not been studied; however, in the event that kidney transplant patients receiving Nulojix exhibit signs and symptoms of altered administered drugs that are known to be efficacy or adverse events associated with cometabolized by CYP450, the clinician should be aware of potentially altered CYP450 metabolism of these drugs.
	7.2 Use with mycophenolate mofetil	In a pharmacokinetic substudy of Studies 1 and 2, the plasma concentrations of MPA were measured in 41 patients who received fixed MMF doses of 500–1500 mg twice daily with either 5 mg/kg of Nulojix or cyclosporine. The mean dose-normalized MPA C_{max} and AUC_{0-12} were approximately 20% and 40% higher, respectively, with Nulojix coadministration than with cyclosporine coadministration. Clinicians should be aware that there is also a potential change of MPA exposure after crossover from cyclosporine to Nulojix or from Nulojix to cyclosporine in patients concomitantly receiving MMF.
Somatropin (rDNA origin) (Genotropin) human growth hormone	Highlights of prescription	*Inhibition of 11 β-hydroxysteroid dehydrogenase type 1*: may require the initiation of glucocorticoid replacement therapy. Patients treated with glucocorticoid replacement

(*continued*)

TABLE 10-3 (*Continued*)

TP Name (Brand Name) Pharmacologic Class Year of Labeling	Item	Labeling Language
2011	information: drug interactions	therapy for previously diagnosed hypoadrenalism may require an increase in their maintenance doses (§7.1; §7.2). *Glucocorticoid replacement*: should be carefully adjusted (§7.2) *Cytochrome P450-metabolized drugs*: monitor carefully if used with somatropin (§7.3) *Oral estrogen*: larger doses of somatropin may be required in women (§7.4) Insulin and/or oral/injectable hypoglycemic agents: may require adjustment (§7.5)
	7 Drug interactions	
	7.1 11 β-hydroxysteroid dehydrogenase type 1	The microsomal enzyme 11βHSD-1 is required for conversion of cortisone to its active metabolite, cortisol, in hepatic and adipose tissue. GH and somatropin inhibit 11βHSD-1. Consequently, individuals with untreated GH deficiency have relative increases in 11βHSD-1 and serum cortisol. Introduction of somatropin treatment may result in inhibition of 11βHSD-1 and reduced serum cortisol concentrations. As a consequence, previously undiagnosed central (secondary) hypoadrenalism may be unmasked and glucocorticoid replacement may be required in patients treated with somatropin. In addition, patients treated with glucocorticoid replacement for previously diagnosed hypoadrenalism may require an increase in their maintenance or stress doses after initiation of somatropin treatment; this may be especially true for patients treated with cortisone acetate and prednisone because conversion of these drugs to their biologically active metabolites depend on the activity of 11βHSD-1.
	7.3 Cytochrome P450-Metabolized drugs	Limited published data indicate that somatropin treatment increases CYP450-mediated antipyrine clearance in humans. These data suggest that somatropin administration may alter the clearance of compounds known to be metabolized by CYP450 liver enzymes (e.g., corticosteroids, sex steroids, anticonvulsants, cyclosporine). Careful monitoring is advisable when somatropin is administered in combination with other

		drugs known to be metabolized by CYP450 liver enzymes. However, formal drug interaction studies have not been conducted.
	7.4 Oral estrogen	In patients on oral estrogen replacement, a larger dose of somatropin may be required to achieve the defined treatment goal. (§2.2)
	7.5 Insulin and/or oral/injectable hypoglycemic agents	In patients with diabetes mellitus requiring drug therapy, the dose of insulin and/or oral/injectable agent may require adjustment when somatropin therapy is initiated. (§5.4)
2. Immunosuppressive effect		
Infliximab (Remicade) TNF blocker 2011	7 Drug Interactions 7.3 MTX and other concomitant medications	Specific drug interaction studies, including interactions with MTX, have not been conducted. Concomitant MTX use may decrease the incidence of antiinfliximab antibody production and increase infliximab concentrations.
Golimumab (Simponi) TNF blocker 2011	7 Drug interactions 7.1 Methotrexate	For the treatment of RA, Simponi should be used with MTX. (§14.1) Because the presence or absence of concomitant MTX did not appear to influence the efficacy or safety of Simponi in the treatment of PsA or AS, Simponi can be used with or without MTX in their treatment. (§12.3)
	12 Clinical pharmacology 12.3 Pharmacokinetics	Patients with RA, PsA, and AS treated with Simponi, 50 mg and MTX had approximately 52%, 36%, and 21% higher mean steady-state trough concentrations of golimumab, respectively, compared with those treated with Simponi 50 mg without MTX. The presence of MTX also decreased antigolimumab antibody incidence from 7% to 2%. (§6.1) For RA, Simponi should be used with MTX. In the PsA and AS trials, the presence or absence of concomitant MTX did not appear to influence clinical efficacy and safety parameters. (§14.1) Patients who developed anti-Simponi antibodies generally had lower steady-state serum trough concentrations of Simponi.

(*continued*)

TABLE 10-3 (*Continued*)

TP Name (Brand Name) Pharmacologic Class Year of Labeling	Item	Labeling Language
Adalimumab (Humira) TNF blocker 2012	7 Drug interactions	
	7.1 Methotrexate	Humira has been studied in RA patients taking concomitant MTX. Although MTX reduced the apparent adalimumab clearance, the data do not suggest the need for dose adjustment of either Humira or MTX ($12.3)
	12 Clinical pharmacology	
	12.3 Pharmacokinetics	In RA patients receiving 40 mg Humira every other week, adalimumab mean steady-state trough concentrations of approximately 5 μg/mL and 8–9 μg/mL, were observed without and with MTX, respectively. MTX reduced adalimumab apparent clearance after single and multiple dosing by 29% and 44%, respectively, in patients with RA. Adalimumab mean steady-state trough concentrations were slightly higher in psoriatic arthritis patients treated with 40 mg Humira every other week (6–10 μg/mL and 8.5–12 μg/mL, without and with MTX, respectively) compared to the concentrations in RA patients treated with the same dose.
Certolizumab pegol (Cimzia) TNF blocker 2012	6 Adverse reactions	
	Immunogenicity	Antibody formation was associated with lowered drug plasma concentration and reduced efficacy. In patients receiving the recommended Cimzia dosage of 200 mg every other week with concomitant MTX, the ACR20 response was lower among antibody positive patients than among antibody-negative patients (Study RA-I, 48% versus 60%; Study RA-II 35% versus 59%, respectively). In Study RA-III, too few patients developed antibodies to allow for meaningful analysis of ACR20 response by antibody status. In Study RA-IV (monotherapy), the ACR20 response was 33% versus 56%, antibody-positive versus antibody-negative status, respectively. (§12.3)

	12 Clinical Pharmacology 12.3 Pharmacokinetics: drug interaction studies	Methotrexate pharmacokinetics is not altered by concomitant administration with Cimzia in patients with RA. The effect of methotrexate on Cimzia pharmacokinetics was not studied. However, methotrexate-treated patients have lower incidence of antibodies to Cimzia. Thus therapeutic plasma levels are more likely to be sustained when Cimzia is administered with methotrexate in patients with RA.
3. Substrate of the therapeutic enzyme		
Glucarpidase (Voraxaze) Carboxypeptidase 2012	Highlights of prescription information: drug interactions	Leucovorin is a substrate for Voraxaze. Do not administer leucovorin within 2 h before or after a dose of Voraxaze. (§7.1) Other potential exogenous substrates of Voraxaze include reduced folates and folate antimetabolites. (§7.2)
	7 Drug interactions 7.1 Use of Voraxaze with Leucovorin	Leucovorin is a substrate for Voraxaze. Do not administer leucovorin within 2 h before or after a dose of Voraxaze. No dose adjustment is recommended for the continuing leucovorin regimen because the leucovorin dose is based on the patient's pre-Voraxaze methotrexate concentration. (§12.3)
	7.2 Other substrate interference	Other potential exogenous substrates of Voraxaze may include reduced folates and folate antimetabolites.
	12 Clinical pharmacology 12.3 Pharmacokinetics: drug interactions	In a study of cancer patients receiving a high-dose methotrexate ($\geq$1 g/m^2) and leucovorin rescue regimen, intravenous administration of 50 Units/kg Voraxaze 2 h before leucovorin reduced (6S)-leucovorin AUC_{0-3h} by 33% and C_{max} by 52% and also reduced its active metabolite, (6S)-5-methyltetrahydrofolate, $AUC_{0\text{-}3h}$ by 92% and C_{max} by 93%. (§7.1)
4. Mechanisms to be elucidated		
Trastuzumab (Herceptin) HER2/neu receptor antagonist 2010	7 Drug interactions	In Study 5, the mean serum trough concentration of trastuzumab was consistently elevated approximately 1.5-fold, when administered in combination with paclitaxel compared to trough concentrations of trastuzumab when administered in combination with an anthracycline and cyclophosphamide.
Palifermin		Heparin may increase systemic exposure (§7)

(*continued*)

TABLE 10-3 *(Continued)*

TP Name (Brand Name) Pharmacologic Class Year of Labeling	Item	Labeling Language
(Kepivance) Mucocutaneous epithelial human growth factor 2011	Highlights of prescription information: drug interactions	
	7 Drug interactions	*In vitro* and *in vivo* data showed that palifermin interacts with unfractionated as well as low molecular weight heparins. Heparin coadministration resulted in a fivefold increase in palifermin systemic exposure. Avoid coadministration of palifermin with heparin. If heparin is used to maintain an intravenous line, rinse the line with saline before and after Kepivance administration. (§12.3)
	12 Clinical pharmacology	
	12.3 Pharmacokinetic: drug interactions	The potential PK interaction between palifermin and heparin was evaluated in a single-dose study in 27 healthy subjects receiving palifermin (60 μg/kg) coadministered with and without therapeutic levels of unfractionated heparin. This coadministration resulted in a fivefold increase in palifermin AUC and an 80% decrease in the mean CL. There was no significant effect of palifermin on heparin activity with respect to aPTT. The clinical relevance of this observed increase in palifermin systemic exposure is unclear. (§7)

[a]Data from Ref. 26.

ACR20: American College of Rheumatology 20 Response; ANC: Absolute Neutrophil Count; aPTT: activated Partial Thromboplastin Time; AS: Ankylosing Spondylitis; 11βHSD-1: 11β-hydroxysteroid dehydrogenase type 1; CHC: Chronic Hepatitis C; CL: Clearance; CYP: Cytochrome P450; GH: Growth Hormone; HCV: Hepatitis C Virus; HER2: Human Epidermal Growth Factor Receptor 2; IL: Interlukin; IMDH: Inosine Monophosphate Dehydrogenase; INF: Interferon; MMF: Mycophenolate Mofetil; MPA: Mycophenolic Acid; 6MTITP: 6-Methylthioionosine Monophosphate; MTX: Methotrexate; NRTI: Nucleoside/Nucleotide Reverse Transcriptase Inhibitor; P-gp: P-glycoprotein; PK: Pharmacokinetics; PsA: Psoriatic Arthritis; RA: Rheumatoid Arthritis; TFN: Tumor Necrosis Factor.

amount of cytokine and/or cytokine modulator).[20,21,28] However, complex signaling may be involved in TP-DDI and it is not currently possible to ensure that cell culture systems provide the appropriate mix of cell types required for meaningful evaluation *in vitro*. Presently, *in vitro* and animal studies have limited value in the qualitative and quantitative projection of clinical interactions because translation of results from *in vitro* to *in vivo* and from animal to human has been inconsistent, necessitating *in vivo* drug interaction studies. Research that increases our understanding of CYP enzyme or transporter regulation by cytokines may increase the utility of *in vitro* studies. Standardized and optimal assay conditions and the identification of suitable positive controls to monitor assay performance are needed. Until such information is available, it is important to carry out *in vivo* risk-based drug interaction studies for TP-DDI assessments. Many TPs cannot be administered to healthy subjects, and cytokines have a different expression levels between healthy subjects and patients, therefore clinical studies (including TP-DDI studies) are conducted in patients. One efficient study design is the "cocktail" approach. Other *in vivo* study designs allow evaluation of the effect of the TP on substrates for specific CYP enzymes and transporters in targeted patient populations, as described in the next section. Table 10-3 provides an extensive list of labeling examples on PK-based TP-DDI.

Recent BLA review examples include tocilizumab (labeling based on *in vitro* and *in vivo* TP-DDI data information) and rilonacept (labeling without TP-DDI studies).

- If a TP will be used with other drug products (D or TP) as a combination therapy, studies should evaluate the effect of each product on the other. The studies should assess effects on PK and, when appropriate, PD of either drug. This evaluation is particularly important when a drug used in combination has a narrow therapeutic range such as chemotherapeutic agents.

  Examples include cancer therapies in which multiple drugs are combined, such as bevacizumab in combination with intravenous 5-fluorouracil-based chemotherapy including an IV bolus of IFL (irinotecan, 5-FU, and leucovorin) or FOLFOX-4 (oxaliplatin, 5-FU, and leucovorin) for the treatment of first- or second-line metastatic colorectal cancer, and trastuzumab in combination with doxorubicin, cyclophosphamide, and either paclitaxel or docetaxel or in combination with docetaxel and carboplatin for adjuvant treatment of HER2 overexpressing node positive or node negative (estrogen receptor/progesterone receptor [ER/PR] negative or with one high risk feature) breast cancer.[26]

- If there are known mechanisms for interactions or prior experience suggests potential PK or PD interactions, appropriate evaluations for possible interactions should be conducted. Some interactions between drugs and TPs are based on mechanisms other than CYP enzyme or transporter modulation.

  For example, methotrexate, an immunosuppressive agent may alter the clearance of concomitantly administered TPs through the reduction of antibodies formed against TP. Methotrexate decreased the clearance of infliximab

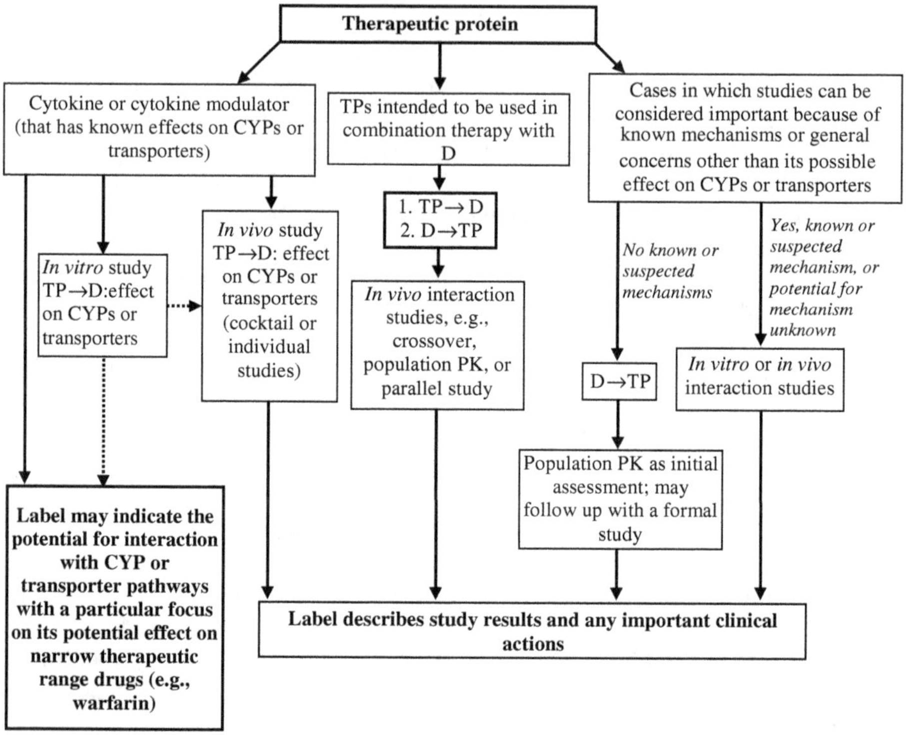

FIGURE 10-1 Summary of the types of studies that have been used during drug development to evaluate therapeutic protein (TP)–small molecule drug (D) interactions. Evaluation includes the effect of TP on D (TP → D) and the effect of D on TP (D → TP). Broken lines: limited use of *in vitro* studies for informing *in vivo* study design or labeling; CYP: cytochrome P450; PK: pharmacokinetics. (Modified from Ref. 13.)

(a tumor necrosis factor α [TNF-α] blocker), which was associated with the decreased incidence of anti-infliximab antibody formation that affects infliximab clearance. At 8 weeks after a maintenance dose of 3–10 mg/kg of infliximab, median infliximab serum concentrations ranged from approximately 0.5 to 6 μg/mL; however, infliximab concentrations were not detectable (<0.1 μg/mL) in patients who tested positive for antibodies to infliximab.[26,27] The incidence of developing anti-infliximab antibodies was 7% and 21%, respectively, with or without concomitant administration of methotrexate at 12 weeks after last infliximab infusion of a 3 mg/kg dose.[26,27] Methotrexate may also affect the clearance of other TNF-α blocking agents. For example, it decreased the clearance of adalimumab by 29% after single dosing and by 44% after multiple dosing, which may result from reduced immunogenicity incidence; patients treated with concomitant methotrexate had a lower incidence of anti-adalimumab antibody formation than patients on adalimumab monotherapy (1% versus 12%).[26]

Other examples include heparin's effect on palifermin (increased exposure) and paclitaxel's effect on trastuzumab (increased exposure)[26] (Table 10-3), for which mechanisms have not been fully elucidated. In postmarketing studies, coadministration of palifermin with heparin resulted in a 5-fold increase in the systemic exposure of palifermin.[26,29] In clinical studies, administration of paclitaxel in combination with trastuzumab resulted in a 1.5-fold increase in trastuzumab serum levels.[26]

The recently approved glucarpidase label includes information about an interaction between methotrexate and leucovorin involving glucarpidase.[26] Glucarpidase is a carboxypeptidase enzyme indicated for the treatment of toxic plasma methotrexate concentrations (>1 μmole/L) in patients with delayed methotrexate clearance due to impaired renal function. In a study of cancer patients receiving a high-dose methotrexate ($\geq$1 g per m^2) and leucovorin rescue regimen, intravenous administration of glucarpidase 2 h before leucovorin reduced (6S)-leucovorin AUC_{0-3h} by 33% and C_{max} by 52%, and also reduced its active metabolite, (6S)-5 methyltetrahydrofolate, AUC_{0-3h} by 92% and C_{max} by 93%.[26] This interaction is anticipated because both methotrexate and leucovorin are substrates of glucarpidase.

In addition to PK interactions, PD interactions can also occur between TP and drug or TP and TP. For example, an increased risk of serious infections has been observed in clinical studies of concurrent therapy for the treatment of rheumatoid arthritis, for eample, etanercept and anakinra, or etanercept and abatacept, without added benefit. Therefore, a warning is stated in the label of a newly approved TNF-α blocker, golimumab, about its risk in combination use with another biologics such as abatacept or anakinra. The labeling states that "use of Simponi (golimumab) with abatacept or anakinra is not recommended" (Table 10-4).[26]

Some PD interactions with TP-DDI have resulted in increased toxicity with combination chemotherapy. It is stated in the Vectibix labeling under Section 10.5 "Warnings and Precautions" that "Vectibix (panitumumab) is not indicated for use in combination with chemotherapy. In an interim analysis of Study 2, the addition of Vectibix to the combination of bevacizumab and chemotherapy resulted in decreased overall survival and increased incidence of NCI CTC grade 3–5 (87% vs 72%) adverse reactions" (Table 10-4).[26]

As described in Figure 10-1, when there have been known mechanisms or prior experience suggesting potential PK or PD interactions, as illustrated by examples discussed earlier, appropriate *in vitro* and/or *in vivo* assessment for possible interactions have been requested and conducted.

10.3 TYPES OF STUDY DESIGNS

In vitro methods to evaluate effects of a drug on CYP enzymes and transporters are discussed in detail in other publications and on the FDA Drug Development and Drug Interaction website.[4–8] A white paper from the International Transporter Consortium provides recommendations on *in vitro* models to evaluate transporter-based interactions, including specific decision trees to evaluate the following transporters:

TABLE 10-4 Examples of Labeling Language for Therapeutic Protein–Drug Interactions (Pharmacodynamics Based)

TP Names (Brand Names) Pharmacological Class; Year of Labeling	Item	Labeling Language
1. Growth factors (Palifermin and G-CSF) and chemotherapy		
Palifermin (Kepivance) Mucocutaneous Epithelial Human Growth Factor 2011	Highlights of prescription information: drug interactions	Myelotoxic chemotherapy (§7; §14)
	7. Drug interactions	Do not administer Kepivance within 24 h before, during infusion of, or within 24 h after administration of myelotoxic chemotherapy. (§2.1; §14) In a clinical trial, administration of Kepivance within 24 h of chemotherapy resulted in increased severity and duration of oral mucositis.
Filgrastim (Neupogen) Granulocyte colony-stimulating factor (G-CSF) 2012	Precautions	The safety and efficacy of Neupogen given simultaneously with cytotoxic chemotherapy have not been established. Because of the potential sensitivity of rapidly dividing myeloid cells to cytotoxic chemotherapy, do not use Neupogen in the period 24 h before through 24 h after the administration of cytotoxic chemotherapy (see Dosage and Administration).
Pegfilgrastim (Neulasta) G-CSF 2007	Precautions	Neulasta should not be administered in the period between 14 days before and 24 h after administration of cytotoxic chemotherapy (see Dosage and Administration) because of the potential for an increase in sensitivity of rapidly dividing myeloid cells to cytotoxic chemotherapy
2. Growth factors and myeloablative therapy		
Oprelvekin (Neumega)	Warnings	*Increased toxicity after myeloablative therapy*: Neumega is not indicated after myeloablative chemotherapy. In a randomized, placebo-controlled Phase II study, the

Thrombopoietic growth factor (TGF, IL-11) 2009		effectiveness of Neumega was not demonstrated (see Clinical Studies, Study in Patients Following Myeloablative Chemotherapy). In this study, a statistically significant increased incidence in edema, conjunctival bleeding, hypotension, and tachycardia was observed in patients receiving Neumega as compared to placebo. The following severe or fatal adverse reactions have been reported in postmarketing use in patients who received Neumega after bone marrow transplantation: fluid retention or overload (e.g., facial edema, pulmonary edema), capillary leak syndrome, pleural and pericardial effusion, papilledema and renal failure.
3. Increased toxicities with concurrent therapy (psychotropic, nephrotoxic, myelotoxic, cardiotoxic or hepatotoxic drugs)		
Aldesleukin (Proleukin) Human recombinant IL-2 2011	Drug interactions	Proleukin may affect central nervous function. Therefore, interactions could occur after concomitant administration of psychotropic drugs (e.g., narcotics, analgesics, antiemetics, sedatives, tranquilizers). Concurrent administration of drugs possessing nephrotoxic (e.g., aminoglycosides, indomethacin), myelotoxic (e.g., cytotoxic chemotherapy), cardiotoxic (e.g., doxorubicin) or hepatotoxic (e.g., methotrexate, asparaginase) effects with Proleukin may increase toxicity in these organ systems. The safety and efficacy of Proleukin in combination with any antineoplastic agents have not been established. In addition, reduced kidney and liver function secondary to Proleukin treatment may delay elimination of concomitant medications and increase the risk of adverse events from those drugs.
Peginterferon α-2b (PegIntron) INF-α 2011	7. Drug interactions 7.3 Use with ribavirin (nucleoside analogues)	Hepatic decompensation (some fatal) has occurred in cirrhotic HIV/HCV co-infected patients receiving combination antiretroviral therapy for HIV and INF-α and ribavirin. Adding treatment with INF-α alone or in combination with ribavirin may increase the risk in this patient subset. Patients receiving interferon with ribavirin and NRTIs should be closely monitored for treatment-associated toxicities, especially hepatic decompensation and anemia. Discontinuation of NRTIs should be considered as medically appropriate (see Individual NRTI Product Information). Dose reduction or discontinuation of interferon, ribavirin, or both should also be considered if worsening

(continued)

TABLE 10-4 (*Continued*)

TP Names (Brand Names) Pharmacological Class; Year of Labeling	Item	Labeling Language
		clinical toxicities are observed, including hepatic decompensation (e.g., Child-Pugh < 6).
4. Increased infections with concurrent therapy		
Etanercept (Enbrel) TNF blocker 2009	Precaution: drug interactions	In a study in which patients with active RA were treated for ≤ 24 weeks with concurrent Enbrel and anakinra therapy, a 7% rate of serious infections was observed, which was higher than that observed with Enbrel alone (0%). Two percent of patients treated concurrently with Enbrel and anakinra developed neutropenia (ANC $< 1 \times 10^9$/L In a study of patients with Wegener's granulomatosis, the addition of Enbrel to standard therapy (including cyclophosphamide) was associated with a higher incidence of noncutaneous solid malignancies. The use of Enbrel in patients receiving concurrent cyclophosphamide therapy is not recommended (see Warnings: Malignancies and Adverse reactions: Malignancies). Patients in a clinical study who were on established therapy with sulfasalazine, to which Enbrel was added, were noted to develop a mild decrease in mean neutrophil counts compared to groups treated with either Enbrel or sulfasalazine alone. The clinical significance of this observation is unknown.
Infliximab (Remicade) TNF blocker 2011	Highlights of prescription information: drug interactions	Use with anakinra or abatacept increased risk of serious infections (§7.1)
	7 Drug interactions 7.1 Use with anakinra or abatacept	An increased risk of serious infections was seen in clinical studies of other TNF-α-blocking agents used in combination with anakinra or abatacept, with no added clinical benefit. Because of the nature of the adverse reactions seen with these

		combinations with TNF-blocker therapy, similar toxicities may also result from the combination of anakinra or abatacept with other TNF-α-blocking agents. Therefore, the combination of Remicade and anakinra or abatacept is not recommended. (§5.9; §5.10)
	7.2 Use with tocilizumab	The use of tocilizumab in combination with biological DMARDs, such as TNF antagonists, including Remicade, should be avoided because of the possibility of increased immunosuppression and increased risk of infection.
Adalimumab (Humira) TNF blocker 2012	Highlights of prescription information: drug interactions	Abatacept increased risk of serious infection (§5.1; §5.11; §7.2) Anakinra increased risk of serious infections (§5.1; §5.7; §7.2)
	7 Drug interactions 7.2 Biologic products	In clinical studies in patients with RA, an increased risk of serious infections has been seen with the combination of TNF blockers with anakinra or abatacept, with no added benefit; therefore, use of Humira with abatacept or anakinra is not recommended in patients with RA. (§5.7; 5.11) A higher rate of serious infections has also been observed in patients with RA treated with rituximab who received subsequent treatment with a TNF blocker. There is insufficient information to provide recommendations regarding the concomitant use of Humira and other biologic products for the treatment of RA, PsA, AS, Crohn's disease, and plaque psoriasis.
Golimumab (Simponi) TNF blocker 2010	Highlights of prescription information: drug interactions	Abatacept increased risk of serious infection (§5.1; §5.5; §7.2) Anakinra increased risk of serious infections (§5.1; §5.6; §7.2)
	7 Drug interactions 7.2 Biologic products for RA, PsA, and/or AS	An increased risk of serious infections has been seen in clinical RA studies of other TNF blockers used in combination with anakinra or abatacept, with no added benefit; therefore, use of Simponi with abatacept or anakinra is not recommended. (§5.6) A higher rate of serious infections has also been observed in RA patients treated with rituximab who received subsequent treatment with a TNF blocker. There is insufficient

(*continued*)

TABLE 10-4 (*Continued*)

TP Names (Brand Names) Pharmacological Class; Year of Labeling	Item	Labeling Language
		information to provide recommendations regarding the concomitant use of Simponi and other biologic products approved to treat RA, PsA, or AS.
Certolizumab pegol (Cimzia) TNF blocker 2012	Highlights of prescription information: drug interactions	Use with biological DMARDs increased risk of serious infections (§5.8; §7.1)
	7 Drug interactions	
	7.1 Use with anakinra, abatacept, rituximab, and natalizumab	An increased risk of serious infections has been seen in clinical studies of other TNF-blocking agents used in combination with anakinra or abatacept, with no added benefit. Formal drug interaction studies have not been performed with rituximab or natalizumab. Because of the nature of the adverse events seen with these combinations with TNF-blocker therapy, similar toxicities may also result from the use of Cimzia in these combinations. There is not enough information to assess the safety and efficacy of such combination therapy. Therefore, the use of Cimzia in combination with anakinra, abatacept, rituximab, or natalizumab is not recommended. (§5.8)
Canakinumab (Ilaris) IL-1β blocker 2009	7 Drug interactions	
	7.1 TNF and IL-1 blockers	An increased incidence of serious infections and an increased risk of neutropenia have been associated with administration of another IL-1 blocker in combination with TNF inhibitors in another patient population. Use of Ilaris with TNF inhibitors may also result in similar toxicities and is not recommended because this may increase the risk of serious infections. (§5.1) The concomitant administration of Ilaris with other drugs that block IL-1 has not been studied. Based on the potential for pharmacological interactions between Ilaris and a recombinant IL-1ra, concomitant administration of Ilaris and other agents that block IL-1 or its receptors is not recommended

Rilonacept (Arcalyst) IL-1β blocker 2008	7 Drug interactions 7.1 TNF and IL-1 blockers	Specific drug interaction studies have not been conducted with Arcalyst. Concomitant administration of another drug that blocks IL-1 with a TNF-blocking agent in another patient population has been associated with an increased risk of serious infections and an increased risk of neutropenia. The concomitant administration of Arcalyst with TNF-blocking agents may also result in similar toxicities and is not recommended. (§5.1) The concomitant administration of Arcalyst with other drugs that block IL-1 has not been studied. Based on the potential for pharmacologic interactions between rilonacept and a recombinant IL-1ra, concomitant administration of Arcalyst and other agents that block IL-1 or its receptors is not recommended.
5. Increased myelotoxicity INF-α-2b and myelosuppressive agents)		
INF-α-2b (Intron A) INF-α 2011	Warnings: drug interactions	Caution should be exercised when administering Intron A therapy in combination with other potentially myelosuppressive agents such as zidovudine.
Interferon alfa-2a (Roferon-A) Alpha interferon 2006	Precautions: drug interactions	Caution should be exercised when administering Roferon-A in combination with other potentially myelosuppressive agents. Synergistic toxicity has been observed when Roferon-A is administered in combination with zidovudine (AZT) (see Warnings: Bone Marrow Toxicity). The neurotoxic, hematotoxic or cardiotoxic effects of previously or concurrently administered drugs may be increased by INF. Interactions could occur following concurrent administration of centrally acting drugs. Use of Roferon-A in conjunction with IL-2 may potentiate risks of renal failure.

(*continued*)

TABLE 10-4 (*Continued*)

TP Names (Brand Names) Pharmacological Class; Year of Labeling	Item	Labeling Language
6. Increased toxicity with combination chemotherapy		
Panitumumab (Vectibix) EGFR antagonist 2009	Highlights of prescribing information: warnings and precautions	Increased toxicity with combination chemotherapy. Vectibix is not indicated for use in combination with chemotherapy (§5.3)
	5 Warnings and precautions	
	5.3 Increased toxicity with combination chemotherapy	Vectibix is not indicated for use in combination with chemotherapy. In an interim analysis of Study 2, the addition of Vectibix to the combination of bevacizumab and chemotherapy resulted in decreased overall survival and increased incidence of NCI-CTC grade 3–5 (87% vs 72%) adverse reactions. (§14)
7. Others		
Etanercept (Enbrel) TNF blocker 2009	Precautions: drug interactions	In a study of patients with Wegener granulomatosis, the addition of Enbrel therapy to standard therapy (including cyclophosphamide) was associated with a higher incidence of noncutaneous solid malignancies. The use of Enbrel in patients receiving concurrent cyclophosphamide therapy is not recommended (see Warnings: Malignancies and Adverse Malignancies).
Somatropin	7. Drug interactions	

(Genotropin) Human growth hormone 2011		
	7.2 Pharmacologic glucocorticoid therapy and supraphysiologic glucocorticoid treatment	Pharmacologic glucocorticoid therapy and supraphysiologic glucocorticoid treatment may attenuate the growth-promoting effects of somatropin in children. Therefore, glucocorticoid replacement dosing should be carefully adjusted in children receiving concomitant somatropin and glucocorticoid treatments to avoid both hypoadrenalism and an inhibitory effect on growth.
Incobotulinumtoxi A (Xeomin) Acetylcholine release inhibitor and neuromuscular blocker 2011	Highlights of prescribing information: drug interactions	Concomitant treatment of Xeomin and aminoglycoside antibiotics, spectinomycin, or other agents that interfere with neuromuscular transmission (e.g., tubocurarine-like agents), or muscle relaxants, should be observed closely because the effect of Xeomin may be potentiated. (§7)
	7 Drug interactions	Coadministration of Xeomin and aminoglycoside antibiotics or other agents interfering with neuromuscular transmission (e.g., tubocurarine-type muscle relaxants) should be performed only with caution because these agents may potentiate the effect of the toxin. Use of anticholinergic drugs after administration of Xeomin may potentiate systemic anticholinergic effects. The effect of administering different botulinum toxin products at the same time or within several months of each other is unknown. Excessive neuromuscular weakness may be exacerbated by administration of another botulinum toxin before the resolution of the effects of a previously administered botulinum toxin. Excessive weakness may also be exaggerated by administration of a muscle relaxant before or after administration of Xeomin.
Pegloticase (Krystexxa) Recombinant uricase 2012	7 Drug interactions	Because antipegloticase antibodies appear to bind to the PEG portion of the drug, there may be potential for binding with other PEGylated products. The impact of anti-PEG antibodies on patients' responses to other PEG-containing therapeutics is unknown.
Abobotulinumtoxin A	7 Drug interactions	

(*continued*)

TABLE 10-4 (*Continued*)

TP Names (Brand Names) Pharmacological Class; Year of Labeling	Item	Labeling Language
(Dysport) Neuromuscular blocker 2012		Patients treated concomitantly with botulinum toxins and aminoglycosides or other agents interfering with neuromuscular transmission (e.g., curare-like agents) should be observed closely because the effect of the botulinum toxin may be potentiated. Use of anticholinergic drugs after administration of Dysport may potentiate systemic anticholinergic effects such as blurred vision. The effect of administering different botulinum neurotoxin products at the same time or within several months of each other is unknown. Excessive weakness may be exacerbated by another administration of botulinum toxin before the resolution of the effects of a previously administered botulinum toxin. Excessive weakness may also be exaggerated by administration of a muscle relaxant before or after administration of Dysport.

[a]Data from Ref. 26.

ANC: Absolute Neutrophil Count; AS: Ankylosing Spondylitis; AZT: Azidothymidine; DMARD: Disease-Modifying Anti-Rheumatic Drugs; EGFR: Epidermal Growth Factor Receptor; G-CSF: Granulocyte colony-stimulating factor; HCV: Heptitis C Virus; IL: Interlukin; INF: Interferon; NCI-CTC: National Cancer Institute Common Toxicity Criteria; NRTI: Nucleoside/Nucleotide Reverse Transcriptase Inhibitor; PEG: Polyethylene Glycol; PsA: Psoriatic Arthritis; RA: Rheumatoid Arthritis; TGF: Thrombopoietic Growth Factor.

P-glycoprotein (P-gp), breast cancer resistance protein (BCRP), organic anion transporting polypeptide (OATP), organic anion transporter (OAT), and organic cation transporter (OCT). These models, when used with other *in vitro* and *in vivo* human data, can inform clinical study design and provide helpful information on drug labels.[30–32] However, at the present time, there are no well–validated *in vitro* models that can help predict TP-DDI. This knowledge gap was acknowledged at a AAPS/NBC TP-DDI workshop held in 2010.[12] Additional research is ongoing to further explore the utility of *in vitro* models for TP-DDI assessments.

An important consideration for most TP-DDI studies is the general need to conduct these studies in patient populations instead of healthy subjects. There are often concerns with the toxicity of the TP or drug in healthy subjects, and PK characteristics and/or target protein expression levels may differ between patients and healthy subjects. The need to conduct studies in patient populations has an impact on the feasibility of study designs.

Table 10-5 lists study designs that have been used in the *in vivo* evaluation of drug interactions involving TP products in human subjects. In studies with crossover designs, each subject serves as his or her own control, minimizing variability and allowing smaller sample size with increased statistical power. Many TP-DDI studies have been conducted using a one-way crossover

TABLE 10-5 Examples of Study Designs Used to Assess Therapeutic Protein and Small Molecule Drug Interactions[a]

Study Designs[b]	Therapeutic Protein (TP) and Drug (D) Names	Study Purposes[c]
One-way crossover (sequential)		
1. D; D + TP (×8)	1. Trastuzumab on paclitaxel	1. TP → D
2. D; TP; TP + D	2. Bevacizumab on irinotecan (and SN38)	2. TP → D
3. D; D+TP (x63.	3. Bevacizumab on carboplatin (or paclitaxel)	3. TP → D
4. D; D+TP (x4.	4. Peginterferonα-2b on methadone	4. TP → D
5. TP + low D; TP + high D	5. 5-FU on cetuximab	5. D → TP
6. D; TP + D	6. Tocilizumab and probe drugs (simvastatin, omeprazole, or dextromethorphan)	6. TP → D
Population pharmacokinetics	1. Cisapride, paclitaxel, doxorubicin, gemcitabine, irinotecan on cetuximab	1. D → TP
	2. Bolus-IFL on bevacizumab	2. D → TP
	3. Comedications on tocilizumab	3. D → TP

(*continued*)

TABLE 10-5 (*Continued*)

Study Designs[b]	Therapeutic Protein (TP) and Drug (D) Names	Study Purposes[c]
Parallel group (within study)		
1. Group A: D; D + TP; D + TP; TP Group B: TP; TP; TP; TP+D	1. Cetuximab on irinotecan Irinotecan on cetuximab	1. TP → D[d,] D → TP[e]
2. Group A: D Group B: TP Group C: D1 + TP Group D: D2 + TP	2. Methotrexate, cyclophosphamide on rituximab	2. D1 → TP D2 → TP TP → D1
Cross-study comparison (historical control)	1. Basiliximab or muromonab-CD3 and cyclosporine or tacrolimus	1. TP → D
	2. IFN-α and theophylline (patients vs. healthy)	2. TP → D
Others	IL-6 or TNF-α and a cocktail (evaluation of the correlation of IL-6 level with enzyme activities)	TP → D

[a]Data from Refs. 13,14,26,33–38.
[b]Study design: the administration of D or TP was listed sequentially; for example, "D; TP; TP + D" indicates that D alone was given followed by TP alone, which was followed by TP and D given together; "D; D + TP (×4)" indicates that D alone was given followed by D and TP given together for 4 cycles (days or weeks).
[c]Study purpose: TP → D denotes a study in which the effect of TP on D is examined; D → TP denotes a study in which the effect of D on TP is examined.
[d]Comparison of D exposure was made between Groups A (first D) and Group B (TP + D).
[e]This portion of study is a one-way crossover (sequential) design.
5-FU:5-Fluorouracil; IL-6: Interlukin-6; INF-α: Interferon-α; TNF-α: Tumor Necrosis Factor-α.

(sequential) design. This study design is often used for evaluation of the effect of TP on concomitant chemotherapy agents (TP → D) in the oncology clinical setting. In the evaluation of TP-DDI between bevacizumab and carboplatin, patients received carboplatin first and bevacizumab was then added to the regimen. PK information collected in both occasions was evaluated to determine the effect of bevacizumab on carboplatin (Table 10-5). This within-study crossover design may not be feasible, however, for the assessment of the effect of chemotherapy agents on TP (D → TP). Many TP products, such as mAbs, have long half-lives, nonlinear PK because of a component of saturable target-mediated clearance, and potential for formation of antiproduct antibodies (APAs) that may alter their own clearance. These unique characteristics of TP produce unavoidable sequence effects, which preclude the use of crossover designs for the assessment of the effect of chemotherapy agents on TP in the combination therapy (D → TP). Alternative study designs, such as parallel group study or population PK analysis, have therefore been used for such assessments.

Population PK studies have been conducted to evaluate drug interactions, and sponsors have been encouraged to include population PK sample collection (and PD, if appropriate) and data analysis in late phase clinical trials to evaluate patient factors and concomitant medications that may contribute to the variability in systemic exposure and/or drug response for small molecular drugs.[2,4–8] Population PK analyses have also been used to determine the effect of concomitant medications on PK and PD of TPs. Table 10-5 includes examples of the use of a population PK approach for evaluating the effect of a drug on a TP (D → TP). For example, the tocilizumab label states that population PK analyses did not detect any effect of methotrexate, nonsteroidal anti-inflammatory drugs, or corticosteroids on tocilizumab clearance.[26] In most cases, the concentrations of concomitantly administered drugs are not determined in these clinical trials, and therefore the studies do not provide information on the effect of a TP on the exposure of concomitantly used drugs (TP → D).

Table 10-5 also shows examples of parallel study design used in the assessment of the TP → D or D → TP effects. In addition to parallel design studies, some assessments were made using a cross-study evaluation, which may involve comparisons of data between healthy subjects and patients or comparisons of data between combination therapy and monotherapy. Data from such analyses should be considered as "hypothesis-generating" and the results generally need to be confirmed.

Finally, Table 10-5 presents information collected in association studies, such as the evaluation of correlation between cytokine levels and enzyme activities. Association studies provide mechanistic understanding of the effects of cytokines on CYP enzyme or transporter activities and predict effects of various cytokine modulators on the activities of specific CYP enzymes or transporters.

10.4 LABELING IMPLICATIONS

When a drug product's optimal dose for an indication has been selected based on assessment of the overall relation of risk-to-benefit, it is relatively straightforward to determine dosing adjustments needed for patients with specific intrinsic and extrinsic factors that alter their PK profiles: dosing adjustments for individuals with altered PK profiles are generally made to achieve the same systemic exposure as has been shown to be safe and effective for patients without altered PK profiles.[4–8,12] The principles that guide recommendations for drug dosing adjustments have been used in the interpretation of the results from TP-DDI studies. Details of the drug interaction results have been included in the pharmacokinetics subsection (12.3) of the clinical pharmacology section of the labeling.[2,39]

Tables 10-3 and 10-4 provide examples of TP-DDI information (PK and PD based, respectively) in current product labels. Although some TP-DDI information is included in TP labels, unlike small molecule drugs, where some drug interactions have resulted in specific labeling language in the sections titled "Highlights,"

"Contraindications," and "Dosage and Administration," most TP labels include descriptive information about drug interactions in the sections called "Drug interactions," "Clinical Pharmacology," and "Warnings and Precautions," suggesting lesser perceived clinical significance of these interactions with TPs. In particular, the current labels do not recommended specific dosing adjustments for TP-DDI in the "Dosage and Administration" section.

The label for tocilizumab, an interleukin 6 (IL-6) receptor inhibitor approved in the United States for rheumatoid arthritis and systemic juvenile idiopathic arthritis, includes results of studies that demonstrate an effect of tocilizumab on the PK of CYP2C19 or CYP3A substrates.[24–26] CYP enzymes in the liver are downregulated by infection and inflammation stimuli, including cytokines such as IL-6. Inhibition of IL-6 signaling in rheumatoid arthritis patients treated with tocilizumab may restore CYP enzyme activities to higher levels than those not treated with tocilizumab leading to increased metabolism of drugs that are CYP enzyme substrates. The tocilizumab labeling states that the effect of tocilizumab on CYP enzymes may be clinically relevant for CYP450 substrates with a narrow therapeutic index, where the dose is individually adjusted. It further recommends that

> Upon initiation or discontinuation of tocilizumab, in patients being treated with these types of medicinal products, therapeutic monitoring of effect (e.g., warfarin) or drug concentration (e.g., cyclosporine or theophylline) should be performed and the individual dose of the medicinal product adjusted as needed.[26]

The development of physiologically based PK (PBPK) models linking the *in vitro* CYP suppression data for tocilizumab to the *in vivo* CYP alternation (measured via changes in systemic exposure of probe substrates, such as simvastatin) helps explain the association between IL-6 levels and CYP suppression.[40] PBPK models have been shown to be a useful tool in clinical pharmacology evaluations, including drug–drug interactions involving small molecule drugs.[41,42]

Several TP labels included language regarding the potential for drug interactions (either at the time of market approval or at their labeling updates). These examples include rilonacept (IL-1β blocker), golimumab (TNF-α blocker), balatacept (IL-2 modulator), infliximab (TNF-α blocker), and ustekinumab (anti-IL12/IL23 mAb) (Table 10-3). Labeling language for these cytokine modulators indicates that increased levels of cytokines (such as IL-1, IL-6, IL-10, TNF-α, IFN-α) during inflammation can alter the expression or activity of CYP enzymes, and that these cytokine modulators may normalize the activity of CYP enzymes, thus affecting systemic exposure of CYP enzyme substrates. The labels include a general recommendation to monitor the effect on drug concentration and to adjust the dose as needed for CYP enzyme substrates with a narrow therapeutic index (such as warfarin, cyclosporine, theophylline). However, no specific TP-DDI study results have been included in the labels of these TPs. In some cases, postmarketing drug interaction studies that provide useful information are added to the labels at a later date.[29]

10.5 CONCLUSION

Information about drug–drug interactions is critical to public health. The use of therapeutic proteins has increased in recent years, and they are often used in combination with small molecule drugs in critical therapeutic areas. It is anticipated that as more data are collected and the understanding of potential therapeutic protein–drug interactions increases, we will be better able to assess the potential for TP-DDIs during the development of therapeutic proteins.

REFERENCES

1. Huang SM, Temple R. Is this the drug or dose for you? Impact and consideration of ethnic factors in global drug development, regulatory review, and clinical practice. *Clin Pharmacol Ther* **84**, 287–94 (2008).
2. US Food and Drug Administration. Clinical pharmacology guidances. Available at www.fda.gov/Drugs/GuidanceComplianceRegulatoryInformation/Guidances/ucm064982.htm.
3. US Food and Drug Administration, US Department of Health and Human Services, Center for Drug Evaluation and Research, Center for Biologics Evaluation and Research (CBER). Draft guidance for industry: drug interaction studies–study design, data analysis, implications for dosing, and labeling. Silver Spring, MD, 2012. Available at http://www.fda.gov/Drugs/GuidanceComplianceRegulatoryInformation/Guidances/ucm064982.htm
4. Huang SM, Temple R, Throckmorton DC, Lesko LJ. Drug interaction studies: study design, data analysis, and implications for dosing and labeling. *Clin Pharmacol Ther* **81**, 298–304 (2007).
5. Huang SM, Strong JM, Zhang L, et al. New era in drug interaction evaluation: US Food and Drug Administration update on CYP enzymes, transporters, and the guidance process. *J Clin Pharmacol* **48**, 662–70 (2008).
6. Zhang L, Zhang YD, Huang SM. Scientific and regulatory perspectives on metabolizing enzyme-transporter interplay and its role in drug interactions–challenges in predicting drug interactions. *Mol Pharm* **6**, 1766–74 (2009).
7. Zhang L, Zhang YD, Zhao P, Huang SM. Predicting drug-drug interactions: an FDA perspective. *AAPS J* **11**, 300–06 (2009).
8. Zhang L, Reynolds KS, Zhao P, Huang SM. Drug interactions evaluation: an integrated part of risk assessment of therapeutics. *Toxicol Appl Pharmacol* **243**, 134–45 (2010).
9. US Food and Drug, Administration. Drug Development and Drug Interaction. Available at www.fda.gov/Drugs/DevelopmentApprovalProcess/DevelopmentResources/DrugInteractionsLabeling/ucm080499.htm.
10. Mahmood I, Green MD. Drug interaction studies of therapeutic proteins or monoclonal antibodies. *J Clin Pharmacol* **47**, 1540–54 (2007).
11. Seitz K, Zhou H. Pharmacokinetic drug-drug interaction potentials for therapeutic monoclonal antibodies: reality check. *J Clin Pharmacol* **47**, 1104–18 (2007).
12. Girish S, Martin SW, Peterson MC, et al. AAPS workshop report: strategies to address therapeutic protein-drug interactions during clinical development. *AAPS J* **13**, 405–16 (2011).

13. Huang SM, Zhao H, Lee JI, et al. Therapeutic protein-drug interactions and implications for drug development. *Clin Pharmacol Ther* **87**, 497–503 (2010).

14. Lee JI, Zhang L, Men AY, et al. CYP-mediated therapeutic protein-drug interactions: clinical findings, proposed mechanisms and regulatory implications. *Clin Pharmacokinet* **49**, 295–310 (2010).

15. Lloyd P, Zhou H, Theil FP, et al. Highlights from a recent BIO survey on therapeutic protein-drug interactions. *J. Clin Pharm.* 52(11):1755–1763, 2012 (Epub Dec 13, 2011).

16. Zhou H, Mascelli MA. Mechanisms of monoclonal antibody-drug interactions. *Annu Rev Pharmacol Toxicol* **51**, 359–72 (2011).

17. Kraynov E, Martin SW, Hurst S, et al. How current understanding of clearance mechanisms and pharmacodynamics of therapeutic proteins can be applied for evaluation of their drug-drug interaction potential. *Drug Metab Dispos* **39**, 1779–83 (2011).

18. Le VM, Gripon P, Stieger B, Fardel O. Down-regulation of organic anion transporter expression in human hepatocytes exposed to the proinflammatory cytokine interleukin 1beta. *Drug Metab Dispos* **36**, 217–22 (2008).

19. Le VM, Lecureur V, Moreau A, et al. Differential regulation of drug transporter expression by hepatocyte growth factor in primary human hepatocytes. *Drug Metab Dispos* **37**, 2228–35 (2009).

20. Morgan ET, Goralski KB, Piquette-Miller M, et al. Regulation of drug-metabolizing enzymes and transporters in infection, inflammation, and cancer. *Drug Metab Dispos* **36**, 205–16 (2008).

21. Morgan ET. Impact of infectious and inflammatory disease on cytochrome P450-mediated drug metabolism and pharmacokinetics. *Clin Pharmacol Ther* **85**, 434–38 (2009).

22. Petrovic V, Teng S, Piquette–Miller M. Regulation of drug transporters during infection and inflammation. *Mol Interv* **7**, 99–111 (2007).

23. Vee ML, Lecureur V, Stieger B, Fardel O. Regulation of drug transporter expression in human hepatocytes exposed to the proinflammatory cytokines tumor necrosis factor-alpha or interleukin-6. *Drug Metab Dispos* **37**, 685–93 (2009).

24. Schmitt C, Kuhn B, Zhang X, et al. Disease-drug-drug interaction involving tocilizumab and simvastatin in patients with rheumatoid arthritis. *Clin Pharmacol Ther* **89**, 735–40 (2011).

25. Zhang X, Schmitt C, Grange S, K, et al. Disease-drug interaction studies of tocilizumab with cytochrome P450 substrates in vitro and in vivo. *Clin Pharm Ther* **85**(ser. 1), 59 (2009).

26. US Food and Drug Administration. Drugs@FDA. Available at www.accessdata.fda.gov/scripts/cder/drugsatfda/index.cfm.

27. Maini RN, Breedveld FC, Kalden JR, et al. Therapeutic efficacy of multiple intravenous infusions of anti-tumor necrosis factor alpha monoclonal antibody combined with low-dose weekly methotrexate in rheumatoid arthritis. *Arthritis Rheum* **41**, 1552–63 (1998).

28. Dickmann LJ, Patel SK, Rock DA, et al. Effects of interleukin-6 (IL-6) and an anti-IL-6 monoclonal antibody on drug-metabolizing enzymes in human hepatocyte culture. *Drug Metab Dispos* **39**, 1415–22 (2011).

29. US Food and Drug Administration. Postmarket Requirements and Commitments: Introduction. Available at www.fda.gov/Drugs/GuidanceComplianceRegulatoryInformation/Post-marketingPhaseIVCommitments/default.htm.

30. Giacomini KM, Huang SM, Tweedie DJ, et al. Membrane transporters in drug development. *Nat Rev Drug Discov* **9**, 215–36 (2010).
31. Huang SM, Zhang L, Giacomini KM. The International Transporter Consortium: a collaborative group of scientists from academia, industry, and the FDA. *Clin Pharmacol Ther* **87**, 32–36 (2010).
32. Huang SM, Woodcock J. Transporters in drug development: advancing on the critical path. *Nat Rev Drug Discov* **9**, 175–76 (2010).
33. Folprecht G, Lutz MP, Schoffski P, et al. Cetuximab and irinotecan/5-fluorouracil/folinic acid is a safe combination for the first-line treatment of patients with epidermal growth factor receptor expressing metastatic colorectal carcinoma. *Ann Oncol* **17**, 450–56 (2006).
34. Leyland-Jones B, Gelmon K, Ayoub JP, et al. Pharmacokinetics, safety, and efficacy of trastuzumab administered every three weeks in combination with paclitaxel. *J Clin Oncol* **21**, 3965–71 (2003).
35. Denlinger CS, Blanchard R, Xu L, et al. Pharmacokinetic analysis of irinotecan plus bevacizumab in patients with advanced solid tumors. *Cancer Chemother Pharmacol* **65**, 97–105 (2009).
36. Breedveld F, Agarwal S, Yin M, et al. Rituximab pharmacokinetics in patients with rheumatoid arthritis: B-cell levels do not correlate with clinical response. *J Clin Pharmacol* **47**, 1119–28 (2007).
37. Delbaldo C, Pierga JY, Dieras V, et al. Pharmacokinetic profile of cetuximab (Erbitux) alone and in combination with irinotecan in patients with advanced EGFR-positive adenocarcinoma. *Eur J Cancer* **41**1739–45 (2005).
38. Lu JF, Bruno R, Eppler S, et al. Clinical pharmacokinetics of bevacizumab in patients with solid tumors. *Cancer Chemother Pharmacol* **62**, 779–86 (2008).
39. US Food and Drug Administration, US Department of Health and Human Services, Center for Drug Evaluation and Research, Center for Biologics Evaluation and Research (CBER). Guidance for Industry: Labeling for Human Prescription Drug and Biological Products—Implementing the New Content and Format Requirements. Rockville, MD, 2006. Available at http://www.fda.gov/Drugs/GuidanceComplianceRegulatoryInformation/Guidances/ucm065010.htm
40. Kenny J. Cytokine mediated regulation of human cytochrome P450—can we translate in vitro evaluation to clinical relevance? Paper presented at the AAPS/National Biotechnology Conference, San Francisco, May 17, 2011.
41. Huang SM, Rowland M. The role of physiologically based pharmacokinetic modeling in regulatory review. *Clin Pharmacol Ther* **91**, 542–49 (2012).
42. Zhao P, Zhang L, Grillo JA, et al. Applications of physiologically based pharmacokinetic (PBPK) modeling and simulation during regulatory review. *Clin Pharmacol Ther* **89**, 259–67 (2011).

CHAPTER 11

DISEASE–DRUG–DRUG INTERACTION ASSESSMENTS FOR TOCILIZUMAB—A MONOCLONAL ANTIBODY AGAINST INTERLEUKIN-6 RECEPTOR TO TREAT PATIENTS WITH RHEUMATOID ARTHRITIS

XIAOPING ZHANG and BARBARA J. BRENNAN

11.1 INTRODUCTION

Tocilizumab (TCZ, Actemra) is a recombinant humanized monoclonal antibody of the immunoglobulin G (IgG) subclass that selectively binds to human interleukin 6 (IL-6) receptors. TCZ binds to both membrane-bound and soluble forms of IL-6R, blocking signal transduction pathways through competitive inhibition of IL-6 binding.[1] TCZ 8 mg/kg given intravenously (IV) is currently approved for the treatment of rheumatoid arthritis (RA) in over 106 countries, including the United States,[2] Canada,[3] Australia,[4] European countries,[5] and Japan.[6] IL-6 receptor blockade was considered as a mechanism for pharmacologic treatment of RA due to the central role of IL-6 in inflammatory processes. IL-6 is a pleiotropic cytokine with numerous activities, including effects on the immune response and inflammation (e.g., interaction with neutrophils in the synovium), bone metabolism, and hematopoiesis.[7,8] Elevated IL-6 levels have been found in the serum and synovial fluid of RA patients, and inflammatory cell infiltration has been correlated with IL-6 levels in synovial tissues.[9] Furthermore, IL-6 levels correlate with disease activity in RA patients and improvement of the disease after treatment with disease-modifying antirheumatic drugs (DMARDs) is accompanied by a reduction in serum IL-6.[10]

Drug–Drug Interactions for Therapeutic Biologics, First Edition.
Edited by Honghui Zhou and Bernd Meibohm.

A systematic review of TCZ for RA has been reported.[11] Ten published clinical trials (seven Phase III, three Phase II) for TCZ were retrieved. Compared with methotrexate (MTX) monotherapy, TCZ 8 mg/kg IV monotherapy had higher ACR20 ($p < 0.001$), ACR50 ($p < 0.002$), and ACR70 ($p < 0.001$) response rates at week 24. TCZ 8 mg/kg IV plus oral MTX had a higher ACR20 response rate than oral MTX plus placebo in patients with RA who failed to respond to MTX or anti–tumor necrosis factor therapy ($p < 0.001$). Patients receiving TCZ 8 mg/kg had less radiographic progression as assessed by Genant-modified Sharp score (85% had no progression progression) than the control group (67% had no progression) ($p < 0.001$). In the all exposure population composed of 4009 RA patients (3577 received treatment for at least 6 months, 3309 for at least one year; 2954 received treatment for at least 2 years and 2189 for 3 years), the rate of serious infections was 4.7 events/100 patient-years of exposure in all TCZ exposure groups containing 4009 patients (3577 received TCZ for at least 6 months, 3309 for at least one year, 2595 for at least 2 years and 2189 for 3 years).[2] A greater frequency of neutropenia, thrombocytopenia, hyperlipidemia, and transaminitis was observed with TCZ compared with placebo.

Following 4 and 8 mg/kg administration of TCZ to RA patients, serum TCZ concentration-time profiles were adequately described by a 2-compartment disposition pharmacokinetic model with parallel linear and nonlinear elimination kinetics. At low TCZ concentrations, the target-mediated elimination (e.g., via IL-6R) is the predominant elimination pathway. As the concentration of TCZ increases, the target is gradually saturated, and linear clearance by the nonspecific reticuloendothelial system becomes a predominant elimination pathway. Total TCZ clearance is therefore increased with time as TCZ concentration decreases after administration. As a result, the terminal half-life of TCZ increases as TCZ concentration decreases during a dosing interval. The half-life of TCZ was up to 11 to 13 days for therapeutic dose of 4 and 8 mg/kg, respectively, every 4 weeks at steady state.[12]

Several biomarkers of TCZ activity have been routinely measured in clinical pharmacology studies as well as in Phase III studies of TCZ, including IL-6 plasma levels, soluble IL-6 receptor plasma levels, and C-reactive protein (CRP). In RA patients, mean plasma IL-6 concentrations increase significantly after TCZ IV infusion and remain elevated for at least 4 weeks after a single 10 mg/kg dose and return to baseline after approximately 8 weeks.[13] This increase in IL-6 is thought to occur as a result of IL-6R blockade inhibiting clearance of IL-6 from the serum.[14,15] Similarly, increases in the soluble IL-6 receptor are also observed after administration of TCZ. CRP is an acute phase protein that is elevated in the serum of RA patients and used as a marker of inflammation. As a result of IL-6R blockade by TCZ, CRP serum levels are reduced to within the normal range by approximately 7 days postinfusion.[13]

The role of proinflammatory cytokines, such as interleukins, interferons, and tumor necrosis factor α (TNF-α), in regulation of cytochrome P450 (CYP) mediated drug metabolism has been well described.[16–18] There are several examples of altered drug metabolism in patients with acute or chronic inflammatory conditions, such as influenza and cancer. Furthermore, several reports have described the ability of interferon to reduce the metabolic activity of several CYP enzymes in clinical drug–drug interactions studies.[19–21]

IL-6 has been reported to down-regulate several CYP isoforms *in vitro*, with CYP3A4 being the most sensitive.[22–24] RA patients have elevated mean IL-6 levels (~50–60 pg/mL) as compared to approximately 5 pg/mL or less in healthy subjects.[25] Treatment with TCZ blocks the IL-6 signaling pathway and therefore has the potential to reverse the downregulation of the CYP system and restore metabolic activity. In this situation, a decrease from pretreatment baseline concentrations would be expected for drugs that are substrates of CYP upon initiation of treatment with TCZ.

11.2 PRECLINICAL EVALUATION

As a first step to investigate the potential for disease–drug–drug interactions (DDDIs) resulting from reduced IL-6 signaling, an *in vitro* assessment in human hepatocytes was performed.[26] The effects of TCZ and IL-6 on hepatic drug metabolizing enzyme expression levels were investigated. The hepatocytes were obtained as cells seeded on 24 well plates. Expression levels of mRNA for CYP3A4, 2D6, 1A2, 2B6, 2C19, and 2C9 were evaluated in the presence and absence of IL-6 at a low concentration of 0.5 ng/mL and at a high concentration of 12.5 ng/mL. Expression levels were also assessed in the presence of 12.5 ng/mL IL-6 plus 250 μg/mL of TCZ. All incubations lasted 72 h.

Results of the hepatocyte experiment are provided in Table 11-1. IL-6 had a profound impact on CYP3A4 expression, particularly at the high concentration of 12.5 ng/mL. No changes were apparent in the other isoforms at the low IL-6 concentration of 0.5 ng/mL; however, there was suppression of all isoforms at the high concentration of 12.5 ng/mL. The addition of TCZ at a concentration of 250 μg/mL to the incubation was able to completely prevent the IL-6-mediated downregulation of CYP3A4. A similar effect was apparent for the other CYP isoforms to a lesser extent than for CYP3A4. These *in vitro* findings supported that a clinical

TABLE 11-1 Meanm RNA Expression Levels Relative to Control in Human Hepatocytes[a,b]

IL-6 (ng/mL)	TCZ (μg/mL)	CYP Isozyme 3A4	2D6	1A2	2B6	2C19	2C9
0	0	**1.00 (0)**	1.00 (0)	1.00 (0)	1.00 (0)	1.00 (0)	1.00 (0)
0.5	0	**0.79 (0.14)**	1.69 (0.29)	1.03 (0.31)	1.08 (0.36)	3.76 (2.02)	1.15 (0.49)
12.5	0	**0.06 (0.07)**	0.67 (0.28)	0.18 (0.13)	0.28 (0.12)	0.43 (0.29)	0.58 (0.25)
12.5	250	**0.95 (0.82)**	0.68 (0.42)	0.76 (0.23)	0.92 (0.34)	0.96 (0.87)	1.44 (0.63)

[a]From Ref. 26.
[b]Standard deviations are in parentheses.
TCZ: tocilizumab.

disease–drug–drug interaction between TCZ and drugs that are metabolized via CYP enzymes should be evaluated in RA patients to assess the clinical relevance of the *in vitro* findings.

11.3 CLINICAL DDDI EVALUATIONS

Following the *in vitro* assessment, three clinical DDDI assessments were performed in patients with RA. Because the mechanism underlying the potential clinical DDDI is a function of the increased IL-6 levels in RA patients, it was essential that the clinical studies were performed in RA patients as opposed to healthy subjects. The effect of TCZ on the pharmacokinetics (PK) exposure of omeprazole (a sensitive CYP2C19 substrate with some CYP3A4 involvement),[26] dextromethorphan (a sensitive CYP2D6 substrate with some CYP3A4 involvement),[26] and simvastatin (a sensitive CYP3A4 substrate),[27] was therefore studied in RA patients. In addition to the studies that were planned with a reasonable mechanistic hypothesis, supported by *in vitro* data suggesting a clinical DDDI could be likely, the effect of TCZ on methotrexate (MTX) PK was also evaluated clinically because MTX is a commonly administered concomitant medication for the treatment of RA.[29] IL-6 has been reported *in vitro* to suppress mRNA of some anion transporters such as MRP2 that are involved in MTX elimination.[28] Elevated IL-6 levels present in RA patients could therefore be associated with downregulation of these transporters. If true, TCZ may reverse this IL-6-induced downregulation of transporters as it does for CYP3A4. Therefore DDDI for TCZ and MTX in RA patients was also evaluated.[29] The potential for other common concomitant medications in RA patients to alter the PK of TCZ was evaluated by population PK analysis.[12]

The first clinical DDDI study was a two-part study that evaluated the effect of TCZ on the PK of omeprazole (OME) and dextromethorphan (DEX) in 18 and 13 patients with RA, respectively. Patients diagnosed with RA based on the 1987 American College of Rheumatology (ACR) criteria[30] with CRP > 1.5 mg/dL, were enrolled. DMARDs or nonsteroidal anti-inflammatory drugs (NSAIDs) had been stabilized for at least 2 weeks before receiving the study drug on day 1. More criteria for inclusion and exclusion from the study can be found elsewhere.[26] All patients who participated in the OME evaluation were genotyped for CYP2C19 metabolizer status.

The second clinical DDDI study was a two-part study evaluated the effect of TCZ on the PK of simvastatin (SIM) in 12 patients in Group 1 and effect of TCZ on the PK of methotrexate (MTX) in 12 patients in Group 2. Patients diagnosed with RA based on the 1987 ACR criteria[30] with CRP > 1.5 mg/dL, were enrolled. MTX was given for at least 12 weeks immediately before day 1, of which the last 8 weeks were at a stable dose between 10 and 25 mg/week. Patients on other DMARDs were excluded. NSAIDs and oral corticosteroids (CSs) were stabilized for at least 2 weeks before receiving the study drug on day 1. More criteria for inclusion and exclusion from the study can be found elsewhere.[27]

11.3.1 Omeprazole DDDI Evaluation

The OME DDDI study was an open-label, two period, single-dose crossover study in 18 patients with RA.[13,26] Of the 18 patients, 5 were classified as CYP2C19 poor metabolizers (PMs), 8 were extensive metabolizers (EMs), and 5 were considered intermediate metabolizers (IMs). All patients received a single dose of OME 10 mg orally on day 1, followed by a single 1-h IV infusion of 8 mg/kg of TCZ on day 8 and then another single oral 10 mg dose of OME 7 days after the TCZ infusion. Seven days was selected as the appropriate time period necessary to evaluate changes in CYP-mediated metabolism as a result of inhibition of IL-6 signal transduction. Serial blood samples were collected after each OME dose for measurement of plasma OME concentrations.

The mean (%CV) OME PK parameters in the absence and presence of TCZ are provided in Table 11-2 for each of the CYP2C19-metabolizer groups.

As expected, plasma exposure of OME at baseline was lower in the EM group as compared to the IM or PM groups. One week after administration of TCZ, plasma OME concentrations were lower than those measured at baseline. The geometric mean ratio of $AUC_{0-\infty}$ and C_{max} and associated 90% confidence intervals for each group are provided in Table 11-2. The decrease was more pronounced in the EM group as compared to the IM or PM group, as expected, because patients with the EM phenotype should be more sensitive to an inflammation-associated reduction in CYP2C19 activity. In the EM group, OME $AUC_{0-\infty}$ was reduced by approximately 28% as compared to an approximate 12% decrease in both the IM and PM groups.

11.3.2 Dextromethorphan DDDI Evaluation

The DDDI study with DEX was an open-label, single-dose cross-over study conducted in RA patients. Thirteen RA patients received a single 30-mg dose of DEX alone and a single 30-mg dose of DEX 1 week after receiving a single 1-h IV infusion of TCZ 8 mg/kg. Blood samples were collected for evaluation of DEX plasma exposure. DEX is metabolized primarily by CYP2D6 to dextrorphan, which then undergoes further metabolism via CYP3A4. Therefore, both DEX and dextrorphan plasma concentrations were measured to assess effects on both CYP2D6 and CYP3A4.

After administration of TCZ, plasma exposure to DEX was essentially unchanged, as shown in Table 11-3, with $AUC_{0-\infty}$ geometric mean ratio of approximately 1 and associated 90% confidence intervals for $AUC_{0-\infty}$ between the equivalence limits of 80% and 125%. However, exposure to the major metabolite, dextrorphan, a CYP3A4 substrate, was approximately 30% lower in the presence of TCZ, suggesting a potential effect of TCZ on CYP3A4 activity. Although this study provided some insight into the potential for alteration of CYP3A4-mediated metabolism by TCZ, DEX is not considered a sensitive CYP3A4 substrate. Therefore a subsequent DDDI study was planned with simvastatin to fully characterize the effect of TCZ on CYP3A4 activity in RA patients.

TABLE 11-2 Mean Omeprazole (OME) Pharmacokinetics Parameters and Ratio of Geometric Least Squares Mean[a]

	OME (PM)		OME (IM)		OME (EM)	
	OME Day 1	OME + TCZ Day 15	OME Day 1	OME + TCZ Day 15	OME Day 1	OME + TCZ Day 15
$AUC_{0-\infty}$ (ng·h/mL)	2723 (74)	1904 (56)	1923 (65)	1207 (79)	1099 (110)	789 (124)
AUC ratio	–	0.88 [0.78–0.99]	–	0.88 [0.77–1]	–	0.72 [0.66–0.79]
C_{max} (ng/mL)	612 (36)	564 (30)	586 (46)	555 (58)	417 (52)	212 (97)
C_{max} ratio	–	0.98 [0.83–1.15]	–	0.94 [0.82–1.08]	–	0.69 [0.55–0.87]

[a]Coefficient of variation ([standard deviation/mean]100%) is in parentheses; 90% confidence interval is in brackets.

$AUC_{0-\infty}$: area under plasma concentration-time curve from zero to infinity; C_{max}: maximum plasma concentration; EM: extensive metabolizer; IM: intermediate metabolizer; PM: poor metabolizer; TCZ: tocilizumab.

TABLE 11-3 Mean Dextromethorphan (DEX) and Dextrorphan Pharmacokinetics Parameters and Ratio of Geometric Least Squares Mean[a]

	DEX PK		Dextrorphan PK	
	DEX Day 1	DEX + TCZ Day 15	DEX Day 1	DEX + TCZ Day 15
$AUC_{0-\infty}$ (ng·h/mL)	21.9 (136)	21.3 (126)	30.2 (55)	20.8 (45)
AUC ratio	–	0.95 [0.82–1.1]	–	0.71 [0.59–0.84]
C_{max} (ng/mL)	2.76 (115)	2.17 (109)	7.45 (58)	4.36 (46.7)
C_{max} ratio	–	0.79 [0.66–0.96]	–	0.63 [0.53–0.75]

[a]Coefficient of variation ([standard deviation/mean]100%) is in parentheses; 90% confidence interval is in brackets.

$AUC_{0-\infty}$: area under plasma concentration-time curve from zero to infinity; C_{max}: maximum plasma concentration; TCZ: tocilizumab.

11.3.3 Simvastatin DDDI Evaluation

To further elucidate and characterize the effect of IL-6 blockade with TCZ on the metabolic activity of CYP3A4, a second DDDI study was undertaken to examine the impact of TCZ on the pharmacokinetics of SIM, which is considered a sensitive substrate of CYP3A4. SIM is almost exclusively metabolized by CYP3A4 and is recommended as a sensitive probe substrate by the FDA.[31] This study was performed in 12 RA patients and he study design is illustrated in Figure 11-1.

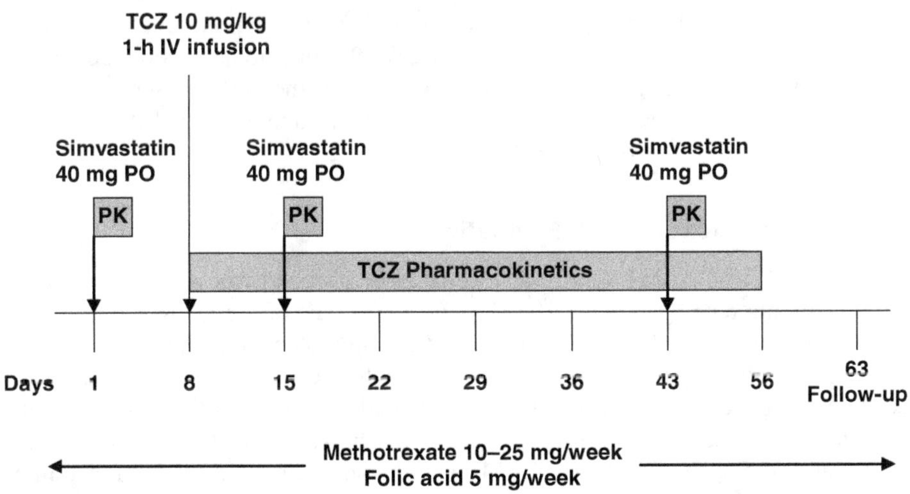

FIGURE 11-1 Design of simvastatin disease–drug–drug interaction study. PK: pharmacokinetics; TCZ: tocilizumab. PO: per oral. IV: intravenous.

Briefly, a single 40 mg dose of SIM was administered alone, followed by a single 10 mg/kg infusion of TCZ after a 7-day washout period (day 8). In this study a 10-mg/kg dose of TCZ was used to represent steady-state exposure of the 8 mg/kg every 4 weeks regimen. One week (day 15) after the TCZ infusion, another single 40-mg dose of SIM was administered to evaluate the peak effect of TCZ on CYP3A4 activity. In this study, the time course of the potential effect on CYP3A4 activity was also evaluated by administration of an additional 40-mg dose of SIM 5 weeks after the TCZ infusion (day 43), to determine if the effect on CYP3A persisted. Blood samples were collected serially for pharmacokinetic assessment of SIM and its major metabolite, simvastatin acid, which is also a CYP3A4 substrate.

The study included male and female RA patients aged 18 to 75 years. Inclusion and exclusion criteria for the study were similar to the previous drug–drug interaction study.[26,27]

Mean $AUC_{0\text{–last}}$ and C_{max} for SIM and simvastatin acid are provided in Table 11-4. The SIM $AUC_{0\text{–last}}$ in RA patients in this study was 102 ng·h/mL compared to values ranging from 11 to 25 ng·h/mL in studies in healthy subjects reported in the literature.[32–35] One week after administration of TCZ the $AUC_{0\text{–last}}$ and C_{max} of SIM were significantly reduced by approximately 57%. This effect persisted even 5 weeks after the TCZ infusion, with an approximate reduction from baseline in $AUC_{0\text{–last}}$ of 40%. The findings were similar for simvastatin acid. Plasma concentrations were higher in RA patients at baseline as compared to literature reports in healthy adults ($AUC_{0\text{–last}}$ of 35 ng·h/mL in RA patients vs. 18 to 26 ng·h/mL in healthy adults). The $AUC_{0\text{–last}}$ of simvastatin acid was reduced by approximately 40% and 20%, 1 week and 5 weeks after TCZ administration, respectively. The effect of TCZ on CYP3A4 activity 5 weeks post dosing was less than 1 week post dosing because SIM $AUC_{0\text{–last}}$ and C_{max} were higher on day 43 than on day 15 (Table 11-4).

These findings are consistent with the pattern of IL-6 receptor blockade by TCZ as evidenced by the time course of CRP (Figure 11-2). The maximum effect of TCZ in normalizing CRP levels was observed 1 week after infusion and corresponded to the maximum decrease in SIM levels. At 5 weeks after the infusion, serum concentrations of CRP represented only one-third of the baseline level, and the effect on SIM exposure was less pronounced.

11.3.4 Methotrexate DDDI Evaluation

The effect of TCZ on MTX PK was evaluated as a second part of the SIM DDDI study. Patients receiving MTX 10–25 mg/week as part of their RA therapy were enrolled in the study. All patients had serial PK samples collected for MTX, 1 week before TCZ administration and then again after the TCZ infusion, at similar times as for SIM. Results of the clinical DDDI evaluation confirmed that no significant changes were observed in MTX exposure in the presence of TCZ with geometric mean ratio of C_{max} and $AUC_{0'last}$ for MTX with and without TCZ coadministration of 1.0 (0.92–1.09) and 1.10 (1.03–1.18), respectively.[29] Despite that IL-6 has been reported *in vitro* to suppress mRNA to downregulate the levels of MRP2,[28] which

TABLE 11-4 Mean Simvastatin (SIM) and Simvastatin Acid Pharmacokinetics Parameters and Ratio of Geometric Least Squares Mean

	Simvastatin			Simvastatin Acid		
	SIM Day 1	SIM + TCZ Day 15	SIM + TCZ Day 43	SIM Day 1	SIM + TCZ Day 15	SIM + TCZ Day 43
$AUC_{0\text{–last}}$ (ng·h/mL)	102 (43)	42.9 (41)	63.2 (54)	35.1 (72)	22.4 (75)	28 (63)
AUC ratio	–	0.43 [0.34–0.55]	0.6 [0.46–0.76]	–	0.61 [0.49–0.76]	0.8 [0.64–1]
C_{max} (ng/mL)	36.1 (60)	14.0 (45)	22.4 (82)	5.37 (67)	3.45 (81)	3.7 (58)
C_{max} ratio	–	0.43 [0.33–0.55]	0.61 [0.47–0.78]	–	0.59 [0.47–0.74]	0.7 [0.56–0.88]

[a]Coefficient of variation ([standard deviation/mean]100%) is in parentheses; 90% confidence interval is in brackets.

$AUC_{0\text{–last}}$: area under plasma concentration-time curve from zero to the last measurable concentration; C_{max}: maximum plasma concentration; TCZ: tocilizumab.

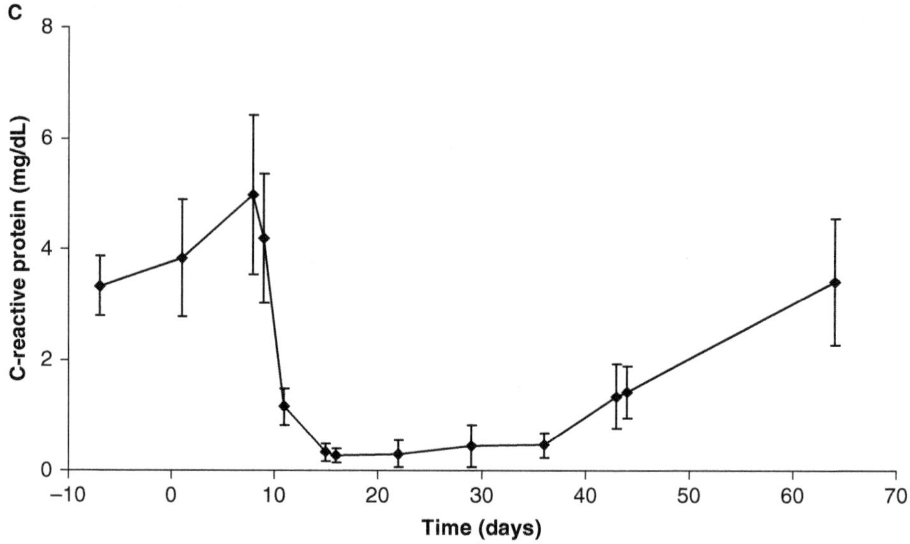

FIGURE 11-2 Mean (±SE) C-reactive protein concentration vs. time.

is one of the main efflux transporters involved in the MTX elimination,[36] the absence of DDDI may suggest that elevated levels of IL-6, which as present in RA patients, have no effect on the expression and/or activity of MRP2 *in vivo* (expression of MRP2 was not measured in this study). Alternatively, other transporters might have compensated for MRP2 action.

11.3.5 Additional Drug–Drug Interaction Evaluations

The potential for drug–drug interactions with common agents coadministered for the treatment of RA was evaluated using population PK analysis with concentration-time data collected from RA patients enrolled in four Phase III clinical trials.[12] In this analysis, the concomitant use of RA-specific treatments–including MTX, chloroquine and derivatives, azathioprine, leflunomide, glucocorticoids, folic acid and derivatives, nonsteroidal anti-inflammatory agents, cyclooxygenase inhibitors, and analgesics (paracetamol, tramadol, codeine, and derivatives)—did not alter the PK of TCZ.

11.4 LABELING

The findings of these clinical DDDI evaluations have been incorporated into TCZ prescribing information globally to provide important guidance to clinicians caring for RA patients and managing comorbidities that require additional medications. As an example, we provide an excerpt from the US prescribing information:[2]

Interactions with CYP450 Substrates

Cytochrome P450s in the liver are down-regulated by infection and inflammation stimuli including cytokines such as IL-6. Inhibition of IL-6 signaling in RA patients treated with tocilizumab may restore CYP450 activities to higher levels than those in the absence of tocilizumab leading to increased metabolism of drugs that are CYP450 substrates. In vitro studies showed that tocilizumab has the potential to affect expression of multiple CYP enzymes including CYP1A2, CYP2B6, CYP2C9, CYP2C19, CYP2D6 and CYP3A4. Its effects on CYP2C8 or transporters is unknown. In vivo studies with omeprazole, metabolized by CYP2C19 and CYP3A4, and simvastatin, metabolized by CYP3A4, showed up to a 28% and 57% decrease in exposure one week following a single dose of ACTEMRA, respectively. The effect of tocilizumab on CYP enzymes may be clinically relevant for CYP450 substrates with narrow therapeutic index, where the dose is individually adjusted. Upon initiation or discontinuation of ACTEMRA, in patients being treated with these types of medicinal products, therapeutic monitoring of effect (e.g., warfarin) or drug concentration (e.g., cyclosporine or theophylline) should be performed and the individual dose of the medicinal product adjusted as needed. Prescribers should exercise caution when ACTEMRA is coadministered with CYP3A4 substrate drugs where decrease in effectiveness is undesirable, e.g., oral contraceptives, lovastatin, atorvastatin, etc. The effect of tocilizumab on CYP450 enzyme activity may persist for several weeks after stopping therapy.

11.5 DISCUSSION

The results of the clinical DDDI evaluations performed with TCZ confirm the longstanding hypothesis that clinically relevant drug interactions can occur as a result of reversal of an inflammatory process that suppresses CYP metabolic enzyme activity, hence "normalizing" activity. However, unlike direct drug–drug interactions, disease–drug–drug interactions are relevant only for CYP substrates with a narrow therapeutic index for which doses are individually adjusted. For those drugs, monitoring of drug effects or concentrations should be performed within 2 weeks upon initiation of TCZ therapy or several weeks after stopping the therapy. From these evaluations, CYP3A4 appears to be the most sensitive isozyme, based on *in vitro* and *in vivo* findings.

When considering the DDDI observed with simvastatin, $AUC_{0-\infty}$ was approximately sixfold higher in RA patients at baseline compared to healthy subjects (105 vs. 17.3 ng·h/mL). This observation confirmed that CYP3A4 activity is suppressed in RA patients by RA-associated inflammation. Upon initiation of TCZ (1 week postinfusion), SIM $AUC_{0-\infty}$ was reduced by approximately twofold (105 vs. 44.9 ng·h/mL), indicating that suppressive effect of inflammation on CYP3A4 was reversed and CYP3A4 activity was increased approaching "normalization." After stopping TCZ treatment (5 weeks postinfusion), SIM $AUC_{0-\infty}$ increased compared to the exposure at week 1 (65.2 vs 44.9 ng·h/mL), as the effect of the single-dose TCZ administration decreased with decreasing TCZ concentration. The clinical relevance of these changes can be revealed when compared to the effect of rifampin (a

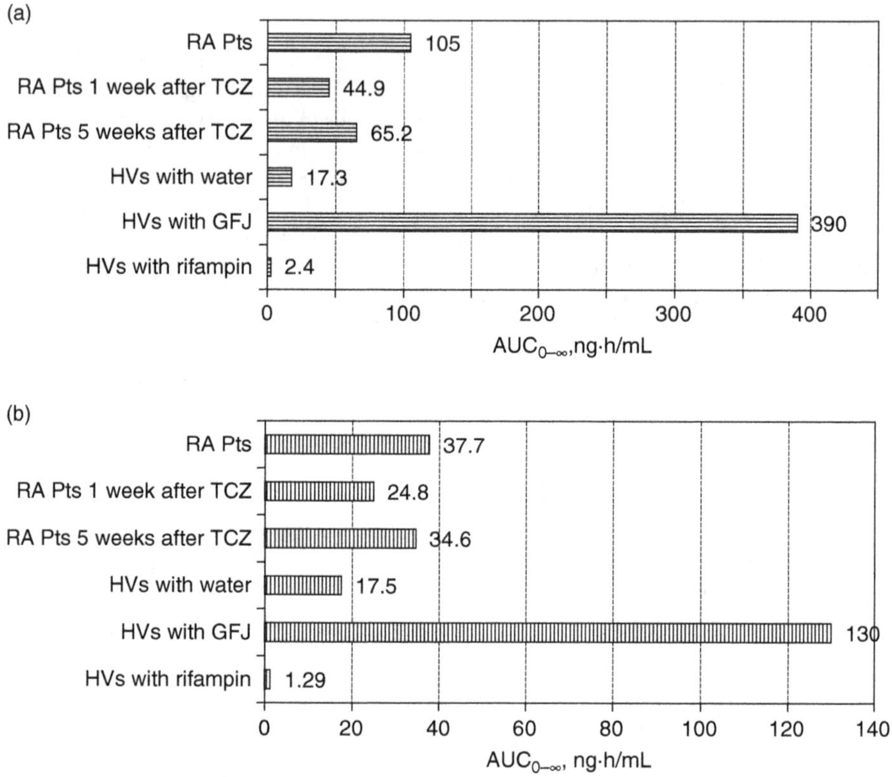

FIGURE 11-3 Comparison of (a) simvastatin and (b) simvastatin acid metabolite. AUC change in rheumatoid arthritis patients (RA Pts) and in RA patients treated with tocilizumab (TCZ) rifampin, and grapefruit juice (GFJ) compared to healthy volunteers (HVs). Simvastatin was administered with water alone, 24 h after the last dose of 200 mL GFJ (3 times a day for 3 days) and 5 days after 600 mg rifampin daily treatment.

typical CYP3A4 inducer[34] and grapefruit juice (GFJ; a CYP3A4 inhibitor).[37] SIM $AUC_{0-\infty}$ was decreased by 87% after 5 daily doses of 600 mg rifampin,[34] and SIM $AUC_{0-\infty}$ was increased by ninefold after 3 days dosing of 200 mL GFJ three times daily.[37] The decrease in SIM $AUC_{0-\infty}$ upon initiation of TCZ (1 week post-dosing) was much less than the change by rifampin (Figure 11-3a) and the lincrease in SIM $AUC_{0-\infty}$ increase after stopping of TCZ treatment (5 weeks postinfusion) was also much less than the change by GFJ (Figure 11-3b). A similar conclusion can also be drawn for the $AUC_{0-\infty}$ of SIM acid metabolite (Figure 11-3b). Interpretation of the findings of the TCZ DDDI assessments in the context of the effects of other well-known DDI perpetrators supports the recommendation that "the effect of tocilizumab on CYP enzymes may be clinically relevant for CYP450 substrates with narrow therapeutic index, where the dose is individually adjusted."[2]

REFERENCES

1. Nishimoto N, Kishimoto T. Humanized antihuman IL-6 receptor antibody, tocilizumab. In: Chernajovsky Y, Nissim A, ed. Therapeutic antibodies. Handbook of experimental pharmacology 181. Berlin, Springer, 2008: 151–60.
2. Genentech. Package insert for Actemra. South San Francisco, CA, USA, 2012.
3. Hoffmann-La Roche. Package insert for Actemra. Mississauga, Ontario, Canada, 2010.
4. Roche Products. Package insert for Actemra. Dee Why, Australia, 2010.
5. RoActemra. Package insert for Actemra. Welwyn, Garden City, UK, 2010.
6. Chugai Pharmaceutical. Package insert for Actemra. Tokyo, Japan, 2008.
7. Hushaw LL, Sawaqed R, Sweis G, et al. Critical appraisal of tocilizumab in the treatment of moderate to severe rheumatoid arthritis. *Ther Clin Risk Manag* **6**, 143–52 (2010).
8. Okuda Y. Review of tocilizumab in the treatment of rheumatoid arthritis. *Biologics* **2**(1), 75–82 (2008).
9. Matsumoto T, Tsurumoto T, Shindo H. Interleukin-6 levels in synovial fluids of patients with rheumatoid arthritis correlated with the infiltration of inflammatory cells in synovial membrane. *Rheumatol. Int* **26**(12), 1096–100 (2006).
10. Milman N, Karsh J, Booth RA. Correlation of a multi-cytokine panel with clinical disease activity in patients with rheumatoid arthritis. *Clin Biochem* **43**(16–17), 1309–14 (2010).
11. Navarro-Millán I, Singh JA. Systematic review of tocilizumab for rheumatoid arthritis: a new biologic agent targeting the interleukin-6 receptor. *Clin Thera* **34**(4), 788–802 (2012).
12. Frey N, Grange S, Woodworth T. Population pharmacokinetic analysis of tocilizumab in patients with rheumatoid arthritis. *J Clin Pharmacol* **50**(7), 754–66 (2010).
13. Zhang X, Peck R Clinical pharmacology of tocilizumab for the treatment of patients with rheumatoid arthritis. *Expert Rev Clin Pharmacol* **4**(5), 539–58 (2011).
14. Nishimoto N, Terao K, Mima T, et al. Mechanisms and pathologic significances in increase in serum interleukin-6 (IL-6) and soluble IL-6 receptor after administration of an anti-IL-6 receptor antibody, tocilizumab, in patients with rheumatoid arthritis and Castleman disease. *Blood* **112**(10), 3959–64 (2008).
15. Uchiyama Y, Yoshida H, Koike N, et al. Anti-IL-6 receptor antibody increases blood IL-6 level via the blockade of IL-6 clearance, but not via the induction of IL-6 production. *Int Immunopharmacol* **8**(11), 1595–601 (2008).
16. Morgan ET. Impact of infectious and inflammatory disease on cytochrome P450-mediated drug metabolism and pharmacokinetics. *Clin Pharmacol Ther* **85**(4), 434–38 (2009).
17. Renton KW. Regulation of drug metabolism and disposition during inflammation and infection. *Expert Opin Drug Metab Toxicol* **1**(4), 629–40 (2005).
18. Morgan ET, Goralski KB, Piquette-Miller M, et al. Regulation of drug-metabolizing enzymes and transporters in infection, inflammation, and cancer. *Drug Metab Dispos* **36**(2), 205–16 (2008).
19. Williams SJ, Farrell GC. Inhibition of antipyrine metabolism by interferon. *Br J Clin Pharmacol* **22**, 610–12 (1986).

20. Williams SJ, Baird-Lambert JA, Farrell GC. Inhibition of theophylline metabolism by interferon. *Lancet* **8565**, 939–41 (1987).
21. Islam M, Frye RF, Richards TJ, et al. Differential effect of IFNa- 2b on the cytochrome P450 enzyme system: a potential basis of IFN toxicity and its modulation by other drugs. *Clin Cancer Res* **8**, 2480–87 (2002).
22. Jover R, Bort R, Gómez-Lechón MJ, Castell JV. Down-regulation of human CYP3A4 by the inflammatory signal interleukin-6: molecular mechanism and transcription factors involved. *FASEB J* **16**, 1799–01 (2002).
23. Warren GW, van Ess PJ, Watson AM, et al. Cytochrome P450 and antioxidant activity in interleukin-6 knockout mice after induction of the acute phase response. *J Interferon Cytokine Res* **21**, 821–26 (2001).
24. Ward LD, Howlett GJ, Hammacher A, et al. Use of a biosensor with surface plasma response detection for the determination of binding constants: measurement of interleukin-6 binding to the soluble interleukin-receptor. *Biochemistry* **34**, 2901–07 (1995).
25. Robak T, Gladalska A, Stepien H, Robak E. Serum levels of interleukin-6 type cytokines and soluble interleukin-6 receptor in patients with rheumatoid arthritis. *Mediators Inflamm* **7**, 347–53 (1998).
26. Zhang X, Schmitt C, Grange S, et al. Disease-drug interaction studies of tocilizumab with cytochrome P450 substrates in vitro and in vivo. *Clin Pharmacol Ther* **85**, S59 (2009).
27. Schmitt C, Kuhn B, Zhang X, et al. Disease-drug-drug interaction involving tocilizumab and simvastatin in patients with rheumatoid arthritis. *Clin Pharmacol Ther* **89**(5), 735–40 (2011).
28. Hartmann G, Cheung AK, Piquette-Miller M. Inflammatory cytokines, but not bile acids, regulate expression of murine hepatic anion transporters in endotoxemia. *J Pharmacol Exp Ther* **303**(1), 273–81 (2002).
29. Schmitt C, Kuhn B, Banken L, et al. Tocilizumab has no relevant effect on methotrexate pharmacokinetics in patients with rheumatoid arthritis. *International Journal of Clinical Pharmacology and Therapeutics.* **50**(3), 218–23 (2012).
30. Arnett F, Edworthy SM, Bloch DA, et al. The American Rheumatism Association 1987 revised criteria for the classification of rheumatoid arthritis. *Arthritis Rheum* **31**(3), 315–24 (1988).
31. US Food and Drug Administration, US Department of Health and Human Services, Center for Drug Evaluation and Research, Center for Biologics Evaluation and Research (CBER). Draft guidance for industry: drug interaction studies—study design, data analysis, implications for dosing, and labeling. Silver Spring, MD, 2012. Available at www.fda.gov/downloads/Drugs/GuidanceComplianceRegulatoryInformation/Guidances/UCM292362.pdf.
32. Najib NM, Idkaidek N, Adel A, et al. Pharmacokinetics and bioequivalence evaluation of two simvastatin 40 mg tablets (Simvast and Zocor) in healthy human volunteers. *Biopharm Drug Dispos* **24**, 183–89 (2003).
33. Lilja JJ, Neuvonen M, Neuvonen PJ. Effects of regular consumption of grapefruit juice on the pharmacokinetics of simvastatin. *Br J Clin Pharmacol* **58**, 56–60 (2004).
34. Kyrklund C, Backman JT, Kivistö KT, et al. Rifampin greatly reduces plasma simvastatin and simvastatin acid concentrations. *Clin Pharmacol Ther* **68**, 592–97 (2000).

35. Kantola T, Kivistö KT, Neuvonen PJ. Erythromycin and verapamil considerably increase serum simvastatin and simvastatin acid concentrations. *Clin Pharmacol Ther* **64**, 177–82 (1998).
36. Elsby R, Fox L, Stresser D, et al. In vitro risk assessment of AZD9056 perpetrating a transporter-mediated drug-drug interaction with methotrexate. *Eur J Pharm Sci* **43**(1–2), 41–9 (2011).
37. Lilja JJ, Kivistö KT, Neuvonen PJ. Duration of effect of grapefruit juice on the pharmacokinetics of the CYP3A4 substrate simvastatin. *Clin Pharmacol Ther* **68**(4), 384–90 (2000).

CHAPTER 12

DRUG–DRUG INTERACTIONS FOR ETANERCEPT—A FUSION PROTEIN

JOAN KORTH-BRADLEY

The treatment of immune-mediated inflammatory diseases, such as rheumatoid arthritis, psoriasis, and psoriatic arthritis has been changed with the introduction of biological agents in the 1990s. Previously available biological agents, such as insulin, were replacements for endogenous proteins that were missing in patients with diabetes, while newer agents such as etanercept were constructs not found in nature. In this case study of drug–drug interaction assessments for etanercept, a fusion protein targeting tumor necrosis factor α (TNF-α) used in the treatment of several immune-mediated inflammatory diseases, background information about etanercept will be discussed, followed by a general assessment on the likelihood of drug interactions with fusion proteins such as etanercept and finally the results of specific etanercept–drug interaction investigations will be discussed.

12.1 ETANERCEPT BACKGROUND

Etanercept is a human tumor necrosis factor receptor (TNFR) p75 Fc fusion protein produced by recombinant DNA technology in a Chinese hamster ovary (CHO) mammalian cell expression system. It is a dimer of a chimeric protein genetically engineered by fusing the extracellular ligand binding domain of human tumor necrosis factor receptor 2 (TNFR2/p75) to the Fc domain of human immunoglobulin G1 (IgG1). The Fc component of etanercept contains the hinge, CH2 and CH3 regions, but not the CH1 region of IgG1. Etanercept has a molecular weight of approximately 150 kDa. It causes immune suppression by binding tumor necrosis factor, which is increased in patients with inflammatory diseases.[1]

Initially approved for the treatment of rheumatoid arthritis (RA), etanercept has subsequently received approval for the treatment of juvenile rheumatoid arthritis

Drug–Drug Interactions for Therapeutic Biologics, First Edition.
Edited by Honghui Zhou and Bernd Meibohm.

(JRA), psoriatic arthritis (PsA), psoriasis, and ankylosing spondylitis (AS). It was also investigated for the treatment of congestive heart failure (CHF), without success. Initially, the approved dose was 25 mg administered by subcutaneous injection twice weekly. The dose for children was 0.4 mg/kg twice weekly. Subsequently, 50 mg administered weekly was also approved, as was 0.8 mg/kg administered once weekly in children.

After subcutaneous injection, etanercept is slowly absorbed into systemic circulation, with maximal concentration reached in 2 to 3 days. Dose proportionality has not been formally evaluated, but saturation of clearance is not apparent over the dosing range studied, and systemic exposure appears to be linear. Bioavailability after subcutaneous administration is 58%.[2] Bioequivalence of absorption was shown after administration of either liquid or reconstituted lyophilized etanercept to healthy volunteers.[3] The volume of distribution of etanercept is small. After administration of single subcutaneous etanercept doses in healthy subjects, the apparent volume of distribution was 12 ± 6 L.[4] In 3 subjects with RA, repeated samples of synovial fluid showed etanercept concentrations were similar to those observed in serum. Etanercept is slowly cleared from the body with a terminal half-life of 3 to 4 days. It is assumed that after binding of etanercept to TNF, the etanercept–TNF complex is then metabolized through peptide and amino acid pathways with either recycling of amino acids or elimination in bile and urine.[4] There was elimination of radioactivity observed in urine after administration of radiolabeled etanercept to patients and healthy subjects,[5] but almost no intact etanercept was observed in the collected urine.[6]

The potential impact of various demographic and disease characteristics on the pharmacokinetics of etanercept has been evaluated. A study of 30 healthy Japanese volunteers who received single subcutaneous 10-, 25-, or 50-mg doses of etanercept showed comparable pharmacokinetics with a similar group of healthy American volunteers.[6] A small pharmacokinetic study ($n = 6$) was performed in subjects with end stage renal disease requiring dialysis. The pharmacokinetics in these subjects was similar to subjects with psoriasis not requiring dialysis.[7] The pharmacokinetic profile of etanercept in subjects with RA showed parameters similar to those observed in healthy subjects.[8] Pharmacokinetics in subjects with AS were similar to those observed in subjects with RA.[9] The pharmacokinetics of etanercept were studied in 11 subjects with chronic CHF after repeated etanercept doses of 12 mg/m^2.[10] The pharmacokinetic results in subjects with CHF were similar to those observed in healthy subjects and in subjects with RA.

Several population pharmacokinetic analyses have been performed to further characterize etanercept pharmacokinetics. Lee and co-workers fit a one-compartment model to concentration data collected from 77 subjects who had received either 25- or 50-mg doses twice weekly for 24 weeks.[11] Significant covariates for apparent clearance were sex and race (white vs. nonwhite). White women had mean (95% confidence intervals) clearance estimates of 0.117 (0.108–0.130) L/h, and white men had clearance estimates of 0.138 (0.118–0.163) L/h; nonwhite patients had mean clearances that were 138% (104–160%) compared with white patients. Zhou and co-workers fit a two-compartment model to concentration data collected from 10 studies, including 265 healthy subjects and patients with JRA and RA.[9] One of the studies had data after IV administration. The final model showed that age (<17 years) and body weight (<60 kg) were significant covariates for apparent clearance, which was lower compared with other subjects. The clearance in children (4 to 17 years of age) was lower than clearance in adults; however, when clearance values were adjusted for body weight, clearance was comparable with that of adults. A modest effect of the presence of RA on absorption was observed, and nonwhite individuals had larger central volumes of distribution, which disappeared if weight normalized. Finally, Nestorov's group fit a one-compartment model to data collected from three studies in 1077 patients with psoriasis.[12] Gender, weight, and time were significant covariates for apparent clearance and body weight for apparent volume of distribution. In contrast to what was observed in the previous analysis in patients with RA, in this analysis, the population mean (95% confidence interval) for clearance in women (0.148 L/h, 0.137–0.161 L/h) was higher than in men (0.129 L/h, 0.123–0.136 L/h). The finding of lower clearance with lower body weight may counteract the discrepancy. Apparent clearance increased over the first 4 weeks. With the exception of dosing for children <60 kg, none of the covariates identified is considered clinically significant or needing specific dosing.

12.2 MECHANISMS OF DRUG INTERACTIONS

Formal drug interaction studies were not performed during the initial clinical development of etanercept. It was believed that etanercept has little potential for causing drug interactions, or being influenced by administration of either inducers or inhibitors of important metabolic enzymes for other medications and thus drug interaction studies would not provide information about the metabolic fate of etanercept.

Parenteral administration excludes potential for absorption interactions with either food or other concomitantly administered medications. The volume of the injection is 1 mL when etanercept is administered to adults. It is unlikely that patients would want to combine other medications for simultaneous injection. Such combinations would not be recommended owing to potential physicochemical problems as well as simply too much volume.

Etanercept is a biological product. Methods to assess the extent of protein binding for small, synthetic molecules are not capable of adequately assessing protein binding for large molecules that are proteins themselves. It is believed there is relatively low potential for displacement reactions between etanercept and other medications that are highly protein bound.

As noted previously, it is assumed that after binding of etanercept to TNF, the complex is then metabolized through peptide and amino acid pathways with either recycling of amino acids or elimination in bile and urine. The cytochrome P450 system (CYP) and other commonly used metabolic pathways are not involved in the metabolism of etanercept, so there is little potential for metabolic drug interactions. The role of transporters in etanercept pharmacokinetics has not been studied.

Finally, etanercept is not believed to change biliary or urinary production and flow, so interactions associated with elimination are also not likely to occur. Information about metabolism and excretion, and thus potential for interaction, is obtained by studying the pharmacokinetics of drug in patients with isolated end organ dysfunction. In the case of etanercept, no evidence was obtained to suggest that either renal or hepatic failure alters elimination of the drug. A study of patients with severe sepsis received a single intravenous dose of etanercept to test whether etanercept had activity in this condition.[13] Serial serum concentrations were measured and, although not reported in the paper, no difference in concentrations in different subjects with different degrees of acute hepatic impairment or acute renal impairment.[5] As noted earlier, a study of subjects with chronic renal failure showed no difference in pharmacokinetics.

In the approximately 15 years since the development of etanercept, many more therapeutic proteins have been studied and guidelines for evaluation of the potential for drug interactions have been developed.[14] In addition to the various mechanisms by which drug interactions may occur already discussed, the importance of cytokine induction of important CYP enzymes has become better understood. Etanercept is not a cytokine and so again would not be expected to directly alter CYP activity. A potential indirect effect could be caused through inhibition of TNF-α by etanercept, but there have been no reports in the literature of interactions of this type. In conclusion, etanercept has a low likelihood of causing a drug interaction or being affected by other medications.

12.3 PHARMACODYNAMIC DRUG INTERACTIONS

Although etanercept may have a low potential for interactions from a pharmacokinetic perspective, pharmacodynamic interactions may occur as a result of its activity. It is always reasonable to evaluate the potential for interactions among medications that are commonly administered for the same indications. Rather than performing pharmacokinetic studies, researchers can collect safety and efficacy data as well as pharmacokinetic data in Phase II and III trials; the results are reviewed carefully for evidence of interactions.[15] This type of investigation was the basis for the recommendations that etanercept not be used concomitantly with anakinra,[16] abataept,[17] cyclophosphamide,[18] or sulfasalazine.[19] In each case, the patients receiving combination therapy had more serious adverse reactions, usually infections, than those subjects in the control group without benefit of improved efficacy.

In addition to too much immunosupression, etanercept may interfere with the desired response to vaccines. Patients exposed to varicella virus should temporarily discontinue etanercept and consider prophylactic treatment with varicella zoster immune globulin.[20] A study of patients with PsA showed that patients treated with etanercept or placebo had similar responses to pneumococcal vaccine.[21] Approximately 20% of patients failed to show a twofold response to any of the antigens. It was concluded that there was no increased risk of poor response with etanercept treatment given alone or with methotrexate.

12.4 RESULTS OF DRUG INTERACTION STUDIES WITH ETANERCEPT

Formal drug interaction studies have been performed to evaluate the interactions of etanercept with methotrexate, digoxin, and warfarin. Etanercept concentrations were also measured in clinical trials for patients with RA who were also treated with anakinra.

There was no mutual pharmacokinetic interaction between etanercept and warfarin.[22] In a nonrandomized, three-period study, 12 healthy male subjects received a single oral 25-mg dose of warfarin after an overnight fast, followed by twice-weekly 25-mg subcutaneous doses of etanercept for seven doses. The last dose of etanercept was administered concurrently with a second dose of warfarin. Serial blood samples for plasma warfarin concentration measurement and international normalized ratio (INR) assessment were collected before and up to 144 h after dose administration. Serial blood samples for serum etanercept concentration measurement were collected before and up to 60 h after the sixth dose and 264 h after the seventh dose. Etanercept did not affect the pharmacokinetics and pharmacodynamics of warfarin. All ratios of maximum serum concentration (C_{max}) and area under the serum concentration versus time curve (AUC) for pharmacokinetics (R- and S-enantiomers of warfarin) and INR fell within the confidence interval of 0.8 to 1.25. Warfarin also did not cause a clinically significant alteration in the pharmacokinetics of etanercept. In conclusion, coadministration of etanercept and warfarin would not be expected to change the pharmacokinetics of either medication; therefore, no dosage adjustment is needed in cases in which warfarin and etanercept are coadministered.

There was also no effect of coadministration of etanercept on the pharmacokinetics of digoxin.[23] In a crossover, open-label, nonrandomized, three-period study, 12 healthy male subjects received loading oral doses of digoxin 0.5 mg every 12 h on day 1 and 0.25 mg every 12 h on day 2, followed by a daily maintenance dose of 0.25 mg for a total of 27 days. Etanercept was administered as a twice weekly 25-mg subcutaneous dose beginning on day 9 and continuing up to day 37 for a total of 9 doses. All ratios of C_{max} and AUC for pharmacokinetics of digoxin fell within the confidence interval of 0.8 to 1.25. Although not considered clinically relevant, the mean C_{max} and AUC of etanercept were 4.2% and 12.5% lower, respectively, when etanercept was given with digoxin than when administered alone. There were no clinically relevant changes in the electrocardiogram parameters, and adverse events did not increase when both drugs were combined. In conclusion, there is no clinically relevant interaction between etanercept and digoxin, and both drugs can be safely coadministered without the need for a dosage adjustment.

The results from the population pharmacokinetic analysis demonstrated that the coadministration of methotrexate had no significant effect on the pharmacokinetics of etanercept in subjects with RA.[24] Using the population pharmacokinetic model previously described, which included data from healthy subjects and patients with RA and JRA, nine Bayesian-derived individual pharmacokinetic parameters were estimated for 100 individuals, 50 receiving methotrexate and etanercept and 50 receiving etanercept alone. The patients had been randomly selected from the TEMPO trial[25] and each had had a single serum measurement collected after 24 weeks of treatment. The predicted weight-normalized CL values were 0.071 ± 0.013 and 0.072 ± 0.011 for the etanercept-treated patients and for the combination treated patients, respectively. Thus no dose adjustment of etanercept dosage is needed when coadministered with methotrexate.

The study of combined treatment of etanercept and anakinra mentioned previously[16] also collected blood samples for the measurement of anakinra and etanercept concentrations in plasma before dosing and then at 4, 12 and 24 weeks after the start of treatment. Although there was large intersubject variability, there appeared to be no difference in either anakinra or etanercept concentrations between those on combination therapy and those receiving etanercept alone.

There are references in the literature to an interaction between paclitaxel and etanercept, where coadministration of paclitaxel causes an increase in etanercept concentrations of 1.5-fold,[14,26] but the specifics of the interaction are not well described in the literature or in the prescribing information for any of the commercially available forms of paclitaxel.[27,28]

12.5 CONCLUSIONS

In conclusion, much more is known now about the potential for drug interaction involving therapeutic proteins than when etanercept was being developed. However, a systematic review of potential causes for drug interactions and careful review of safety and efficacy of studies where combinations of medications are used has been successful in identifying pharmacodynamic interactions so that potentially unsafe combinations of treatments can be avoided, even when pharmacokinetic interactions are unlikely.

ACKNOWLEDGMENTS AND CONFLICTS OF INTEREST

The author is an employee of Pfizer Inc. who comarkets Enbrel (etanercept) along with Amgen. The statements or opinions expressed in this paper are her own and do not necessarily represent those of Pfizer.

REFERENCES

1. Alldred A. Etanercept in rheumatoid arthritis. *Expert Opin Pharmacother* **2**, 1137–48 (2001).
2. Lebsack ME, Hanna RK, Lange MA, et al. Absolute bioavailability of TNF receptor fusion protein following subcutaneous injection in healthy volunteers. *Pharmacotherapy* **17**, 1118–19 (1997).

3. Sullivan JT, Ni L, Sheelo C, Salfi M, Peloso PM. Bioequivalence of liquid and reconstituted etanercept subcutaneous injections. *J Clin Pharmacol* **46**, 654–61 (2006).
4. Korth-Bradley JM, Rubin ASR, Hanna RK, et al. The pharmacokinetics of etanercept in healthy volunteers. *Ann Pharmacother* **34**, 161–64 (2000).
5. European Medicinal Agency. Enbrel: EPAR - Scientific Discussion. London, 18 Oct 2006. Available at www.ema.europa.eu/docs/en_GB/document_library/EPAR_-_Scientific_Discussion/human/00262/WC500027358.pdf. Accessed 15 Mar 2013.
6. Kawai S, Sekino H, Yamashita N, et al. The comparability of etanercept pharmacokinetics in healthy Japanese and American subjects. *J Clin Pharmacol* **46**, 418–23 (2006).
7. Don BR, Spin G, Nestorov I, et al. The pharmacokinetics of etanercept in patients with end-stage renal disease on haemodialysis. *J Pharm Pharmacol* **57**, 1407–13 (2010).
8. Kremer JM, Spencer-Green GT, Hanna RK, et al. ENBREL (etanercept) pharmacokinetics in patients with rheumatoid arthritis. *Arthritis Rheum* **43**, S229 (2000).
9. Zhou H, Buckwalter M, Mayer P, et al. Population-based pharmacokinetics of the soluble TNFr etanercept: a clinical study in 43 patient with ankylosing spondylitis compared with post hoc data from patients with rheumatoid arthritis. *Int J Clin Pharmacol Ther* **42**, 267–76 (2004).
10. Soran O, Feldman AM, Schneider R, et al. The pharmacokinetics of etanercept in patients with heart failure. *Br J Clin Pharmacol* **51**, 197–99 (2001).
11. Lee H, Kimko HC, Rogge M, et al. Population pharmacokinetic and pharmacodynamic modeling of etanercept using logistic regression analysis. *Clin Pharm Ther* **73**, 348–65 (2003).
12. Nestorov I, Zitnik R, Ludden T. Population pharmacokinetic modeling of subcutaneously administered etanercept in patients with psoriasis. *J Pharmacokin Pharmacodyn* **31**, 463–90 (2004).
13. Fisher CJ, Agosti JM, Opal SM et al. Treatment of septic shock with the tumor necrosis factor receptor: Fc fusion protein. *N Engl J Med* **334**, 1697–02 (1996).
14. Huang S-M, Zhao H, Reynold SK et al. Therapeutic protein-drug interactions and implications for drug development. *Clin Pharm Ther* **87**, 497–503 (2010).
15. Van Roon EN, van den Bernt PMLA, Jansen TLThA, et al. An evidence-based assessment of the clinical significance of drug-drug interactions between disease-modifying antirheumatic drugs and non-antirheumatic drugs according to rheumatologists and pharmacists. *Clin Ther* **31**, 1737–46 (2009).
16. Genovese MC, Cohen S, Moreland L, et al. Combination therapy with etanercept and anakinra in the treatment of patients with rheumatoid arthritis who have been treated unsuccessfully with methotrexate. *Arthritis Rheum* **50**, 1412–19 (2004).
17. Weinblatt M, Schiff M, Goldman A, et al. Selective costimulation modulation using abatacept in patients with active rheumatoid arthritis while receiving etanercept: a randomized clinical trial. *Ann Rheum Dis* **66**, 228–34 (2007).
18. The Webener's Granulomatosis Etanercerpt Trial (WGET) Research Group. Etanercept plus standard therapy for Wegener's granulomatosis. *N Eng J Med* **352**, 351–61 (2005).
19. Combe B, Codreanu C, Fiocco U, et al. Efficacy, safety and patient-reported outcomes of combination etanercept and sulfasalazine versus etanercept alone in patients

with rheumatoid arthritis: a double-blind randomized 2-year study. *Ann Rheum Dis* **68**, 1146–52 (2009).

20. Amgen, Pfizer. Package insert for Enbrel for subcutaneous injection. Thousand Oaks, CA 2012.
21. Mease PJ, Ritchlin CT, Martin RW, et al. Pneumococcal vaccine response in psoriatic arthritis patients during treatment with etanercept. *J Rheumatol* **31**, 1356–61 (2004).
22. Zhou H, Patat A, Parks V, et al. Absence of a pharmacokinetic interactions between etanercept and warfarin. *J Clin Pharmacol* **44**, 543–50 (2004).
23. Zhou H, Parks V, Patat A, et al. Absence of a pharmacokinetic interactions between etanercept and digoxin. *J Clin Pharmacol* **44**, 1244–51 (2004).
24. Zhou H, Mayer PR, Wajdula J, Fatenejad S. Unaltered etanercept pharmacokinetics with concurrent methotrexate in patients with rheumatoid arthritis. *J Clin Pharmacol* **44**, 1235–43 (2004).
25. Klareskog L, van der Heijde D, de Jager JP, et al. Therapeutic effect of the combination of etanercept and methotrexate compared with each treatment alone in patients with rheumatoid arthritis: double blind, randomized, controlled trial. *Lancet* **363**, 675–81 (2004).
26. Mahood I, Green MD. Drug interaction studies of therapeutic proteins and monoclonal antibodies. *J Clin Pharmacol* **47**, 1540–54 (2007).
27. Celegene Corp. Package insert for Abraxane. Summit, NJ, 2012.
28. Bristol-Myers Squibb.Package insert for Taxol. Princeton, NJ, 2011.

CHAPTER 13

DRUG INTERACTIONS OF CYTOKINES AND ANTICYTOKINE THERAPEUTIC PROTEINS

J. GREG SLATTER, LARRY C. WIENKERS, and LESLIE J. DICKMANN

If it were not for the great variability among individuals, medicine might as well be a science and not an art.

Sir William Osler, 1892[1]

13.1 INTRODUCTION

The suppression of drug metabolism by proinflammatory cytokines and the relief of suppression by disease treatment are a current issue in the development of drug interaction labels for new biologic products. This topic has been the subject of a number of recent reviews that capture the history of related work as well as current regulatory and pharmaceutical industry thinking.[1–12] This chapter provides an overview of some recent data developed in industrially relevant preclinical models and outlines the pros and cons of these models with respect to the generation of actionable data–that is data that are robust enough to preclude the need for a clinical drug–disease interaction study, and/or a class label for a new cytokine or anticytokine therapeutic agent.

Because small molecule clearance mechanisms are suppressed by inflammation, proinflammatory cytokine treatment and/or active inflammatory disease may afford patient populations with impaired drug clearance.[6] As such, the effective treatment of inflammatory disease would be expected to normalize or "desuppress" drug clearance. When clearance is mediated by cytochromes P450 (CYP), suppression works in the same direction as CYP inhibition, leading to increased systemic exposure. In contrast, CYP desuppression increases CYP activity (similar to CYP induction), leading to lower drug concentrations. The caveat with desuppression is that the upper limit of the effect is the clearance

Drug–Drug Interactions for Therapeutic Biologics, First Edition.
Edited by Honghui Zhou and Bernd Meibohm.

observed in healthy patients. The clinical literature that have accurately measured the magnitude of these effects are limited. Where clinical pharmacokinetic effects have been characterized, several drug labels appropriately note that small molecule clearance changes are possible when anti-inflammatory biologic agents or proinflammatory cytokines are coadministered with narrow therapeutic index small molecules that are cleared mainly by CYP enzymes (labels are reviewed in Ref. 5). Many pharma/biotech pharmacokinetics and drug metabolism labs are now discussing best practices for *in vitro* experiment design and interpretation, meaningful *in vitro* to *in vivo* correlations (IVIVC), and the design of suitable clinical investigations involving therapeutic proteins.[13,14]

13.2 CLINICAL RELEVANCE OF CYTOKINE-MEDIATED SUPPRESSION AND DESUPPRESSION OF ADME ENZYMES

When proinflammatory cytokine-mediated effects on CYPs are compared to induction and inhibition of CYPs by agents such as ketoconazole and rifampin, it is so far apparent that the clinical magnitude of these effects are relatively subtle. In many cases, small study populations have made it difficult to discern a clinically measurable effect on transcriptionally variable CYP isoenzymes. The magnitude of known clinical changes of therapeutic protein DDI, along with useful data on dosimetry and necessary mechanistic details are compiled in three useful tables published in Ref. 4.

13.2.1 Treatment with Interferons and Related Proinflammatory Agents

Clinical interest in the suppression of drug metabolism by inflammatory disease rose after the 1978 account of side effects in asthmatic children on theophylline during an influenza outbreak. Theophylline half-life increases averaged about two-fold,[15] and this highlights that subtle disease-mediated changes in oral clearance can be clinically significant in narrow therapeutic index drugs. Shedlofsky et al.[16] showed that changes in theophylline and antipyrine clearance in healthy subjects could be elicited with lipopolysaccharide (LPS) endotoxin. Interferon-mediated suppression of CYP in influenza was proposed as a potential mechanism in a clinical study in healthy subjects that compared antipyrine clearance during influenza (78% of healthy value) and after administration of interferon A (85% of healthy value) or interferon B (85% of healthy value).[17]

Over the time period when disease-related CYP effects were first linked to interferons, these cytokines were also of interest as drug candidates. Type I interferons (α2a, α2b, β1a, β2b) are a key component of polytherapy of Hepatitis B and Hepatitis C viral infections (HBV and HCV), autoimmune diseases, and cancers, by virtue of their ability to elicit a host immune response. As a consequence, CYP suppression by interferons has been investigated largely clinically because combination therapy entails risk of higher exposure to concomitant, narrow therapeutic index CYP substrates.[18] *In vitro* studies and a variety of preclinical *in vivo* studies have also been undertaken to understand the mechanism of CYP suppression, relative to other cytokines such as the interleukins. Flaman et al.[18] recently examined the mechanism of CYP3A4 suppression by interferon α in human hepatoma (HepG2) cells and demonstrated a time-dependent and physiologically relevant concentration-dependent decrease in CYP3A4 transcription at nontoxic concentrations of interferon α. They proposed that suppression was transcriptional and involved an interferon-mediated JAK1-STAT1 signaling cascade that is independent of ERK1/2 and p38 MAP kinases. Aitken and Morgan[19] investigated a type II interferon, interferon γ, and showed that it mediated isoform-specific effects on CYPs in human hepatocytes, with CYP1A2, 2A6, 2B6, and 3A4 activities being selectively suppressed in a nitric oxide–independent manner. Isoenzyme-specific upregulation and downregulation of CYPs by interferon α2b has also been observed at therapeutic concentrations in human hepatocyte coculture with Kupffer cells.[20] Kupffer cells are a contributor to cytokine-related signaling in the liver, and there is current interest in whether data from human hepatocyte/Kupffer cell cocultures are any more informative than hepatocytes alone, in terms of recapitulating *in vivo* acute phase response effects on CYPs.[20,21]

A variety of clinical studies on therapeutic interferons have revealed that probe substrate clearances were generally in the range of 85% to 40% relative to control values. Oral clearance decreases were generally less than twofold to threefold (50% to 33% of control values) and covered a variety of major CYP isoenzymes.[4] Accordingly, there are relatively few drug interactions listed in the product labels of interferons. Label warnings include use of interferon α2a and α2b with theophylline, the latter doubling theophylline levels. In 18 peginterferon α2b–naive chronic hepatitis C subjects on stable methadone maintenance therapy, mean methadone AUC increased by about 16% after 4 weeks of peginterferon α2b treatment and was about twofold higher in two subjects. In 22 HCV patients, peginterferon α2b caused a modest increase in clearance of a CYP2C8/9 probe substrate (mean increase 28%), a variable increase in clearance of a CYP2D6 probe substrate (mean increase 66% with 13 subjects increased, 5 subjects decreased, and 4 subject unchanged) and with no significant effect on CYP1A2 and CYP3A4 probe substrates.[5] With these counterintuitive increases in CYP substrate clearance, it is apparent that suppression of CYPs by the interferons used for HCV therapy is not a straightforward problem. Data from Nakai et al.[22] may explain the subtle effects and counterintuitive data because the authors have shown that suppression of drug-metabolizing enzymes increases with the progression of liver fibrosis in chronic HCV patients. Accordingly, when interferons are used successfully in the treatment of HCV infection, sustained virologic responses over time may reduce hepatic inflammation and therefore mitigate the suppressive effects of the proinflammatory cytokine treatments. This highlights the temporal efficacy-related aspects of this drug–disease interaction that make the quantitative *in vivo* extrapolation of *in vitro* data tenuous.

13.2.2 Treatment of Inflammatory Conditions through the Mitigation of Cytokine-Mediated Signaling

It is now well established that inflammation contributes to pathology in a wide range of diseases and that inflammation can also contribute to altered drug pharmacokinetics and pharmacodynamics.[11,23] For example, drug–disease interactions of this sort have been described for HIV patients whose oral midazolam clearance was 50% (a twofold decrease) or 82% relative to that of healthy subjects.[24,25] Losartan oral clearance was unaffected by rheumatoid arthritis, but its active metabolite AUC was 40% of control in subjects with active arthritis.[26] In other examples, antipyrine clearance was impaired in sepsis,[27] and clozapine clearance was impaired in infection.[28,29]

Haas et al.[30] showed an inverse correlation of serum interleukin 6 (IL-6) with CYP3A activity in postsurgical patients, as measured by the Erythromycin Breath Test. Frye et al.[31] showed inverse correlations of CYP1A2- and CYP2C19-mediated probe substrate metabolism with serum IL-6 in patients with congestive heart failure, and Rivory et al.[32] showed a correlation of the acute phase marker, serum C-reactive protein (CRP), with the Erythromycin Breath Test in cancer patients. Advanced cancer patients with reduced clearance of docetaxel and elevated markers of inflammation experienced more severe adverse effects of chemotherapy.[33] Chen et al.[34] revealed a strong correlation between IL-6 concentrations and increased cyclosporine levels in six bone marrow transplant patients receiving a continuous cyclosporine infusion over the posttransplant interval. On average IL-6 increased to a maximum of 19.8-fold and CRP was elevated to a maximum of 8.8-fold. Cyclosporine levels increased by 3.60-fold at 4.83 days after the maximum IL-6 concentration. The Spearman rank order correlation R^2 for the correlation between $C_{\mathrm{max\ IL\text{-}6}}$ and dose normalized $C_{\mathrm{max\ CsA}}$ was 0.794. For the cyclosporine metabolite AM1, the correlation with $C_{\mathrm{max\ IL\text{-}6}}$ was 0.975.[34] Taken together, these selected studies show how systemic manifestations of inflammation are correlated with CYP status and how CYP suppression could be a contributor to adverse events, especially during acute care polytherapy.

As the studies cited indicate, the magnitude of CYP effects can be mild. Accordingly, there are also studies that failed to reveal significant cytokine-mediated effects on CYPs, such as when arthritis treatment with sodium aurothiomalate, sulfasalazine or penicillamine had no significant effect on the CYP3A marker, urinary 6β-hydroxycortisol in rheumatoid arthritis patients.[35] Similarly, there was no difference in cyclosporine pharmacokinetics in elderly patients with and without arthritis.[36]

To date, the upper magnitude of inflammatory disease treatment effects on CYP-related drug metabolism appears to be in the range of onefold to fourfold, with most examples occurring at the low end of the range. Upper limits of CYP suppression could be better defined with adequately powered studies comparing probe substrate pharmacokinetics and inflammation biomarkers in patients with severe inflammatory disease, measured either against predisease status or age and sex matched healthy subjects. The Chen et al.[34] study on cyclosporine pharmacokinetics in transplant patients stands out as a good example of the value of this approach.

13.2.3 The Tocilizumab–Simvastatin Interaction

So far there is only one published clinical study investigating a drug–disease interaction involving cytokine desuppression by anticytokine therapy, the tocilizumab–

simvastatin interaction.[37] Rheumatoid arthritis patients were treated with the anti-IL6R monoclonal antibody, tocilizumab (Actemra^{TM}),[36] and simvastatin pharmacokinetics were measured before and after tocilizumab administration. Rheumatoid arthritis patients experienced an 2.3 fold and (43% of control) decrease in the plasma AUC of the CYP3A4 substrate simvastatin after an 10 mg/kg tocilizumab dose.[36,37] In comparison, in separate pharmacokinetic studies, the decreases in simvastatin AUC after treatment with the prototypical inducer rifampin were 7.2- and 11.1-fold.[38,39] The view that desuppression normalized but did not induce simvastatin oral clearance in tocilizumab-treated rheumatoid arthritis patients is supported by a simvastatin $AUC_{0-\infty}$ comparison across separate studies (44.5 ± 19 ng·h/mL in tocilizumab-treated rheumatoid arthritis patients versus 29 ng·h/mL (range 8–56 ng∗h/mL) in healthy subjects).[37,39]

13.2.4 Can C-reactive Protein (CRP) and/or IL-6 Be Used to Predict the Clinical Magnitude of CYP Suppression and Desuppression?

Schmitt et al.[37] noted that the plasma concentration versus time profile of the acute phase reaction protein, serum CRP, "paralleled the time course of tocilizumab's effect on simvastatin pharmacokinetics." Production of CRP by hepatocytes and release into the systemic circulation is primarily elicited by IL-6 and to a lesser extent by IL-1β and tumor necrosis factor and is mediated by the transcription factors C/EBPβ and NF-κB.[40] This observation and the other independent correlations of CYP3A activity with CRP and IL-6 discussed earlier,[31,32] may indicate that initial and posttreatment CRP concentrations are a potential quantitative biomarker for overall hepatic inflammation of sufficient magnitude to suppress CYPs. Therefore, although it is widely acknowledged to be a variable inflammation biomarker, large changes in circulating CRP (and/or IL-6) may be inexpensively predictive of the clinical magnitude of suppression and/or desuppression of CYPs in adequately powered studies. Clinical data on other cytokine and anticytokine drugs will be needed to discern whether clinically measurable suppression and desuppression of CYP enzymes is possible in the absence of significant changes in CRP and IL-6. These data will be useful in understanding the relative magnitude of hypothetical "direct" effects of various individual cytokines on CYPs against a possible central role of transcriptional CYP suppression that occurs secondary to inflammation-related hepatocyte exposure to IL-6 and/or IL-1β.

13.3 MECHANISM

13.3.1 Cytokine-Mediated Suppression of CYPs

No single global mechanism has yet been identified that accounts for all inflammation-mediated changes for all drug-metabolizing enzymes. Ashino et al.[41] and others have conducted work related to the premise that blanket downregulation of pro-oxidant CYP oxidative pathways is hepatoprotective against immune-activation-related oxidative stress. Morgan[42] reviewed the complexities of possible adaptive mechanisms for CYP suppression, including the relief of oxidative stress. In this insightful review, he posed some excellent "rational speculation" on why animals that experience inflammation might find it beneficial to suppress hepatic CYP enzymes. Transcriptional effects are the most investigated and may be coordinately regulated by a variety of different nuclear receptors and transcription factors[43] as they are in liver disease.[44] Because a number of cytokines mimic the effects of infection when injected *in vivo* or when incubated with hepatocytes in primary culture, downregulation of many CYP isoenzymes has been associated with elevated levels of a variety of cytokines.[19,42,45] Inflammation-related CYP suppression is mainly transcriptionally mediated, although posttranscriptional mechanisms have also been proposed.[46,47] Consequently, NF-κB-mediated regulation and related crosstalk with the pregnane X receptor (PXR, NR1I2) are often inferred, though an overall mechanism for suppression has not been fully elucidated.[48–52] Ghose et al.[53] proposed that suppression of many CYP isoenzymes may be mediated by inflammation-induced alteration of the subcellular location of retinoid X receptor α (RXRα), which is an obligate partner in heterodimeric nuclear receptors liver X receptor (LXR), farnesoid X receptor (FXR), retinoic acid receptor (RAR), constitutive androstane receptor (CAR), pregnane X receptor (PXR), and peroxisome proliferator activated receptor alpha (PPARα). Gu et al.[54] demonstrated direct interactions of NF-κB P65 with the DNA binding domain of retinoid X receptor alpha (RXRα). Taken together, these studies indicated that effects on RXRα are a potential explanation for cytokine-mediated suppressive effects that are observed across the spectrum of enzymes involved in steroid, lipid, bile acid, and drug metabolism.

A recent genomewide association study on CYP regulation profiled liver tissue from over 427 donors. The study revealed negative correlations between the expression of genes in a gene module related to the acute phase response (APR) and genes in an hepatic steroid and lipid metabolism gene module (see Figure 3 in Ref.[55]). Subtle downregulation of steroid and lipid metabolism, in consort with up-regulation of the APR related genes is apparently a core component of the hepatic response to systemic inflammation. This is illustrated in Figure13-1, for which supplemental data from Yang et al.[55] were used to plot pairwise correlations among mRNA of CYP3A4, PXR, IL-6, IL-1β, and the APR marker CRP. CYP3A4 expression in this population was correlated with many genes under the regulatory control of LXR, FXR, RAR, CAR, PXR, and PPARα. CYP3A4 was also weakly correlated with a variety of other hepatic acute phase markers associated with the APR.[55,56] It is attractive to speculate, based on these gene expression correlations, that hepatic inflammation is a covariate of basal CYP3A4 expression in human populations. If this were true, the dynamic range needed to discern these correlations in this population of liver donors may have been related to individual differences in disease state or suitability for transplant, hepatic inflammation related to the process of death and dying, and/or liver collection and preservation.

13.3.2 A Related Mechanism: Drug Binding to α1-Acid Glycoprotein

There is a second mechanism that alters small molecule pharmacokinetics in inflammatory diseases. Jamali and others have shown that drugs that bind

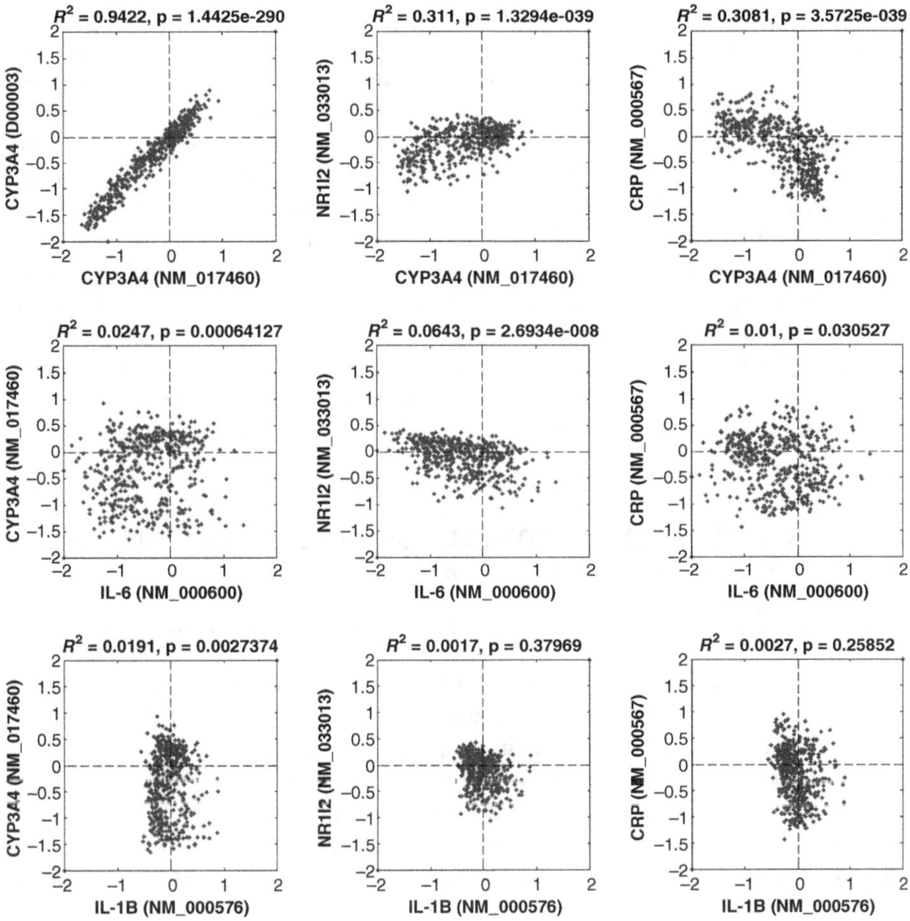

FIGURE 13-1 Gene to gene mRNA expression correlations. CYP3A4, IL-6, and IL1β are correlated with a second CYP3A4 probe, the nuclear pregnane X receptor (PXR; gene name NR1I2) and the acute phase response marker, C-reactive protein (CRP). *Top row, CYP3A4 correlations*: (left) Two unique CYP3A4 probes were present on the array and show the expected high correlation between different probes for the same enzyme across the donor population; (middle) expected positive correlation between CYP3A4 and PXR (NR1I2) expression; (right) negative correlation between CYP3A4 and CRP. *Middle row, IL-6 correlations*: (left) Lack of correlation with CYP3A4; (middle) negative correlation between IL-6 and PXR (NR1i2) expression; (right) lack of correlation between IL-6 and CRP. *Bottom row IL-1β correlations*: (left) Dynamic range of IL-1β expression is narrower than IL-6 ($< \pm$ 10-fold), no correlation with CYP3A4; (middle) no correlation between IL-1β and PXR (NR1i2) expression; (right) no correlation between IL-1β and CRP. Plots prepared by Dr. Yudong He (Amgen Corp), from open access data at sagebase.org and data from Ref. 55. (See the color version of this figure in Color Plates section.)

extensively and exclusively to the acute phase reaction protein α1-acid glycoprotein (AAG) can exhibit inflammatory disease-dependent clearance.[11,23] A recent study on the kinase inhibitor vismodegib illustrates how modern biophysical techniques can be used to assess the role of AAG-binding-dependent pharmacokinetics in patient populations.[57] The AAG drug–disease interaction mainly occurs with basic oral drugs that have intermediate to high extraction ratios, such as verapamil and β-blockers.[58] Most AAG correlations with drug clearance were developed using *in vivo* experiments in rats or mice, for which treatment-mediated relief of experimental inflammation was accomplished with ketoprofen and infliximab.[59,60] In these cases, binding to elevated AAG may work in consort with inflammation-impaired CYP-mediated clearance to increase drug exposure. Accordingly, changes in oral clearance of AAG-binding drugs with rate-limiting CYP-mediated metabolism may be greater than in drugs that do not bind AAG. For example, rheumatoid arthritis resulted in a fourfold increase in total verapamil AUC, relative to control subjects.[58]

13.4 CAN PRECLINICAL MODELS BE USED TO PREDICT CLINICAL SUPPRESSION OR DESUPPRESSION?

13.4.1 Cytokine Effects in Hepatocyte Culture: Studies on IL-6

Human hepatocytes are an *in vitro* model of choice for investigating cytokine effects on CYPs.[21,19,61] IL-6 is the most studied cytokine to date and is known to decrease CYP mRNA and the activities of major human CYP isoenzymes.[19] For the interleukins, virtually all *in vitro* hepatocyte studies conducted before 2010 were conducted at cytokine concentrations that were much higher than human serum cytokine concentrations. For example, in healthy individuals, serum IL-6 concentrations range from 1.3 to 10.3 pg/mL and are increased to 2.6–123 pg/mL in inflammatory diseases.[61]

Our laboratory carried out an analysis of CYP regulation by IL-6 and desuppression of CYP by an anti-IL-6 monoclonal antibody.[61] IL-6 was chosen as a model cytokine to define industrially relevant pros and cons of *in vitro* experimentation on cytokine-mediated suppression of CYPs in primary human hepatocytes. IL-6 was also selected because well-designed clinical and related *in vitro* work became the foundation of the tocilizumab drug–disease–drug interaction label[37] and because of the multiple roles of IL-6 in liver physiology.[62] mRNA expression of diverse drug metabolizing enzymes and CYP1A2 and 3A4 activities were measured.

Cytokines and antibodies from commercial vendors required special characterization that is outside of the realm of standard small molecule *in vitro* drug metabolism protocols. For example: Experiments were benchmarked to physiologically relevant serum cytokine concentrations. Binding affinities of monoclonal antibodies for their targets were characterized and compared to the affinity of the cytokine for its receptor. Lastly, the expression of relevant cytokine receptors on hepatocytes, and the cytokine-mediated generation of acute phase proteins (serum amyloid A and/or CRP) were followed through the course of the experiment.[61,63]

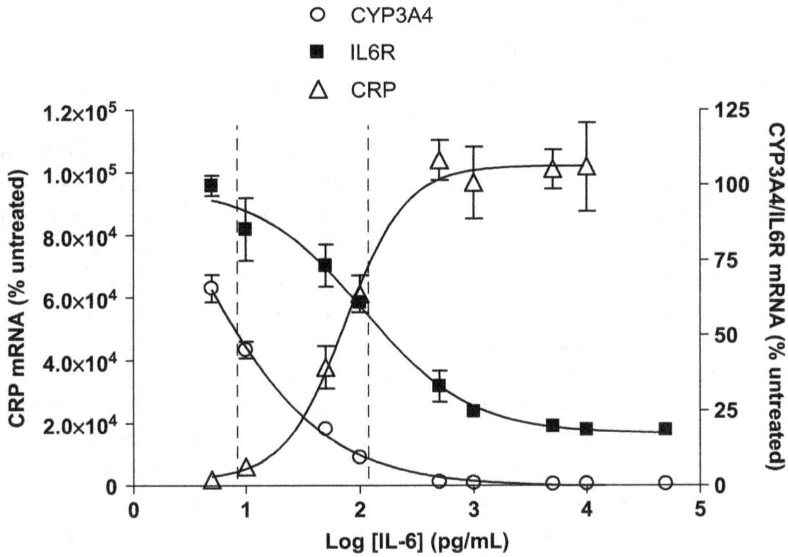

FIGURE 13-2 Single donor data showing the dose-response of human hepatocytes to interleukin 6 (IL-6). mRNA expression of CYP3A4 is compared to mRNA of the acute phase response marker C-reactive protein (CRP) and mRNA expression of the IL-6 receptor. Dashed vertical bars show the normal physiological range of IL-6 and IL-1β serum concentrations in health and disease. Selected data from Ref. 61 were replotted with permission.

Across the spectrum of major drug metabolizing enzymes, CYP3A4 was the most sensitive to suppression by IL-6 and CYP1As were the least sensitive. The EC_{50} values of many CYP isoenzymes were above the range of physiological IL-6 concentrations. The dynamic range of IL-6 responsiveness was determined in human hepatocytes and compared to the development of the hepatocyte APR, as measured by CRP production. In Figure13-2, a single hepatocyte donor dose-response to IL-6 shows that CYP suppression, CRP production, and IL-6 receptor (IL-6R) downregulation all occurred within or near the physiological range of serum IL-6 concentrations. This suggests two intriguing possibilities. First, that the hepatocytes in this model corroborate clinical indications that CYP suppression and the APR are correlated, indicating that CRP could be a bona fide biomarker of CYP suppression and desuppression in a large clinical population. Second, IL-6R downregulation may indicate an adaptive and/or self-limiting response to sustained cytokine exposure, and this could translate into the relatively modest downregulation of the CYP isoenzymes that has been observed clinically to date.

We also investigated the dynamic range of inflammation effects on CYP expression *in vitro* by studying cytokine-mediated CYP suppression during sustained induction pressure. Rifampin and omeprazole were used to induce CYP3A4 and CYP1A2, respectively. Both inducers shifted the IL-6 dose-response curve to the right, affording new EC_{50} values that were well above the physiologically relevant range of IL-6 concentrations. Therefore, at least *in vitro*, CYP suppression by cytokines may not necessarily overcome CYP induction by concomitant administration of a CYP inducer. The effect of CYP induction and suppression in rodents *in vivo* has been studied,[64] but a definitive study on the magnitude of CYP induction in inflammatory disease patients, relative to healthy individuals, could not be found.

No *in vitro* experiments have yet addressed the antibody-mediated abrogation of IL-6-mediated suppression of CYP that is the basis of the tocilizumab–simvastatin interaction. In human hepatocytes, an anti-IL-6 monoclonal antibody shifted the IL-6 dose-response curve for suppression of CYP3A4 activity to the right. Although this expected result indicated that the human hepatocyte model can indeed measure desuppression, the magnitude of the effect was a consequence of a long list of experimental variables, including antibody concentration and the competing affinities of antibodies, cytokine ligands, and their receptors.

Overall, these data illustrated that an individual cytokine can be studied in hepatocytes *in vitro* and can afford quantitative CYP suppression and desuppression data. However, these data represent only a small component of the complicated biology of the APR and/or chronic inflammation in the intact liver. This limits the value of the model for quantitative prediction of clinical drug–disease interactions related to CYP suppression and desuppression.

13.4.2 Is Interleukin-6 Always Involved When Hepatic CYP Suppression Occurs?

IL-6 is secreted by several different cell types, including T-cells, osteoblasts, and macrophages such as Kupffer cells in the liver. Acting via the IL-6 receptor on hepatocytes, IL-6 uniquely mediates diverse hepatic effects that include the APR, hepatic regeneration, and decreases in apoptosis and necrosis.[62] Tocilizumab efficacy data[65] have shown that the IL-6/IL-6R axis is a valid target for the treatment of rheumatoid arthritis. Similarly, the tocilizumab–simvastatin interaction has illustrated that IL-6 is also a key mediator of inflammatory disease manifestations in the liver.[37] Many other cytokines have been shown to cause CYP suppression *in vitro* (interferons, IL-1, IL-2, IL-10, TNF, etc.), yet the diverse roles of IL-6 in the liver may (or may not) render it preeminent among all cytokines as a biomarker or rate-limiting mediator of a clinically measurable effect. As such, cytokine combination experiments *in vitro* might help determine which cytokines are key drivers of hepatic CYP responses to systemic inflammation.

13.4.3 Recapitulating Hepatic Inflammation *in vitro* with Cytokine Combinations

The evaluation of a single cytokine in isolation in hepatocyte culture *in vitro* has little chance of recapitulating the complexity of acute or chronic inflammation in the liver *in vivo*. Many studies have examined the effect of LPS endotoxin,[66,67] and others have systematically dissected components of the APR through the study of individual cytokines.[19,21,68–70] To probe whether two clinically significant cytokines could act additively or synergistically *in vitro*, we examined the effects of

IL-1β alone and in combination with IL-6, on drug-metabolizing enzymes in human hepatocytes.[63,71,72]

It has been shown that IL-1β can suppress hepatic drug-metabolizing enzyme and transporter levels, implying a role of IL-1β in CYP suppression.[19,41] As with IL-6, isoform-specific differences have been observed in CYP suppression in IL-1β-treated hepatocytes *in vitro*.[19,46] However, all studies conducted to date have examined supraphysiologic concentrations of IL-1β relative to physiologically relevant serum concentrations.

Like IL-6, IL-1β alone suppressed mRNA of several CYP isoenzymes, with EC_{50} values that differed by isoenzyme.[63] EC_{50} values of IL-1β were markedly higher than those of IL-6. IL-1β also caused upregulation of CRP and downregulation of IL-6R mRNA. These two events occurred near the physiological range of serum IL-1β concentrations.[73,74] As shown in Figure13-3,[63] IL-1β and IL-6 combinations at concentrations from 1 to 100 pg/mL in human hepatocytes demonstrate that IL-6 was the dominant cytokine in terms of increases in CRP mRNA as well as decreases in CYP3A4 and 2B6 mRNA.

The effects of pairwise or ternary combinations of other cytokines on CYPs, when benchmarked to physiological serum cytokine concentrations, may inform

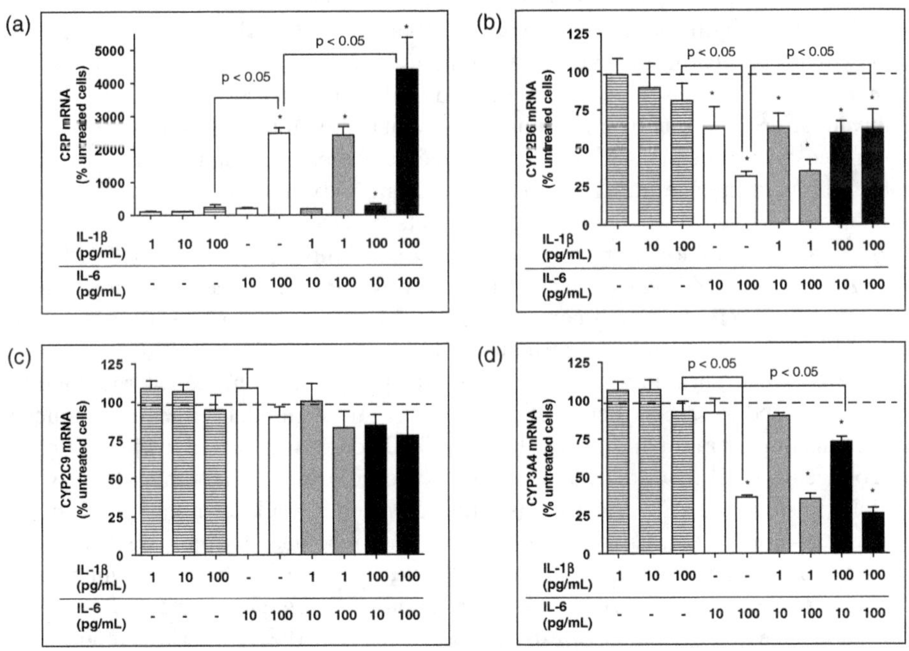

FIGURE 13-3 A comparison of the acute phase reactant (a) C-reactive protein (CRP), (b) CYP2B6, (c) CYP2C9, and (d) CYP3A4 mRNA responses in human hepatocytes treated with interleukin 6 (IL-6) and IL-1β. Asterisk: a p-value <0.05 when compared to untreated cells. Adapted from Ref. 63, Figures 2 and 3 with permission.

which cytokines are dominant in effects on CYP in isolated hepatocytes. These experiments may help in the development of a mechanistic understanding of which cytokines are the most important, even if this approach is certain to fall short of the *in vitro* reconstitution of human disease in liver. Important data in these investigations are the presence of relevant cytokine receptors on hepatocytes, the measurement of CRP and the downregulation of relevant cytokine receptors. Cosgrove et al.[75] developed a cytokine combination matrix and coincubated major cytokines with drugs known to elicit idiosyncrasy with human hepatocytes. The study evaluated synergy among major cytokines in the manifestation of drug-induced hepatocellular toxicity and may represent a useful approach for the systematic assessment of effects of ternary cytokine combinations on CYP suppression.

13.4.4 CYP Suppression and Desuppression in Murine Models of Inflammatory Disease *In Vivo*

There are many published studies on suppression of drug-metabolizing enzymes by cytokines in rodent models. A general inflammatory stimulus from bacterial LPS in preclinical *in vivo* models suppresses CYP probe substrate clearances generally by onefold to twofold.[66] However, there are relatively few studies on disease treatment effects on CYP suppression. Piquette-Miller and Jamali[59] studied nonsteroidal anti-inflammatory drug (NSAID) treatment effects on propranolol pharmacokinetics in severe adjuvant arthritis in rats. They investigated CYP markers, protein binding, and cytokine levels for this basic, AAG-binding drug. Ling and Jamali[60] studied the effect of the anti-TNF monoclonal antibody infliximab on the resolution of early adjuvant arthritis in rats by examining effects on CYPs, inflammation markers, and pharmacokinetics of the AAG-binding molecule verapamil. Ashino et al.[76] studied the effect of a murine anti-IL-6 monoclonal antibody treatment on arthritis in a transgenic mouse arthritis model. The effects of disease and treatment on CYP activity and expression were measured. In the Ashino study, IL-6 was markedly upregulated and CYP3A was the most sensitive CYP isoenzyme to suppression in arthritic rats. CYP3A protein content fell to 60% of that in nonarthritic transgenic rats, and an anti-IL-6 antibody normalized this response.

The collagen antibody-induced arthritis (CAIA) mouse model of rheumatoid arthritis is used to screen therapeutic candidates. In this model, arthritis is induced by initial administration of a monoclonal antibody cocktail directed toward epitopes in collagen type II that are highly autoantigenic, followed by administration of an endotoxin such as LPS.[77] Arthritis is induced within a week of initiation, with the severity of arthritis corresponding to the dose of monoclonal antibody cocktail.[78]

We adopted murine CAIA as a potential fast-turnaround model that might help in the development of *in vitro* to *in vivo* correlations.[79,80] On days 5 and 7 after initiation of the CAIA protocol, serum IL-1β, IL-6, and the APR marker serum amyloid A (SAA) were significantly elevated (Figure13-4), whereas both IL-2 and TNF-α were below limits of quantitation (not shown in the figure). Hepatic CYP mRNA levels were generally downregulated in the CAIA model, although CYP1b1, 2d9, and 3a13 were upregulated, relative to arthritis-naive animals at day 5 of treatment.

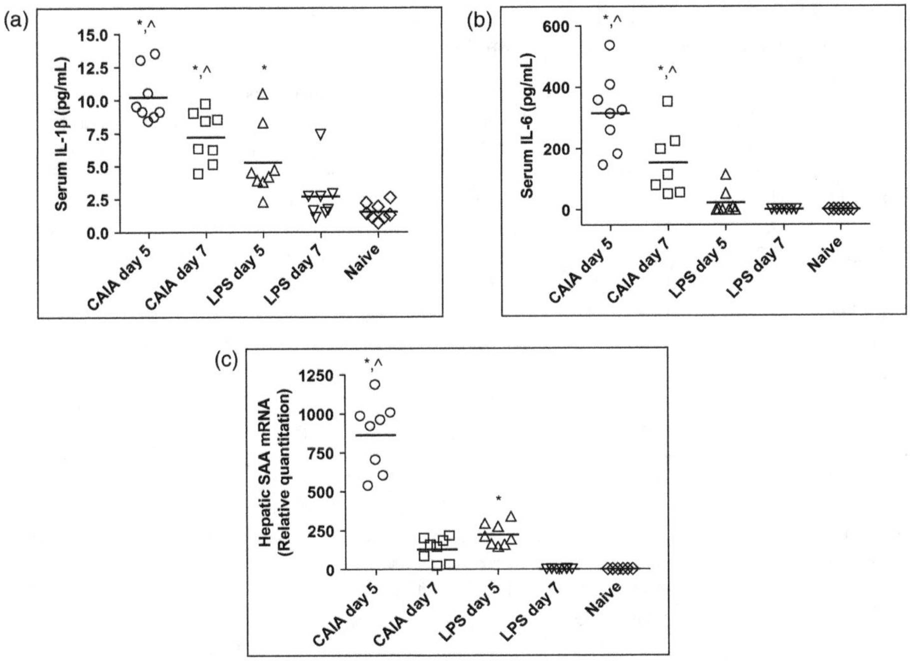

FIGURE 13-4 Serum cytokine and hepatic serum amyloid A (SAA) levels measured in naive, lipopolysaccharide (LPS) treated and collagen antibody-induced arthritis (CAIA). mice (a) Interleuken 1β (IL-1β) protein, (b) IL-6 protein, and (c) SAA mRNA. Serum IL-2 and tumor necrosis factor α were below the limits of quantitation. Adapted from Ref. 79 with permission.

CYP1a2, 2c29, 2b9, and 3a11 mRNA were significantly downregulated to 36%, 29%, 44%, and 45% of naive values at day 5, whereas CYP2e1 levels were unchanged (Table13-1).[79,80] These data were consistent with results obtained using a transgenic mouse model of arthritis where CYP2c29 and CYP3a11 hepatic mRNA were downregulated to 75% and 45%, respectively, compared to wild type animals.[76] Other models are known to afford strain- and isoform-dependent CYP suppression data, such as in a bacterial LPS model of acute inflammation.[67] Taken together, these studies indicate that the direction and magnitude of changes in hepatic CYP enzymes in preclinical inflammation models depend on both the model of inflammation and the mouse strain used.

Intrinsic clearance data obtained using hepatic S9 with phenacetin, bupropion, diclofenac, dextromethorphan, and testosterone CYP probe substrate activities suggests that the maximum decrease in CYP activity in the CAIA model is approximately twofold (Figure13-5).[79,80] With the exception of dextromethorphan, the CAIA day 5 group and the LPS day 5 group had nearly identical profiles. This suggested that the observed decreases in the CAIA day 5 intrinsic clearance may be due to the administration of LPS alone and not the development of CAIA. Indeed, paw swelling in the CAIA model did not correlate with CYP suppression because CAIA

TABLE 13-1 Relative mRNA Levels Compared to Either Naive Animals (CAIA or LPS-Treated Mice) or Untreated Cells (Mouse Hepatocytes)[a]

	CAIA Day 5	CAIA Day 7	Hepatocytes, 10/500[b] IL-1β/IL-6	Hepatocytes 10/100[b] IL-1β/IL-6	LPS Day 5	LPS Day 7
CYP1a2	0.36*,**	0.93*	0.13*	0.30*	0.97	1.28
CYP2b9	0.44*,**	0.77*	0.21*	0.39*	0.77	0.84
CYP2c29	0.29*,**	0.79	0.38*	0.41*	0.56*	0.78*
CYP2d9	2.28*	0.95	0.51*	0.64*	2.47*	1.14
CYP2e1	1.10	0.77	0.02*	0.03*	1.06	0.72
CYP3a11	0.45*,**	0.54*	0.19*	0.38*	0.93	0.65*
CYP3a13	4.37*,**	1.97*,**	0.90	0.88	1.59	0.66
POR	1.84*	1.38	0.46*	0.60	1.87*	0.92
SAA	918*	144*	464*	18.0*	223*	0.98

[a]Adapted from Ref. 79 with permission.
[b]Units of pg/mL.
*$p < 0.05$ when compared to the naive group (CAIA) or untreated cells (hepatocytes);
**$p < 0.05$ when compared to the respective LPS counterpart.
CAIA: collagen antibody-induced arthritis; IL: interleukin; LPS: bacterial lipopolysaccharide; POR: P450 oxidoreductase; SAA: serum amyloid A.

day 5 mice had, on average, a clinical score of 2.6 (out of 16) while the CAIA day 7 mice had an average score of 6.4. The differences observed between day 5 and day 7 for CYP suppression, cytokine markers levels, and disease severity show that this model is time dependent. The CAIA model measured maximum CYP suppression

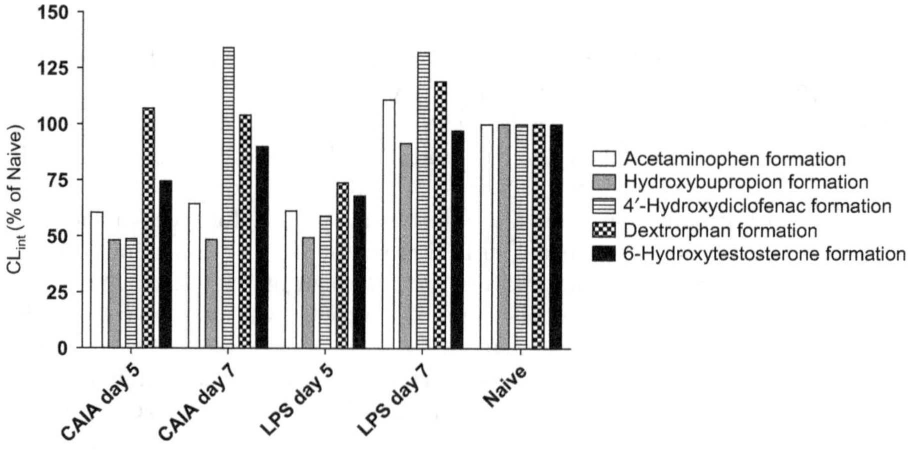

FIGURE 13-5 *In vitro* intrinsic clearance (CL_{int}) of several cytochrome CYP substrates in naive, lipopolysaccharide (LPS) treated and collagen antibody-induced arthritis (CAIA) mice. Values were obtained from pooled S9 fractions ($n = 8$ animals/group). Eadie-Hofstee plots indicated a multiple enzyme system for acetaminophen and 6β-hydroxytestosterone clearance. Changes in CL_{int} reflected changes in both V_{max} and K_m. Adapted from Ref. 79 with permission.

that was in the same general range as other murine models. However, aside for a short experiment cycle time, there were no compelling advantages over other arthritis models.

13.4.5 *In Vivo/In Vitro* Similarities and Differences between the CAIA Model and Cultured Mouse Hepatocytes Treated with a Cytokine Cocktail

Serum cytokine levels from the CAIA study were used as a guide for experiments on CYP suppression in mouse hepatocyte culture. While some CYP isoforms (CYP1a2, 2b9, 2c29, and 3a11) were in agreement in terms of magnitude and direction of suppression in both the *in vitro* and *in vivo* models, several isoforms (CYP2d9, 2e1, 3a13) were not (Table13-1).[79,80] Although this *in vitro/in vivo* comparison was rudimentary, it highlights the difficulty in properly reconstructing a complex inflammatory disease state in an *in vitro* hepatocyte model.

13.4.6 Are *In Vitro* Data Adequate to Preclude the Need for a Clinical DDI Study?

The pros of investigating cytokine-mediated effects on CYP expression in human hepatocytes are that *in vitro* experimentation affords quantitative data showing that ablating a particular inflammatory cytokine might de-suppress/normalize CYP activity. As such, mechanisms can be studied and hypotheses developed using this mainstream industrial technology.[61] However, it is at least technically challenging if not impossible, to accurately re-create an inflammatory disease state in hepatocytes *in vitro* and then treat it. This would be a crucial foundation for a clinically relevant and actionable *in vitro-in vivo* extrapolation. The single cytokine approach in hepatocytes is not likely to quantitatively predict subtle disease treatment effects on CYPs in humans *in vivo*. The diverse variables that are not controlled in hepatocytes include relevant hepatic cytokine concentrations, time-dependent adaptive response mechanisms like cytokine receptor downregulation, the absence of other interacting cytokines, the absence (or unquantified presence) of crucial cell types like Kupffer cells in primary hepatocyte culture, and of course the nuances of patient and disease-specific aspects of inflammatory disease.[61] Other models, such as Kupffer cell coculture may, theoretically, better recapitulate cytokine effects in whole liver. However, CYP3A suppression by IL-1β and IL-6 was not significantly altered by the addition of Kupffer cells in the first report of the application of this model to CYP suppression.[21] As such, human hepatocyte coculture is still in development, and the benefits and limitations of the technique for the study of cytokine-mediated suppression of CYPs have not been established.

Attempts are under way to simulate human pharmacokinetic outcomes based on currently available *in vitro* and *in vivo* data on IL-6. For example, cyclosporine pharmacokinetics and IL-6 concentrations in transplant patients[20] and *in vitro* human hepatocyte data from Aitken and Morgan[19] were used with a simCYP minimal physiologically based pharmacokiuetic (PBPK) model to retrospectively predict a 2-5-fold increase in cyclosporine levels, against an actual 2-7-fold change.[81] Kenny[82] tested *in vitro* drug–drug interaction calculations using human hepatocyte data on IL-6-mediated suppression of CYP3A4 to calculate K_i and k_{inact} values. Population simulations were developed via simCYP using these *in vitro* data, along with clinical IL-6 and simvastatin concentrations from Schmitt et al.[37] to retrospectively simulate the magnitude of the tocilizumab–simvastatin interaction at low and high systemic IL-6 concentrations. They confirmed that IL-6 was a key variable in the interaction because closer predictions were obtained at higher IL-6 serum concentrations. More examples of clinically measurable effects of diverse cytokines and related therapeutic proteins on small molecule pharmacokinetics are needed before the potential merits of predictions based on preclinical *in vitro* experiments can be established for cytokines other than IL-6.

13.5 CURRENT REGULATORY PERSPECTIVE

The tocilizumab–simvastatin interaction was an important event from a regulatory perspective. The clinical observation of increased simvastatin oral clearance after tocilizumab treatment in patients with rheumatoid arthritis, along with some related *in vitro* characterization of IL-6 effects on CYPs in human hepatocytes, afforded a drug interaction label recommending caution for tocilizumab use with narrow therapeutic index CYP substrates.[4] Subsequently, regulatory authorities have requested *in vitro* and/or *in vivo* study data for other cytokine-modulating biologics such as golimumab[5] and for some therapeutic proteins that do not have direct connections to inflammatory disease. A science-driven article by regulatory scientists at the FDA on therapeutic protein–drug interactions (TP-DIs) was published recently,[5] and elements of that paper have reappeared in the revised draft drug interaction guidance that was issued for comment in 2012.[83] For cytokines or cytokine modulators, the new draft guidance advocates clinical investigation in targeted patient populations using probe substrates or probe substrate cocktails. Consistent with the complexities outlined in this chapter, the draft guidance also states that *in vitro* or animal studies have limited value in the qualitative and quantitative projection of clinical interactions because translation of *in vitro* to *in vivo* and animal to human results to date have been inconsistent.[83]

13.6 CLINICAL OPTIONS

Drug–disease interactions associated with an inflammatory disease could be investigated in a compound independent way by comparing CYP probe clearances in adequate numbers of patients with a given disease, against age- and sex-matched healthy controls. This would help establish the absolute upper magnitude of drug-metabolizing enzyme suppression for a given disease state. For disease treatment effects on CYP-mediated metabolism, midazolam AUC could be compared before and after treatment with the biologic agent. Alternatively, if it is thought that

CYP3A4 may not be the most sensitive CYP isoenzyme, a CYP probe cocktail approach may be pursued in a similar fashion.[84] To date, the study conducted by Roche on the tocilizumab–simvastatin interaction has a hallmark design, as it included data on the biomarkers IL-6 and CRP and included two sampling points after the administration of the long half-life antibody.[37] Whether the 80–125% confidence intervals employed in the tocilizumab–simvastatin interaction study to discern a clinically measurable interaction are appropriate for other compounds or diseases is open to discussion. Alternatively, confidence intervals could be set in a case-by-case fashion, possibly based on statistical assessment of treatment-induced changes in inflammatory disease severity biomarkers such as CRP, known population pharmacokinetic norms for narrow therapeutic index victim drugs in health and disease states, or variability in the expression of CYPs of interest.

In most cases, therapeutic protein drug–disease interaction studies will need to be conducted in the target population and are complicated by the long duration of the effects of most therapeutic antibodies. These and other factors may place practical limitations on the number of treatment arms in the study and may exclude some simple DDI study designs. Some study design options for DDI are overviewed in Huang et al.[5] When a cotherapy is likely to be employed extensively in the intended clinical population and plasma concentrations of the concomitant medication are monitored in larger clinical trials, a population pharmacokinetics–based approach may discern a measurable change in exposure to a select victim drug.

Industry scientists and regulators alike are interested in knowing whether *in vitro* data will ever be understood well enough to be used to develop a therapeutic protein drug interaction label that will effectively protect patients and inform healthcare practitioners. New therapeutic cytokines, anticytokine antibodies, and other biologic agents may have direct effects on CYP expression in hepatocytes, or indirect effects via hepatic manifestations of systemic inflammation, or both. It will continue to be difficult to rule out the need for a clinical pharmacokinetic study based on *in vitro* data alone because current *in vitro* models entail risk of false positive or false negative data and expected clinical effects are mild relative to classical CYP inducers and inhibitors.

The alternative is to proceed directly to clinical investigation. Some factors in determining the need for a clinical study might include a lack of direct effect on CYP3A4 in hepatocytes *in vitro* but might also consider whether relevant cytokine receptors are present on hepatocytes and whether large changes in systemic concentrations of inflammatory markers or APR proteins are expected in the target patient population. A risk-based, case-by-case approach is advised.

13.7 CONCLUSIONS

The chapter focused on how inflammatory disease can affect the expression of CYPs and other drug-metabolizing enzymes. In particular, clinical experience with inflammation-related effects on drug metabolism was reconciled with the *in vitro* and *in vivo* models that could be used to model or predict the effects of disease, or cytokine/anticytokine treatments on drug metabolism. Recent work was used to illustrate the pros and cons of preclinical experimentation and to make recommendations with respect to actionable approaches to the development of a robust and informative drug interaction label.

ACKNOWLEDGMENTS

Sonal Patel, Dan Rock, Josh Pearson, Graham Jang, Chris Endres (Amgen PKDM), Yudong He (Amgen Comparative Biology and Safety Sciences), Helen McBride, Kent Miner (Amgen Inflammation research). Raymond Evers (Merck Drug Metabolism) and the *in vitro* LM DDI working group of the IQ Consortium.

DECLARATION OF INTEREST

The authors were employed by Amgen, Inc.

REFERENCES

1. Piquette-Miller M, Grant DM. The art and science of personalized medicine. *Clin Pharmacol Ther* **81**, 311–15 (2007).
2. Kraynov E, Martin SW, Hurst S, et al. How current understanding of clearance mechanisms and pharmacodynamics of therapeutic proteins can be applied for evaluation of their drug-drug interaction potential. *Drug Metab Dispos* **39**, 1779–83 (2011).
3. Zhou H, Mascelli MA. Mechanisms of monoclonal antibody-drug interactions. *Annu Rev Pharmacol Toxicol* **51**, 359–72 (2011).
4. Lee JI, Zhang L, Men AY, et al. CYP-mediated therapeutic protein-drug interactions: clinical findings, proposed mechanisms and regulatory implications. *Clin Pharmacokinet* **49**, 295–310 (2010).
5. Huang SM, Zhao H, Lee JI, et al. Therapeutic protein-drug interactions and implications for drug development. *Clin Pharmacol Ther* **87**, 497–503 (2010).
6. Morgan ET. Impact of infectious and inflammatory disease on cytochrome P450-mediated drug metabolism and pharmacokinetics. *Clin Pharmacol Ther* **85**, 434–38 (2009).
7. Zhou H, Davis HM. Risk-based strategy for the assessment of pharmacokinetic drug-drug interactions for therapeutic monoclonal antibodies. *Drug Discov Today* **14**, 891–98 (2009).
8. Seitz K, Zhou H. Pharmacokinetic drug-rug interaction potentials for therapeutic monoclonal antibodies: reality check. *J Clin Pharmacol* **47**, 1104–18 (2007).
9. Petrovic V, Teng S, Piquette-Miller M. Regulation of drug transporters during infection and inflammation. *Mol Interv* **7**, 99–111 (2007).
10. Mahmood I, Green MD. Drug interaction studies of therapeutic proteins or monoclonal antibodies. *J Clin Pharmacol* **47**, 1540–54 (2007).

11. Kulmatycki KM, Jamali F. Drug disease interactions: role of inflammatory mediators in disease and variability in drug response. *J Pharm Pharm Sci* **16**, 8: 602–25 (2005).

12. Haas CE. Drug-cytokine interactions. In Piscitelli SC, ed *Drug interactions in infectious diseases*. Totawa NJ, Humana Press, 2005: 431–61.

13. Pharmacokinetics and Disposition Expert Working Group Drug-Drug Interaction Task-force, Lyoyd P, Zhou H, Theil FP, Kakkar T, Nestorov I, roberts SA. Highlights from a recent BIO survey on therapeutic protein-drug interactions. *J. clin Pharmacol.* **52**: 1755–63 (2012).

14. Girish S, Martin SW, Peterson MC, et al. AAPS workshop report: strategies to address therapeutic protein-drug interactions during clinical development. *AAPS J* **13**, 405–16 (2011).

15. Chang KC, Bell TD, Lauer BA, Chai H. Altered theophylline pharmacokinetics during acute respiratory viral illness. *Lancet* **1**(8074), 1132–33 (1978).

16. Shedlofsky SI, Israel BC, McClain CJ, et al. Endotoxin administration to humans inhibits hepatic cytochrome P450-mediated drug metabolism. *J Clin Invest* **94**, 2209–14 (1994).

17. Brockmeyer NH, Barthel B, Mertins L, Goos M. Changes of antipyrine pharmacokinetics during influenza and after administration of interferon-alpha and -beta. *Int J Clin Pharmacol Ther* **36**, 309–11 (1998).

18. Flaman AS, Gravel C, Hashem AM, et al. The effect of interferon-α on the expression of cytochrome P450 3A4 in human hepatoma cells. *Toxicol Appl Pharmacol* **253**, 130–36 (2011).

19. Aitken AE, Morgan ET. Gene-specific effects of inflammatory cytokines on cytochrome P450 2C, 2B6 and 3A4 mRNA levels in human hepatocytes. *Drug Metab Disp* **35**, 1687–93 (2007).

20. Chen C, Han YH, Yang Z, Rodrigues AD. Effect of interferon-α2b on the expression of various drug-metabolizing enzymes and transporters in co-cultures of freshly prepared human primary hepatocytes. *Xenobiotica.* **41**:476–85 (2011).

21. Sunman JA, Hawke RL, LeCluyse EL, Kashuba AD. Kupffer cell-mediated IL-2 suppression of CYP3A activity in human hepatocytes. *Drug Metab Dispos* **32**, 359–363 (2004).

22. Nakai K, Tanaka H, Hanada K, et al. Decreased expression of cytochromes P450 1A2, 2E1, and 3A4 and drug transporters Na+-taurocholate-cotransporting polypeptide, organic cation transporter 1, and organic anion-transporting peptide-C correlates with the progression of liver fibrosis in chronic hepatitis C patients. *Drug Metab Dispos* **36**, 1786–93 (2008).

23. Kulmatycki KM, Jamali F. Therapeutic relevance of altered cytokine expression. *Cytokine* **14**, 1–10 (2001).

24. Jetter A, Fätkenheuer G, Frank D, et al. Do activities of cytochrome P450 (CYP)3A, CYP2D6 and P-glycoprotein differ between healthy volunteers and HIV-infected patients? *Antivir Ther* **15**, 975–83 (2010).

25. Jones AE, Brown KC, Werner RE, et al. Variability in drug metabolizing enzyme activity in HIV-infected patients. *Eur J Clin Pharmacol;* **66**, 475–85 (2010).

26. Daneshtalab N, Lewanczuk RZ, Russell AS, Jamali F. Drug-disease interactions: losartan effect is not downregulated by rheumatoid arthritis. *J Clin Pharmacol* **46**, 1344–55 (2006).

27. Carcillo JA, Doughty L, Kofos D, et al. Cytochrome P450 mediated-drug metabolism is reduced in children with sepsis-induced multiple organ failure. *Intensive Care Med* **29**, 980–84 (2003).

28. Haack MJ, Bak ML, Beurskens R, et al. Toxic rise of clozapine plasma concentrations in relation to inflammation. *Eur Neuropsychopharmacol* **13**, 381–85 (2003).
29. Raaska K, Raitasuo V, Neuvonen PJ. Effect of influenza vaccination on serum clozapine and its main metabolite concentrations in patients with schizophrenia. *Eur J Clin Pharmacol* **57**, 705–08 (2001).
30. Haas CE, Kaufman DC, Jones CE, et al. Cytochrome P450 3A4 activity after surgical stress. *Crit Care Med* **31**, 1338–46 (2003).
31. Frye RF, Schneider VM, Frye CS, Feldman AM. Plasma levels of TNF-alpha and IL-6 are inversely related to cytochrome P450-dependent drug metabolism in patients with congestive heart failure. *J Card Fail* **8**, 315–19 (2002).
32. Rivory LP, Slaviero KA, Clarke SJ. Hepatic cytochrome P450 3A drug metabolism is reduced in cancer patients who have an acute-phase response. *Br J Cancer* **87**, 277–80 (2002).
33. Charles KA, Rivory LP, Stockler MR, et al. Predicting the toxicity of weekly docetaxel in advanced cancer. *Clin Pharmacokinet* **45**, 611–22 (2006).
34. Chen YL, Le Vraux V, Leneveu A, et al. Acute-phase response, interleukin-6, and alteration of cyclosporine pharmacokinetics. *Clin Pharmacol Ther* **55**, 649–60 (1994).
35. Beyeler C, Frey BM, Bird HA. Urinary 6 beta-hydroxycortisol excretion in rheumatoid arthritis. *Br J Rheumatol* **36**, 54–58 (1997).
36. Kovarik JM, Koelle EU. Cyclosporin pharmacokinetics in the elderly. *Drugs Aging* **15**, 197–205 (1999).
37. Schmitt C, Kuhn B, Zhang X, et al. Disease-drug-drug interaction involving tocilizumab and simvastatin in patients with rheumatoid arthritis. *Clin Pharmacol Ther* **89**, 735–40 (2011).
38. Kyrklund C, Backman JT, Kivistö KT, et al. Rifampin greatly reduces plasma simvastatin and simvastatin acid concentrations. *Clin Pharmacol Ther* **68**, 592–97 (2000).
39. Chung E, Nafziger AN, Kazierad DJ, Bertino JS Jr. Comparison of midazolam and simvastatin as cytochrome P450 3A probes. *Clin Pharmacol Ther* **79**, 350–61 (2006).
40. Rhodes B, Fürnrohr BG, Vyse TJ. C-reactive protein in rheumatology: biology and genetics. *Nat Rev Rheumatol* **7**, 282–89 (2011).
41. Ashino T, Oguro T, Shioda S, et al. Involvement of interleukin-6 and tumor necrosis factor alpha in CYP3A11 and 2C29 down-regulation by Bacillus Calmette-Guerin and lipopolysaccharide in mouse liver. *Drug Metab Dispos* **32**, 707–14 (2004).
42. Morgan ET. Regulation of cytochrome p450 by inflammatory mediators: why and how? *Drug Metab Dispos* **29**, 207–12 (2001).
43. Zhou C, Verma S, Blumberg B. The steroid and xenobiotic receptor (SXR), beyond xenobiotic metabolism. *Nucl Recept Signal* **7**, e001 (2009).
44. Congiu M, Mashford ML, Slavin JL, Desmond PV. Coordinate regulation of metabolic enzymes and transporters by nuclear transcription factors in human liver disease. *J Gastroenterol Hepatol* **24**, 1038–44 (2009)
45. Renton KW and Nicholson TE. Hepatic and central nervous system cytochrome P450 are down-regulated during lipopolysaccharide-evoked localized inflammation in brain. *J Pharmacol Exp Ther* **294**, 524–30 (2000).
46. Lee, CM, Pohl J, Morgan ET. Dual mechanisms of CYP3A protein regulation by proinflammatory cytokine stimulation in primary hepatocyte cultures. *Drug Metab Dispos* **37**, 865–72 (2009).

47. Ferrari L, Peng N, Halpert JR, Morgan ET. Role of nitric oxide in down-regulation of CYP2B1 protein, but not RNA, in primary cultures of rat hepatocytes. *Mol Pharmacol* **60**, 209–16 (2001).
48. Jover R, Bort R, Gómez–Lechón MJ, Castell JV. Down-regulation of human CYP3A4 by the inflammatory signal interleukin-6: molecular mechanism and transcription factors involved. *FASEB J* **16**, 1799–01 (2002).
49. Zhou C, Tabb MM, Nelson EL, et al. Mutual repression between steroid and xenobiotic receptor and NF-kappaB signaling pathways links xenobiotic metabolism and inflammation. *J Clin Invest* **116**, 2280–89 (2006).
50. Xie W, Tian Y. Xenobiotic receptor meets NF-kappaB, a collision in the small bowel. *Cell Metab* **4**, 177–78 (2006).
51. Zordoky BN, El-Kadi AO. Role of NF-kappaB in the regulation of cytochrome P450 enzymes. *Curr Drug Metab* **10**, 164–78 (2009).
52. Pascussi JM, Gerbal–Chaloin S, Pichard-Gracia L, et al. Interleukin-6 negatively regulates the expression of pregnane X receptor and constitutively activated receptor in primary human hepatocytes. *Biochem Biophys Res Commun* **274**, 707–13 (2000).
53. Ghose R, Zimmerman TL, Thevananther S, Karpen SJ. Endotoxin leads to rapid subcellular re-localization of hepatic RXRalpha: a novel mechanism for reduced hepatic gene expression in inflammation. *Nucl Recept* **2**, 4 (2004).
54. Gu X, Ke S, Liu D, et al. Role of NF-kappaB in regulation of PXR-mediated gene expression: a mechanism for the suppression of cytochrome P-450 3A4 by pro-inflammatory agents. *J Biol Chem* **281**, 17882–889 (2006).
55. Yang X, Zhang B, Molony C, et al. Systematic genetic and genomic analysis of cytochrome P450 enzyme activities in human liver. *Genome Res* **20**, 1020–36 (2010).
56. Slatter JG. Preclinical models for drug interactions with biologics: P450 based DDIs following inflammatory disease treatment with biologics. Paper presented at the Thirteenth Annual Land O' Lakes Conference on Drug Metabolism and Applied Pharmacokinetics, Meerimac, WI, September 13-17, 2010.
57. Giannetti AM, Wong H, Dijkgraaf GJ, et al. Identification, characterization, and implications of species-dependent plasma protein binding for the oral Hedgehog pathway inhibitor vismodegib (GDC-0449). *J Med Chem* **54**, 2592–01 (2011).
58. Mayo PR, Skeith K, Russell AS, Jamali F. Decreased dromotropic response to verapamil despite pronounced increased drug concentration in rheumatoid arthritis. *Br J Clin Pharmacol* **50**, 605–13 (2000).
59. Piquette-Miller M, Jamali F. Influence of severity of inflammation on the disposition kinetics of propranolol enantiomers in ketoprofen-treated and untreated adjuvant arthritis. *Drug Metab Dispos* **23**, 240–45 (1995).
60. Ling S, Jamali F. The effect of infliximab on hepatic cytochrome P450 and pharmacokinetics of verapamil in rats with pre-adjuvant arthritis: a drug-disease and drug-drug interaction. *Basic Clin Pharmacol Toxicol* **105**, 24–29 (2009).
61. Dickmann LD, Patel SK, Rock DA, et al. Effects of interleukin-6 (IL-6) and an anti-IL-6 monoclonal antibody on drug metabolizing enzymes in human hepatocyte culture. *Drug Metab Dispos* **39**, 1415–22 (2011).
62. Taub R. Liver regeneration: from myth to mechanism. *Nat Rev Mol Cell Biol* **5**, 836–47 (2004).

63. Dickmann LJ, Patel SK, Wienkers LC, Slatter JG. Effects of interleukin 1β (IL-1β) and IL-1β /interleukin 6 (IL-6) combinations on drug metabolizing enzymes in human hepatocyte culture. *Curr Drug Metab* **13**, 930–7 (2012).
64. Li-Masters T, Morgan ET. Effects of bacterial lipopolysaccharide on phenobarbital-induced CYP2B expression in mice. *Drug Metab Dispos* **29**, 252–57 (2001).
65. Rose-John S, Waetzig GH, Scheller J, et al. The IL-6/sIL-6R complex as a novel target for therapeutic approaches. **11**, 613–24 (2007).
66. Yang KH, Lee MG. Effects of endotoxin derived from Escherichia coli lipopolysaccharide on the pharmacokinetics of drugs. *Arch Pharm Res* **31**, 1073–86 (2008).
67. Richardson TA, Morgan ET. Hepatic cytochrome P450 gene regulation during endotoxin-induced inflammation in nuclear receptor knockout mice. *J Pharmacol Exp Ther* **314**, 703–09 (2005).
68. Morgan ET, Thomas KB, Swanson R, et al. Selective suppression of cytochrome P-450 gene expression by interleukins 1 and 6 in rat liver. *Biochim Biophys Acta* **1219**, 475–83 (1994).
69. Le Vee M, Lecureur V, Moreau A, et al. Differential regulation of drug transporter expression by hepatocyte growth factor in primary human hepatocytes. *Drug Metab Dispos* **37**, 2228–35 (2009).
70. Assenat, E. Gerbal-Chaloin S, Larrey D, et al., Interleukin 1β inhibits CAR-induced expression of hepatic genes involved in drug and bilirubin clearance. *Hepatology* **40**, 951–60 (2004).
71. Dickmann LJ, Patel SK, Wienkers LC, Slatter JG. *In vitro* effects of cytokines on cytochrome P450 expression in human hepatocyte culture. Paper presented at the 17th North American Regional Meeting of the International Society for the Study of Xenobiotics, Atlanta, GA., October 16–20, 2011.
72. Slatter JG. P450 Based drug-disease interactions following inflammatory disease treatment with biologics: studies in the CAIA murine arthritis model and related primary hepatocyte *in vitro*/*in vivo* correlations. Paper presented at the AAPS National Biotechnology Conference, San Francisco, May 15–18, 2011.
73. Hidaka T, Suzuki K, Kawakami M, et al. Dynamic changes in cytokine levels in serum and synovial fluid following filtration leukocytapheresis therapy in patients with rheumatoid arthritis. *J Clin Apher* **16**, 74–81 (2001).
74. Yilmaz M, Kendirli SG, Altintas D, et al. Cytokine levels in serum of patients with juvenile rheumatoid arthritis. *Clin Rheumatol* **20**, 30–35 (2001).
75. Cosgrove BD, King BM, Hasan MA, et al. Synergistic drug-cytokine induction of hepatocellular death as an *in vitro* approach for the study of inflammation-associated idiosyncratic drug hepatotoxicity. *Toxicol Appl Pharmacol* **237**, 317–30 (2009).
76. Ashino T, Arima Y, Shioda S, et al. Effect of interleukin-6 neutralization on CYP3A11 and metallothionein-1/2 expressions in arthritic mouse liver. *Eur J Pharmacol* **558**, 199–207 (2007).
77. Khachigian LM. Collagen antibody–induced arthritis. *Nat Protoc* **1**, 2512–16 (2006).
78. Chondrex. Adjuvant induced arthritis, 2011. Available at www.chondrex.com/animal-models/adjuvant-induced-arthritis-adjuvants.
79. Dickmann LJ, McBride HJ, Patel SK, et al. Murine collagen antibody induced arthritis (CAIA) and primary mouse hepatocyte culture as models to study cytochrome P450 suppression. *Biochem Pharmacol* **83**, 1682–9 (2012).

80. Dickmann LJ, Patel SK, McBride H, et al. Collagen antibody induced arthritis (CAIA) in mice and primary mouse hepatocyte culture as models to study P450 suppression: *in vivo/in vitro* similarities and differences. Paper presented at the 17th North American Regional Meeting of the International Society for the Study of Xenobiotics, Atlanta, GA, October 16–20, 2011.
81. Machavaram KK, Almond LM, Jamei M, Rostami-Hodjegan A. Predicting the effect of interleukin-6, an inflammatory cytokine, on time-dependent reduction of cyclosporine clearance: an application of the Simcyp Population-Based Simulator to suppression of CYP450 by biologics. Paper presented at the AAPS National Biotechnology Conference. San Francisco, May 16–18, 2011.
82. Kenny J. Cytokine mediated regulation of human cytochrome P450s—can we translate in vitro evaluation to clinical relevance? Paper presented at the AAPS National Biotechnology Conference. San Francisco, May 16–18, 2011.
83. US Food and Drug Administration, US Department of Health and Human Services, Center for Drug Evaluation and Research, Center for Biologics Evaluation and Research (CBER). Draft guidance for industry: drug interaction studies–study design, data analysis, implications for dosing, and labeling. Silver Spring, MD, 2012. Available at www.fda.gov/downloads/Drugs/GuidanceComplianceRegulatoryInformation/Guidances/UCM292362.pdf.
84. Goh BC, Reddy NJ, Dandamudi UB, et al. An evaluation of the drug interaction potential of pazopanib, an oral vascular endothelial growth factor receptor tyrosine kinase inhibitor, using a modified Cooperstown 5+1 cocktail in patients with advanced solid tumors. *Clin Pharmacol Ther* **88**, 652–59 (2010).

CHAPTER 14

DRUG INTERACTIONS FOR GROWTH FACTORS AND HORMONES

YOW-MING C. WANG and TARUNDEEP KAKKAR

14.1 INTRODUCTION

According to the dictionary, the definition of *growth factor* is any of various proteins that promote the growth, organization, and maintenance of cells and tissues. And, the definition of *hormones* is any of various internally secreted compounds that affect the functions of specifically receptive organs or tissues when transported to them by the body fluids. Many of the proteins discussed in other chapters of this book, such as interferons and cytokines, may also be regarded as growth factors or hormones by these definitions. This chapter will discuss the traditional growth factors, primarily hematopoietic growth factors, and hormones, primarily growth hormones and glucagon.

The clinical applications of biotechnology-derived products began with growth factors and hormones in the 1980s. Two decades later, growth factors and hormones ranked the second and the third, respectively, of the top biologic drugs in terms of the U.S. sales[1] only behind therapeutic monoclonal antibodies. The total sales of growth factors and hormones combined accounted for approximately $20 billion (about $10 billion each).

In the growth factor category, epoetin α (recombinant human erythropoietin; rhEPO) approved by the U.S. Food and Drug Administration (FDA) in 1989 represents the first commercially available recombinant growth factor.[2] Filgrastim (recombinant human granulocyte colony stimulating factor, G-CSF) followed closely with an FDA approval in 1991. Subsequently, improvements to these recombinant human growth factors, by hyperglycosylation in the case of darbepoetin α[2,3] and by pegylation in the case of pegfilgrastim[4–6] and methoxy polyethylene glycol-epoetin β,[7,8] resulted in new generations of recombinant proteins with a longer

Drug–Drug Interactions for Therapeutic Biologics, First Edition.
Edited by Honghui Zhou and Bernd Meibohm.

elimination half-life, enabling less frequent dose administration than the first-generation products. Palifermin (recombinant human keratinocyte growth factor; rhKGF) and romiplostim (a thrombopoietin stimulating agent) are growth factors that were approved in the past 10 years. These growth factors made substantial contributions to the therapeutic treatment of several chronic conditions.

Hormones, as a class, consist of a vast range of compounds to cover various ailments, and many have been on the market for decades. Most of the marketed drugs are of synthetic origin, including sex hormones such as estrogens, progesterone, thyroid hormone, and corticosteroids such as betamethasone. In 1960, the FDA approved Lilly's recombinant glucagon, which came more than 30 years after Murlin hypothesized that a hormone with a blood sugar–raising effect was present in the insulin product of low purity in 1923.[9] The first synthetic growth hormone, called somatrem, was developed in 1980s by Genentech in collaboration with a Swedish pharmaceutical company (Kabi), which commercialized the human growth hormone harvested and purified from cadavers.[10] The recombinant growth hormone somatropin, on the market now, is manufactured by an improved method.[10] Today, multiple recombinant human growth hormone and recombinant glucagon products are commercially available globally. Pegvisomant, a growth hormone analog and growth hormone receptor antagonist, was approved by European Medicines Agency (EMA) and the FDA in 2002 and 2003, respectively.

Pharmacokinetics (PK) drug interaction studies were not conducted as part of the clinical drug development program and did not play a role in the regulatory decision for many of the recombinant growth factors and hormone products. The PK drug interaction issue has been deemed unique to the small molecule chemical drugs and not of concern to the biotherapeutic agents because such agents are catabolized by proteolytic enzymes and do not share the metabolic pathways responsible for the *in vivo* disposition of small molecule drugs. Consequently, the understanding of PK drug interactions, or lack thereof, for these products primarily came from clinical experience in patient care over time and research conducted after the approval of marketing authorizations that was done to address these patient care findings.

Administration of two drugs indicated for treating the same disease condition is expected to show pharmacodynamic drug interactions because of additive or synergistic effects; often this type of drug interaction is factored into the study design and monitored during clinical trials and clinical use. For instance, romiplostim administration enables a reduction in the use of corticosteroids in treatment of thrombocytopenia, and the pharmacodynamic synergism of sex hormones is the basis for the use of follitropin α in combination with other agents to stimulate ovulation (e.g., human chorionic gonadotropin, clomiphene citrate) to potentiate the follicular response. This type of "polypharmacy" for additive therapeutic effect is not discussed in this chapter.

For many of the recombinant technology derived hormone products such as gonadotropins (luteinizing hormone, follicle-stimulating hormone, and human chorionic gonadotropin), minimal information is available in the literature and on product labels regarding drug interactions. We have opted to not discuss insulin because many drugs interact with insulin pharmacodynamically; some drugs alter the blood glucose control and some drugs affect insulin sensitivity, which necessitate insulin dose adjustment. On the other hand, few reports on drug interaction involve insulin as a perpetrator even though in *in vitro* systems insulin has been implicated in modulating the expression of hepatic enzymes via its engagement in a multitude of cellular signaling pathways.

The focus of this chapter is on four recombinant growth factors products (erythropoietins, G-CSF, KGF, and thrombopoietins) and three hormone products (glucagon, somatropin, and pegvisomant). We compiled the information related to drug interactions found in scientific publications and regulatory agency approved labeling documents. The referenced U.S. prescribing information (PI) can be found at the Drugs@FDA website, and the European summary of product characteristics (SPC) can be found on the EMA's website for European public assessment reports (EPARs) by keyword search, using the international nonproprietary name (INN). In addition to the postings on regulatory agencies' websites, manufacturers and other interest groups also post up-to-date labeling documents on their product websites. Readers are referred to these online resources for the product labeling information cited in this chapter.

14.2 GROWTH FACTORS

14.2.1 Erythropoiesis-Stimulating Agents

Multiple commercial recombinant erythropoiesis-stimulating agents (ESAs)[2] are produced in Chinese hamster ovary (CHO) cells using recombinant DNA technology. They stimulate erythropoiesis by the same mechanism as the endogenous erythropoietin[11,12] and are indicated for treatment of anemia due to chronic kidney disease or chemotherapy. With the identical amino acid sequence as the natural erythropoietin produced by the kidney, epoetin α has 165 amino acids and a molecular weight (MW) of 30 kDa. Darbepoetin α, a second-generation ESA (MW 37 kDa), has two additional *N*-linked oligosaccharide chains compared to epoetin α, resulting in a prolonged terminal elimination half-life. The third-generation ESA product manufactured by Roche is methoxy polyethylene glycol (PEG)-epoetin β (MW 60 kDa), which differs from epoetin β in that PEG is covalently linked to some of the lysine residues in rhEPO, which also leads to a reduced clearance and a prolonged half-life. The prolonged half-life for these new ESAs afforded greater *in vivo* activity, despite the decrease of *in vitro* potency when compared to their respective first-generation counterpart. Other epoetins (e.g., epoetin β, epoetin δ, epoetin ζ) may have some structural differences from epoetin α and may or may not have differences in the biological activities.

The SPCs of epoetin α and darbepoetin α mentioned a potential for an interaction with substances that are highly bound to red blood cells, such as cyclosporine and tacrolimus, and recommended monitoring of the blood levels of the concomitant medications for assessment of potential dosage adjustment as the hemoglobin rises. Other implications of ESAs on clinical management of patients have been reported where the observed interactions are primarily results of the pharmacodynamic effects of the ESAs. One example is the effect of ESA on the hemoglobin A1C level.[13] Hemoglobin A1C is a measure of the amount of glycosylated hemoglobin in the blood and is a commonly accepted indicator of the long-term glycemic control in diabetic patients.[14,15] But, its level depends on the number of circulating erythrocytes and the duration of red blood cells in circulation because glucose binding to hemoglobin occurs continuously during the life span of a red blood cell.[16] The hemoglobin A1C level decreases on initiation of ESAs and increases on discontinuation of ESA treatment; however, the glucose levels were reported to be stable. These data indicate that caution should be applied in the use of hemoglobin A1C level in patients who are initiating ESA treatment or in those whose ESA doses are being adjusted. Another example occurred in a German multicenter rhEPO trial,[17] where stroke patients receiving rhEPO after tissue plasminogen activator (tPA)-induced thrombolysis appeared to have increased mortality. A subsequent investigation in the mouse cerebral artery occlusion model suggested interaction of t-PA with EPO at the blood–brain interface–that is, promotion of vascular permeability and extracellular matrix breakdown may account for the unfavorable observation. As such, the authors concluded that after t-PA-induced thrombolysis, the use of EPO may not be suitable.[17] This hypothesis will need to be verified with additional studies.

14.2.2 Granulocyte Colony Stimulating Factor

Filgrastim, pegfilgrastim, and lenograstim are three commercially available G-CSF products. Filgrastim is a nonglycosylated recombinant human G-CSF (rhG-CSF) produced by *E. coli* containing 175 amino acids (MW 19 kDa). Pegfilgrastim is the pegylated filgrastim (MW 39 kDa), with a 20 kDa mono-methoxy-polyethylene glycol molecule covalently bound to the *N*-terminal methionyl residue of filgrastim. Lenograstim, a glycosylated rhG-CSF product, produced by CHO cells with 174 amino acids (MW 19 kDa), is approved globally outside the United States. Barring minor differences, these products are indicated for the treatment of neutropenia caused by chemotherapy. Except for pegfilgrastim, they are also approved for the mobilization of the peripheral blood progenitor cells.

Drug interactions were called out in the U.S. PI and EU SPC for filgrastim (with lithium and 5-FU) and pegfilgrastim (with 5-FU). Lithium products are oral medications for the treatment of psychiatric patients with bipolar disorder. Lithium has broad effects on the hematopoietic system and is known to increase neutrophil counts from the research of the 1980s. Mechanistically, lithium dose-dependently reduces the levels of surface CXCR4 protein and mRNA in neutrophils.[18] Because CXCR4 protein plays a critical role in the retention of hematopoietic cells within the bone marrow and in their mobilization into the circulation, a reduction in the CXCR4 is associated with an increased release of neutrophils into the circulation. In addition, lithium induces the production of cytokines, including G-CSF, which contributes to the increase in neutrophils.[19] Preliminary evidence from a small number of patients treated concomitantly with filgrastim and 5-fluorouracil indicates that the severity of neutropenia may be exacerbated, a phenomenon believed to be associated with an increase in the topoisomerase sensitivity in the presence of G-CSF.[20] Furthermore, the product labels indicate that the use of filgrastim or lenograstim is not recommended from 24 h before until 24 h after chemotherapy ends, in view of the sensitivity of rapidly dividing myeloid cells to cytotoxic chemotherapy.

14.2.3 Keratinocyte Growth Factor

Marketed globally, palifermin is a truncated, nonglycosylated form of endogenous human keratinocyte growth factor (KGF) with the first 23 *N*-terminal amino acids deleted to improve protein stability. This recombinant KGF contains 140 amino acids (MW 16.3 kDa). It is a mucocutaneous epithelial growth factor and indicated for treatment of oral mucositis in chemotherapy setting. Palifermin has been shown to bind to heparin *in vitro* because KGF is a heparin-binding member of the fibroblast growth factor family,[21–23] and heparin has been shown to be involved in modulating the KGF binding to its receptor.[21] The potential of palifermin binding to low molecular weight heparin was a concern because heparin is used in the indicated clinical setting. As such, its U.S. PI advises against coadministration with heparin and advises to rinse heparin-containing IV lines with saline before administering palifermin into the line. The FDA's approval letter (December 15, 2004) listed three postmarketing-required studies related to drug interactions between palifermin and heparin: an *in vivo* study in healthy subjects to evaluate the drug interaction of palifermin with heparin, an *in vitro* study to evaluate the drug interaction of palifermin with low molecular weight heparins, and an *in vivo* study in healthy subjects, contingent on the results of the *in vitro* study, to evaluate the drug interaction of palifermin with low molecular weight heparins.

Two clinical studies evaluating the interaction between palifermin and heparin were posted on the ClinicalTrials.gov website, presumably to fulfill the postmarketing requirements. The first one was a single-dose study[24] in 27 healthy subjects receiving palifermin (60 μg/kg) coadministered with and without therapeutic levels of un-fractionated heparin. The primary objective was to evaluate the effect of heparin on the PK of palifermin and the secondary objective was to evaluate the effect (activated partial thromboplastin time, aPTT) of a single dose of palifermin on unfractionated heparin pharmacodynamics (PD) based on the PD measures of $AUCaPTT_{0-6}$ and $AUCaPTT_{0-24}$. As reflected in the drug interactions section of the PI, coadministration of heparin resulted in a fivefold increase in AUC and an 80% decrease in the mean clearance rate of palifermin. This case of drug interaction is one of the early examples to challenge the general perception that drug interactions were not of concern when dealing with biologics products. Although the PD results from this same study showed no significant effect of palifermin on heparin activity with respect to aPTT, the clinical impact of the increased palifermin exposure remains to be investigated.

The second study,[25] presumably conducted to further explore the clinical relevance of the observed PK interaction for the palifermin PD effect, was designed to evaluate the impact of a continuous intravenous (IV) infusion of unfractionated heparin on the multiple-dose pharmacodynamics of palifermin. Forty-four healthy adult subjects were enrolled and randomized into three treatment groups in a ratio of 20 : 15 : 8 receiving either (1) daily palifermin dose of 40 μg/kg for three consecutive days as IV bolus injections and continuous heparin IV infusion, (2) daily palifermin dose of 40 μg/kg for three consecutive days as IV bolus injections, or (3) no treatment (a control group). The PK of palifermin and its PD response were measured, with PD defined as the ratio to baseline of epithelial cell proliferation as assessed by Ki67 staining of buccal mucosal tissue on study day 4. Consistent with the first study, the palifermin exposure was fivefold higher when coadministered with heparin. The PD response was similar with and without heparin, as the geometric mean ratio was 0.863 and the 90% confidence interval was within the predefined interval of 0.7 to 1.43, an equivalence criteria.

Similar to the G-CSF labeling, it is not recommended to administer palifermin within 24 h before or after chemotherapy. Because of the likelihood of coadministration with G-CSF, preclinical studies in murine and nonhuman primate chemotherapy/radiotherapy models explored the possibility of interaction between G-CSF and palifermin and found no evidence of any drug interaction.

14.2.4 Thrombopoiesis-Stimulating Agents

Two recombinant thrombopoiesis-stimulating agents (TSAs) are marketed: recombinant human interleukin 11 (rhIL-11) and romiplostim, a thrombopoietin (TPO) mimetic Fc-fusion protein. These recombinant products are indicated for the treatment of thrombocytopenia as their endogenous counterparts play a key role in platelet production.

Romiplostim, produced by recombinant DNA technology in *E. coli*, is a nonglycosylated Fc-peptide fusion protein (peptibody) that has a specific function to increase platelet production via binding to the TPO receptor,[26,27] despite the lack of amino acid sequence homology to the endogenous TPO. By covalent binding of the TPO mimetic peptides to the human immunoglobulin IgG1 Fc domain, this second-generation TSA has prolonged elimination half-life compared to the first-generation product, which failed to reach the market because of the immunogenicity issue.[26,28] No formal drug interaction studies were conducted in the development program, apart from the expected pharmacodynamic interaction with other medications used to treat thrombocytopenia, such as corticosteroids, danazol, azathioprine, intravenous immunoglobulin (IVIG), and anti-D immunoglobulin.

Oprelvekin, a rhIL-11 produced in *E. coli*, is a nonglycosylated protein containing 177 amino acids (MW 19 kDa). It differs from native IL-11 (178 amino acids) in lacking the amino-terminal proline residue. This alteration has not resulted in measurable differences in bioactivity either *in vitro* or *in vivo*. The endogenous IL-11 binds to IL-11 receptors and exerts pleiotropic activities, including stimulation of hemopoiesis, thrombopoiesis, and many nonhematopoietic activities, such as anti-inflammation. In terms of the potential association with the liver metabolic enzymes, IL-11 has been shown to be involved in modulating the transcriptional expression of aromatase in several tissues[29,30] and has minor effect on other hepatic metabolic enzymes, as shown in the *in vitro* hepatocyte studies.[31] Aromatase is a member of the cytochrome P450 superfamily, which aromatize androgens to form estrogens.[32] Inhibitors of aromatase are clinically used as adjuvants in breast cancer treatment.[32,33] No formal drug interaction studies have been conducted for oprelvekin. Further evaluation of drug interactions related to IL-11 modulation could become important because IL-11 produced by inflammatory cells in the intestine has been implicated in the pathophysiology of cancer, and IL-11 has been proposed as a new target for anticancer therapeutics.[34]

14.3 HORMONES

14.3.1 Human Growth Hormone

Somatropin is the active ingredient of more than 10 recombinant human growth hormone (rhGH) products approved by the FDA and at least three approved by the EMA. The amino acid sequence of somatropin is identical to that of human growth hormone of pituitary origin, which is a single-chain protein of 191 amino acids (MW 20 kDa), including four cysteine residues present as two intrachain disulfides. They are produced in *E. coli* cells or *Saccharomyces cerevisiae* or mammalian cell line by recombinant DNA technology. Somatropin is indicated for the treatment of growth hormone deficiency, growth failures or short stature due to various causes, and wasting and cachexia secondary to HIV.

Growth hormone has been shown to modulate the cytochrome P450 gene expression and activity in primary culture of human hepatocytes, animal models, and humans.[35–38] It affects the PK of small molecule drugs by which the clearance of some drugs is increased, such as antipyrine and theophylline, and for others decreased, such as amobarbital. Much research is still ongoing to elucidate the precise mechanism of the regulatory effect of growth hormone on various CYP450 enzymes.[39] Consequently the labeling of growth hormone products indicates that monitoring is advisable when it is administered in combination with medicinal products known to be metabolized by CYP450 liver enzymes, such as corticosteroids, sex steroids, anticonvulsants, and cyclosporine. The labeling also advises that patients taking insulin for diabetes mellitus should be carefully monitored during treatment with somatropin because an adjustment of the insulin dose may be required.

On the other hand, the treatment effects of growth hormone can be modulated by the coadministered drugs. Patients with coexisting adrenocorticotropic hormone (ACTH) deficiency should have their glucocorticoid replacement dose carefully adjusted to avoid an inhibitory effect on growth. For women taking oral estrogens, a higher dose of growth hormone may be required to achieve the treatment goal.

14.3.1.1 Interaction with Glucocorticoid Therapy Glucocorticoids may cause growth retardation and reduce the effect of exogenously administered rhGH in GH-deficient patients that appears to be mediated by multiple mechanisms.[40] In a study conducted with seven children who were on prednisone therapy, rhGH was given to counteract the impaired linear growth. Based on the change in growth velocity, six of seven children responded to rhGH. One patient who received high doses of prednisone (>0.5 mg/kg per day) throughout the rhGH therapy did not experience an increased growth velocity. The response to rhGH was found to be related to the prednisone dose; no increase in growth velocity at prednisone doses >0.35 mg/kg per day was noted, and the response increased with decreased prednisone dose.[41] Downregulation of the expression of GH receptors and the receptor activities, alteration of the GH binding protein, and regulation of the local IGF-1 secretion and activities are a few of the possible mechanisms reported.[40]

Conversely, treatment effects of glucocorticoids are affected by growth hormone because it inhibits 11β-hydroxysteroid dehydrogenase type 1 (11β-HSD1), an enzyme that converts inactive cortisone to active cortisol.[42] In patients with GH deficiency, 11β-HSD1 activity is enhanced, which is subsequently normalized after they receive rhGH treatment.[42,43] Hence the U.S. PI of GH products indicate that GH-deficient patients being treated with rhGH may require initiation of glucocorticoid therapy and patients treated with glucocorticoid replacement for previously diagnosed hypoadrenalism may require an increase in their maintenance doses when rhGH therapy is started.

14.3.1.2 Interaction with Amobarbital In a clinical study in children amobarbital was used as a probe to detect hepatic cytochrome P450 enzyme-mediated activity because it is eliminated almost completely through oxidation via hepatic cytochrome P450 mixed-function oxidases. Amobarbital PK after single doses at 3 to 5 mg/kg was compared before and after 6 weeks of rhGH treatment (~2 mg dosed intramuscularly 3 times a week) in four GH-deficient children. The results showed that plasma clearance of amobarbital decreased by 50% after rhGH treatment.[44]

14.3.1.3 Interaction with Antipyrine In a clinical study with GH-deficient adults, the antipyrine clearance test was used as a marker for hepatic cytochrome P450 activity. Antipyrine (10 mg/kg) was given orally before and after 3 months of rhGH therapy in which daily SC doses of rhGH were given at 0.125 IU/kg/week for 1 month followed by daily SC doses at 0.25 IU/kg/week. Median antipyrine clearance increased significantly (by 52%) and antipyrine clearance increased for all subjects except for one, who was a smoker with high baseline clearance.[45]

In another clinical study with 12 GH-deficient children, antipyrine clearance was used to detect short-term and long-term rhGH effect on enzyme activity. Antipyrine (18 mg/kg oral) was given before and after either daily rhGH dose at 1U/day for 5–7 days (short-term effect) or 3 times per week treatment of rhGH at 2U/dose for 6 weeks to 12 months (long-term effect) via intramuscular injection. In both cases, the mean clearance did not change statistically relative to baseline.[46]

14.3.1.4 Interaction with Theophylline In two clinical studies in GH-deficient children, the effect of rhGH on PK of theophylline was assessed.[47] In both studies, 4 mg/kg theophylline was administered (either orally or IV as aminophylline) before and after 2 IU rhGH given intramuscularly 3 times a week for 3–5 weeks (first study) or 3 weeks (second study). In the first study in four children, the elimination half-life of theophylline decreased by 55% relative to pretreatment (clearance data not reported). In the second study in eight children, even though the half-life did not change, the apparent clearance of theophylline increased by 28% relative to pretreatment.

14.3.1.5 Interaction with Oral Estrogens The effect of oral estrogen therapy on rhGH treatment was demonstrated in eight hypopituitary GH-deficient women with hypogonadism.[48] Daily doses of 2 mg oral estradiol valerate were administered to all subjects for 8 weeks (over 2 estrogen cycles of 4 weeks each). Starting from the second estrogen cycle, subjects received stepwise increase of rhGH daily dose from 0.5 IU in week 1, to 1 IU in week 2, to 2 IU in week 3. The level of insulin-like growth factor 1 (IGF-1) was measured at baseline before estrogen treatment, during the estrogen-alone treatment phase, and during rhGH treatment at the end of weeks 1, 2, and 3 to evaluate the dose-response relationship of rhGH treatment. Orally administered estrogen reduced the IGF-1 level to below the baseline, which was followed by a weekly stepwise increase in IGF-1 levels owing to increasing rhGH dose, indicating a dose response for rhGH treatment. A daily rhGH dose of 1 IU restored the mean IGF-1 level to approximately the baseline level. Therefore, patients who are on oral estrogen therapy will require a higher dose of rhGH for the desired treatment effect. It is important to note that the same study suggested that the antagonistic effect was dose route dependent, because estrogen administered via transdermal patch did not cause a decrease in IGF-1 levels. The differential effect of estrogen from two routes of administration was also demonstrated using other metabolic endpoints for lipid metabolism and protein metabolism. Although the mechanism of this interaction between estrogen and growth hormone is unclear, estrogen has been implicated in modulating the metabolic action of GH *in vitro* as well as in patients.[49,50]

14.3.1.6 Interaction with Thyroxin In a clinical study the effect of rhGH on biomarkers related to thyroxin (T_4) therapy (owing to central hypothyroidism) was studied in eight GH-deficient adults.[51] These patients were on stable T_4 substitution (0.1–0.2 mg daily dose) for at least 1 year and were studied over four consecutive 4-week periods receiving no rhGH therapy, followed by daily SC doses of 1 IU/m^2, then 2 IU/m^2, and finally 4 IU/m^2 rhGH. At the end of each period, the thyroid hormone levels were measured.

Serum total and free T_4 levels decreased with increasing doses of rhGH; however, the decrease was not statistically significant in both cases. Serum total and free triiodothyronine (T_3) levels, the active thyroid hormone derived from T_4 by deiodinase, increased with increasing doses of rhGH and were significantly greater than no rhGH therapy. The T_3/T_4 ratio exhibited a significant dose-dependent increase, and the reverse T_3, an inactive isomer of T_3 that binds to the thyroid hormone receptor, showed a significant dose-dependent decrease relative to no rhGH dose. In the product label for the GH product Omnitrope it is speculated that in patients receiving replacement therapy with thyroxin mild hyperthyroidism may occur and the label advises to test thyroid function after starting treatment with rhGH and after dose adjustments.

14.3.1.7 Interaction with Insulin The effect of rhGH on insulin dose-response was studied in six normal male subjects in a crossover design[52] with a 72-h washout in between two treatment occasions in which subjects were started on either 0.45% saline or GH (2 μg/kg/h) 12 h before insulin dose-response studies. During each occasion, subjects were administered insulin doses of 0.4, 1.0, 2.0, and 10.0 mU/kg/min sequentially for 2 h, and each insulin dose was given simultaneously with the infusion of glucose as 50% solution to maintain arterial glucose concentrations at 95 mg/dL. The glucose production and utilization were measured before the first insulin infusion and during the last 40 min of each subsequent insulin infusion. The effect of rhGH on insulin receptor function was also determined before the first insulin infusion by measuring the proportion of monocytes in the mononuclear cell preparation and the insulin binding to the monocytes.

In the presence of rhGH, both dose-response curves of insulin vs. glucose production and insulin vs. glucose utilization shifted to the right—that is, the half-maximal response was significantly increased in the presence of rhGH without affecting the maximum response—suggesting that rhGH increases insulin resistance. Treatment with rhGH did not result in differences in the binding of insulin to the receptors on monocytes. Therefore, the insulin resistance effect was postulated to be the result of a postreceptor defect.

14.3.2 Pegvisomant

Pegvisomant (Somavert) is an analog of GH that has been structurally altered to act as a GH receptor antagonist and is indicated for treatment of acromegaly. Pegvisomant is produced in *E. coli* and contains 191 amino acid residues to which several PEG polymers are covalently bound (mostly 4–6 PEG/protein molecule). The predominant molecular weights of pegvisomant are approximately 42, 47, and 52 kDa, with the protein portion being 22 kDa and the PEG being ~5 kDa per unit.

Per the U.S. PI, the treatment regimen for blood sugar control for acromegalic patients with diabetes mellitus who are being treated with insulin and/or oral hypoglycemic agents may require dose reductions after the initiation of therapy with pegvisomant. This is consistent with the report that growth hormone can increase the insulin resistance, as stated in the previous sections.

The effect of growth hormone blockade on insulin sensitivity was studied by Williams *et al.* in seven young adults with type 1 diabetes who were on a combination of short- and long-acting insulins two to four times per day.[53] In this crossover study subjects were given two treatments, either 5 or 10 mg daily SC pegvisomant, over a 3-week period after a run-in period and a 3-week washout in between treatments. At baseline and after each treatment, variable rate insulin infusion was administered overnight to maintain blood glucose at 5 mmol/L. In the morning, this was followed by a two-step hyperinsulinemic euglycemic clamp, using [6,6 $^2H_2$]

glucose and $^2H_5$ glycerol, to measure glucose and glycerol turnover. The investigators showed that overnight insulin requirements significantly decreased for both treatments when compared to baseline. Another study suggested that the decrease in overnight free fatty acid led to an improvement in insulin sensitivity.[54]

The second drug interaction in the U.S. PI stated that in clinical studies patients on opioids often needed higher serum pegvisomant concentrations to achieve appropriate IGF-I suppression compared to patients not receiving opioids. In our literature review, we did not locate any clinical reports of formal drug interaction studies; however, we came across reports of complex interactions between the opioid system and somatostatin, which appears to be an active subject of current research. Morphine and opioid antagonists have been shown to regulate cell growth *in vitro* by interacting with the somatostatin receptors;[55–57] opioids with proliferative effect and opioid antagonists with antiproliferative effect. Analgesic effect associated with octreotide, a peptide analog of somatostatin, administration was reported for a patients treated for acromegaly.[58] The involvement of opioids in regulating the functions and/or secretion of growth hormone has been suggested in clinical setting.[59,60] These research reports were consistent with the advice in the product labeling.

14.3.3 Glucagon

Multiple recombinant glucagon products are available commercially in the United States and one is available in the EU. Recombinant glucagon, a single-chain polypeptide hormone with 29 amino acids (MW 3.5 kDa), is synthesized in *E. coli* and is identical to human glucagon produced by the pancreas. Indicated for the treatment of hypoglycemia and also used in diagnostic testing of the stomach and other digestive organs, it increases blood glucose and relaxes smooth muscle of the gastrointestinal tract.

The labeling of glucagon products refer to interactions with various drugs, and most of them are pharmacology-based: Potential interaction with β-blockers might lead to a greater increase in both pulse and blood pressure, which may require therapy in patients with pheochromocytoma or coronary artery disease. When used with indomethacin, glucagon may lose its ability to raise blood glucose or may even produce hypoglycemia. Coadministration with an anticholinergic drug is not recommended owing to increased gastrointestinal side effects. Glucagon may increase the anticoagulant effect of warfarin. Insulin reacts antagonistically toward glucagon. Summarized in the following sections by product are the results of our literature search, which are consistent with the product labeling.

14.3.3.1 Interaction with Indomethacin and Prostaglandins Some prostaglandins have been demonstrated to evoke the release of glucagon from the α-cells in the pancreas.[61] On the other hand, inhibition of the synthesis of prostaglandins by cyclooxygenase inhibitors, such as indomethacin and acetyl salicylic acids, can inhibit the release of glucagon. The effects of indomethacin on glucose levels and on insulin levels were demonstrated in anesthetized male Wistar rats.[62]

The glucose and insulin levels were monitored over 2 h in rats that were administered (1) glucagon only as IV infusion for 120 min, (2) a pretreatment of indomethacin at 10 mg/kg administered intraperitoneally at 15 and 3 h before the infusions of glucagon, or (3) a pretreatment of indomethacin followed by infusions of saline. The basal glucose levels were similar across the three groups. Indomethacin alone does not affect glucose levels or insulin levels over 2 h. In response to glucagon treatment alone, the glucose level and the insulin level increased and remained elevated over 90 min. On the other hand, glucagon treatment in rats pretreated with indomethacin raised the glucose level and the insulin level above baseline for a short time and then declined to below the basal level after 30 min; peak glucose and insulin levels were lower than in rats not administered indomethacin.

In another set of experiments assessing the effect of indomethacin on hepatic glucose output, 2-^{3}H-glucose was mixed with normal glucose and infused intravenously for 210 min. At 90 min, which is after establishing steady-state, glucagon was intravenously infused for 120 min either alone or with indomethacin. Results showed that indomethacin reduced both glucose production and glucose utilization.

The same research group conducted a study in conscious female dogs with similar design to evaluate the effect of indomethacin or acetyl salicylic acid on glucagon's ability to produce glucose. Pretreatment with indomethacin and salicylate did not alter basal glucose levels but reduced the effect of glucagon in terms of the increase in glucose levels and the glucose production and utilization.[63]

14.3.3.2 Interaction with Warfarin During 9 months of a study in Massachusetts General Hospital, 469 patients on warfarin were monitored to identify drugs that modify the hypoprothrombinemic effect of warfarin.[64] Twenty-four subjects were on glucagon concomitant therapy. Of these, 13 subjects were on glucagon for at least 2 days with a total dose >50 mg of glucagon. In 8 of these, the hypoprothrombinemic effect of warfarin was enhanced, indicating that they became markedly more sensitive to warfarin during glucagon treatment. They had much prolonged prothrombin time and experienced bleeding while on both warfarin and glucagon. The authors postulated that glucagon may potentiate the anticoagulant action of warfarin by acting synergistically to depress hepatic synthesis of vitamin-K-sensitive clotting proteins or by increasing the affinity of warfarin for its receptor sites.

14.3.3.3 Interaction with Aminophylline Glucagon has myocardial inotropic action in experimental animals and in men. In a study conducted in anesthetized dogs the effect of aminophylline and glucagon when coadministered on coronary and systemic circulation was compared to either agent given alone.[65] Comparisons were made on baseline and posttreatment values of multiple coronary and systemic hemodynamic endpoints, including heart rate, coronary sinus blood flow, cardiac output, and pressures. Treatment with aminophylline alone or glucagon alone was delivered as a constant rate infusion at two different infusion rates, one being half of the other. The combination treatment with aminophylline and glucagon were infused at the lower of two infusion rates for each agent. In total, 21 repeated runs of the treatment procedures were made in 10 dogs, and summary statistics were determined for various measurements. The average doses used were 2.7 and 5.5 mg/min for aminophylline and 21 and 42 μg/min for glucagon.

When compared to the treatment effect of glucagon alone at the same dose rate, aminophylline enhanced the cardiovascular effects of glucagon. Furthermore, the effects of the glucagon and aminophylline combined treatment were greater than the glucagon treatment effect as a single agent at twice the dose. This interaction was thought to be the result of an accumulation of cyclic 3′,5′-AMP because glucagon activates adenyl cyclase and aminophylline inhibits 3′,5′ nucleotide phosphodiesterase.

14.4 CONCLUSIONS

Our literature research had limited success in finding drug interactions that involve the discussed growth factor products, whereas information is more readily available for the covered hormone products. We speculate that the target receptor specificity of these growth factors in combination with the specificity of their postreceptor functions may play a role in the limited findings of drug interactions with these products. The observed interactions are mostly consequences of pharmacological effect of either the perpetrator drug or the victim drug. On the other hand, the hormone products we discussed generally have much broader involvement in biological/physiological functions. They can affect not only the cytochrome P450 enzyme system but also other metabolic functions and enzyme systems in addition to pharmacological interactions mediated at the receptor levels. Furthermore, the complexity of the network of signaling pathways for hormones and the intricate interconnectivities with other complex biochemical signaling processes present opportunities for pharmacodynamic interactions. This can explain the higher volume of available drug interaction information in the literature for hormones as compared to growth factors.

Palifermin is the only case of a PK drug interaction we found in which heparin causes an increase in the palifermin systemic exposure; however, the clinical significance of this interaction is yet to be elucidated. Several cases of PK interactions were reported for recombinant hormone products. Many more cases of pharmacodynamic interactions were observed for both recombinant growth factors and hormones. These reported interactions have clear impact on the clinical use of the products in that they are included in the product labeling whether they are caused by PK alteration or PD modulation.

With the advent of the advanced tools for research at the molecular and cellular level we also expect that the empirically identified drug interactions will gradually find the mechanistic explanations. For instance, lithium's effect on neutrophil count was an empirical finding that was called out in the filgrastim label as a potential interaction. This was later mechanistically attributed to its interaction with the chemokine receptor CXCR4 a few years after the initial approval of filgrastim. As growth factor product pipelines in pharmaceutical development rapidly expand, it is very likely that drug interactions may become more common for new growth factors that reach the market. We expect that new reports of drug interaction on currently marketed products will be forthcoming as more and more patients take multiple medications given the increasing average age of the population. In summary, our research indicates that investigations on drug interactions for growth factors and hormone products are important for clinical use of these medications and that this is an area for continued investment by scientists in academia, industry, and regulatory agencies.

REFERENCES

1. Aggarwal S. What's fueling the biotech engine—2009–2010. *Nat Biotechnol* **28**, 1165–71 (2010).
2. American Society of Clinical Oncology. Erythropoiesis-stimulating agents: continued challenges. *J Oncol Pract* **3**, 248–253 (2007).
3. Macdougall IC, Padhi D, Jang G. Pharmacology of darbepoetin alfa. *Nephrol Dial Transplant* **22** (Suppl 4), iv2–iv9 (2007).
4. Molineux G. Pegfilgrastim: using pegylation technology to improve neutropenia support in cancer patients. *Anticancer Drugs* **14**, 259–64 (2003).
5. Renwick W, Pettengell R, Green M. Use of filgrastim and pegfilgrastim to support delivery of chemotherapy: twenty years of clinical experience. *BioDrugs* **23**, 175–86 (2009).
6. Molineux G. The design and development of pegfilgrastim (PEG-rmetHuG-CSF, Neulasta). *Cur Pharmaceut Design* **10**, 1235–44 (2004).
7. Macdougall IC. CERA (Continuous Erythropoietin Receptor Activator): a new erythropoiesis-stimulating agent for the treatment of anemia. *Curr Hematol Rep* **4**, 436–40 (2005).
8. Topf JM. CERA: third-generation erythropoiesis-stimulating agent. *Expert Opin Pharmacother* **9**, 839–49 (2008).
9. Seige K. [Glucagon, its discovery and description and the work of Max Burger.] *Z Gesamte Inn Med* **41**, 568–71 (1986).
10. Somatropin. Human growth hormone history. Available at www.somatropin.net/hgh-history.htm.
11. Lombardero M, Kovacs K, Scheithauer BW. Erythropoietin: a hormone with multiple functions. *Pathobiology* **78**, 41–53 (2011).
12. Chateauvieux S, Grigorakaki C, Morceau F, et al. Erythropoietin, erythropoiesis and beyond. *Bioch Pharmacol* **81**, 1291–303 (2011).
13. Brown JN, Kemp DW, Brice KR. Class effect of erythropoietin therapy on hemoglobin A (1c) in a patient with diabetes mellitus and chronic kidney disease not undergoing hemodialysis. *Pharmacotherapy* **29**, 468–72 (2009).
14. Okada T, Nakao T, Matsumoto H, et al. Influence of erythropoietin-stimulating agent treatment on glycated hemoglobin values in diabetic patients with chronic kidney disease. *Clin Exper Nephrol* **15**, 316–17 (2011).
15. Joy MS, Cefalu WT, Hogan SL, Nachman PH. Long-term glycemic control measurements in diabetic patients receiving hemodialysis. *Am J Kidney Dis* **39**, 297–307 (2002).

16. Inaba M, Okuno S, Kumeda Y, et al. Glycated albumin is a better glycemic indicator than glycated hemoglobin values in hemodialysis patients with diabetes: effect of anemia and erythropoietin injection. *J Am Soc Nephrol* **18**, 896–903 (2007).
17. Zechariah A, ElAli A, Hermann DM. Combination of tissue-plasminogen activator with erythropoietin induces blood-brain barrier permeability, extracellular matrix disaggregation, and DNA fragmentation after focal cerebral ischemia in mice. *Stroke* **41**, 1008–12 (2010).
18. Kim HK, Kim JE, Chung J, P et al. Lithium down-regulates the expression of CXCR4 in human neutrophils. *J Trace Elements Med Biol* **21**, 204–09 (2007).
19. Hager ED, Dziambor H, Hohmann D, et al. Effects of lithium on thrombopoiesis in patients with low platelet cell counts following chemotherapy or radiotherapy. *Biolo Trace Element Res* **83**, 139–48 (2001).
20. Kreuser ED, Wadler S, Thiel E. Biochemical modulation of cytotoxic drugs by cytokines: molecular mechanisms in experimental oncology. *Recent Results Cancer Res* **139**, 371–82 (1995).
21. Berman B, Ostrovsky O, Shlissel M, et al. Similarities and differences between the effects of heparin and glypican-1 on the bioactivity of acidic fibroblast growth factor and the keratinocyte growth factor. *J Biol Chem* **274**, 36132–38 (1999).
22. MacDonald KP, Hill GR. Keratinocyte growth factor (KGF) in hematology and oncology. *Curr Pharm Des* **8**, 395–403 (2002).
23. Rubin JS, Osada H, Finch PW, et al. Purification and characterization of a newly identified growth factor specific for epithelial cells. *Proc Nat Acad Sci U S A* **86**, 802–06 (1989).
24. Clinical Development, Biovitrum AB., An open-label, randomized, 2-part, parallel design study to characterize the effect of heparin on palifermin pharmacokinetics and the effect of palifermin on heparin pharmacodynamics in healthy subjects. NCT00361348, February 27, 2009. Available at http://clinicaltrials.gov/ct2/show/NCT00361348?term=palifermin+heparin&rank=2.
25. Swedish Orphan Biovitrum. An open-label, randomized, parallel-design study to characterize the effect of heparin on palifermin activity in healthy adult subjects. NCT01163097, November 23, 2012. Available at http://clinicaltrials.gov/ct2/show/study/NCT01163097?term=palifermin+heparin&rank=1.
26. Wang YM, Krzyzanski W, Doshi S, et al. Pharmacodynamics-mediated drug disposition (PDMDD) and precursor pool lifespan model for single dose of romiplostim in healthy subjects. *AAPS J* **12**, 729–40 (2010).
27. Wei P. Thrombopoietin factors. *Cancer Treatment Res* **157**, 75–93 (2011).
28. Neumann TA, Foote M. Megakaryocyte growth and development factor (MGDF): an Mpl ligand and cytokine that regulates thrombopoiesis. *Cytokines Cell Molec Ther* **6**, 47–56 (2000).
29. Simpson ER, Michael MD, Agarwal VR, H. et al. Cytochromes P450 11: expression of the CYP19 (aromatase) gene: an unusual case of alternative promoter usage. *FASEB J* **11**, 29–36 (1997).
30. Zhao Y, Agarwal VR, Mendelson CR, Simpson ER. Transcriptional regulation of CYP19 gene (aromatase) expression in adipose stromal cells in primary culture. *J Steroid Biochem Molec Biol* **61**, 203–10 (1997).
31. Guillen MI, Donato MT, Jover R, et al. Oncostatin M down-regulates basal and induced cytochromes P450 in human hepatocytes. *J Pharmacol Exper Ther* **285**, 127–34 (1998).

32. Brodie A. Aromatase and its inhibitors—an overview. *J Steroid Biochem Mol Biol* **40**, 255–261 (1991).
33. Bickenbach KA, Jaskowiak N. Aromatase inhibitors: an overview for surgeons. *J Am Coll Surgeons* **203**, 376–89 (2006).
34. Putoczki T, Ernst M. More than a sidekick: the IL-6 family cytokine IL-11 links inflammation to cancer. *J Leukocyte Biol* **88**, 1109–17 (2010).
35. Liddle C, Goodwin BJ, George J, *et al.* Separate and interactive regulation of cytochrome P450 3A4 by triiodothyronine, dexamethasone, and growth hormone in cultured hepatocytes. *J Clin Endocrinol Metab* **83**, 2411–16 (1998).
36. Jarukamjorn K, Sakuma T, Jaruchotikamol A, et al. Modified expression of cytochrome P450 mRNAs by growth hormone in mouse liver. *Toxicology* **219**, 97–105 (2006).
37. Thangavel C, Dworakowski W, Shapiro BH. Inducibility of male-specific isoforms of cytochrome p450 by sex-dependent growth hormone profiles in hepatocyte cultures from male but not female rats. *Drug Metab Dispos* **34**, 410–19 (2006).
38. Jurgens G, Lange KH, Reuther LO, et al. Effect of growth hormone on hepatic cytochrome P450 activity in healthy elderly men. *Clin Pharmacol Ther* **71**, 162–68 (2002).
39. Kim SK, Novak RF. The role of intracellular signaling in insulin-mediated regulation of drug metabolizing enzyme gene and protein expression. *Pharmacol Ther* **113**, 88–120 (2007).
40. Mushtaq T, Ahmed SF. The impact of corticosteroids on growth and bone health. *Arch Dis Childhood* **87**, 93–96 (2002).
41. Rivkees SA, Danon M, Herrin J. Prednisone dose limitation of growth hormone treatment of steroid-induced growth failure. *J Pediatr* **125**, 322–25 (1994).
42. Moore JS, Monson JP, Kaltsas G, et al. Modulation of 11beta-hydroxysteroid dehydrogenase isozymes by growth hormone and insulin-like growth factor: in vivo and in vitro studies. *J Clin Endocrinol Metab* **84**, 4172–77 (1999).
43. Swords FM, Carroll PV, Kisalu J, et al. The effects of growth hormone deficiency and replacement on glucocorticoid exposure in hypopituitary patients on cortisone acetate and hydrocortisone replacement. *Clin Endocrinol* **59**, 613–20 (2003).
44. Redmond GP, Bell JJ, Perel JM. Effect of human growth hormone on amobarbital metabolism in children. *Clin Pharmacol Ther* **24**, 213–18 (1978).
45. Cheung NW, Liddle C, Coverdale S, et al. Growth hormone treatment increases cytochrome P450-mediated antipyrine clearance in man. *J Clin Endocrinol Metab* **81**, 1999–2001 (1996).
46. Rifkind AB, Saenger P, Levine LS., et al. Effects of growth hormone on antipyrine kinetics in children. *Clin Pharmacol Ther* **30**, 127–32 (1981).
47. Redmond GP. Regulation of theophylline metabolism by growth hormone. *Semin Perinatol* **5**, 374–77 (1981).
48. Wolthers T, Hoffman D M, Nugent AG, et al. Oral estrogen antagonizes the metabolic actions of growth hormone in growth hormone-deficient women. *Am J Physiol Endocrinol Metabol* **281**, E1191–96 (2001).
49. Lopez-Guerra A, Chirino R, Navarro D, et al. Estrogen antagonism on T3 and growth hormone control of the liver microsomal low-affinity glucocorticoid binding site (LAGS). *J Steroid Biochem Molec Biol* **63**, 219–28 (1997).

50. Wolthers T, Hoffman DM, Nugent AG, et al. Oral estrogen antagonizes the metabolic actions of growth hormone in growth hormone-deficient women. *Am J Physiol Endocrinol Metabol* **281**, E1191–96 (2001).

51. Jorgensen JO, Moller J, Laursen T, et al. Growth hormone administration stimulates energy expenditure and extrathyroidal conversion of thyroxine to triiodothyronine in a dose-dependent manner and suppresses circadian thyrotrophin levels: studies in GH-deficient adults. *Clin Endocrinol* **41**, 609–14 (1994).

52. Rizza R A, Mandarino LJ, Gerich J E. Effects of growth hormone on insulin action in man. Mechanisms of insulin resistance, impaired suppression of glucose production, and impaired stimulation of glucose utilization. *Diabetes* **31**, 663–69 (1982).

53. Williams RM, Amin R, Shojaee-Moradie F, et al. The effects of a specific growth hormone antagonist on overnight insulin requirements and insulin sensitivity in young adults with Type 1 diabetes mellitus. *Diabetologia* **46**, 1203–10 (2003).

54. Gormsen LC, Jessen N, Gjedsted J, Gjedde, et al. Dose-response effects of free fatty acids on glucose and lipid metabolism during somatostatin blockade of growth hormone and insulin in humans. *J Clin Endocrinol Metab* **92**, 1834–42 (2007).

55. Hatzoglou A, Kampa M, Castanas E. Opioid-somatostatin interactions in regulating cancer cell growth. *Frontiers Biosci* **10**, 244–56 (2005).

56. Notas G, Kampa M, Nifli AP, et al. The inhibitory effect of opioids on HepG2 cells is mediated via interaction with somatostatin receptors. *Eur JPharmacol* **555**, 1–7 (2007).

57. Hatzoglou A, Ouafik L, Bakogeorgou E, et al. Morphine cross-reacts with somatostatin receptor SSTR2 in the T47D human breast cancer cell line and decreases cell growth. *Cancer Res* **55**, 5632–36 (1995).

58. Pascual J, Freijanes J, Berciano J, Pesquera C. Analgesic effect of octreotide in headache associated with acromegaly is not mediated by opioid mechanisms. Case report. *Pain* **47**, 341–44 (1991).

59. De Marinis L, Mancini A, Valle D, et al. Physiological role of the opioid-cholinergic interaction in growth hormone neuroregulation: effect of sex and food intake. *Metabolism* **46**, 740–44 (1997).

60. Pende A, Musso NR, Montaldi ML, et al. Interaction between morphine, an opioid agonist, and clonidine, an alpha-adrenergic agonist, on the regulation of anterior pituitary hormone secretion in normal male subjects. *Biomed Pharmacother* **41**, 243–47 (1987).

61. Giugliano D, Torella R, D'Onofrio F. Prostaglandins and the alpha-cell. *Prostagland Med* **6**, 283–97 (1981).

62. Ganguli S, Sperling MA, Frame C, Christensen R. Inhibition of glucagon-induced hepatic glucose production by indomethacin. *Am J Physiol* **236**, E358–65 (1979).

63. Miller JD, Ganguli S, Sperling MA. Prostaglandin synthesis inhibitors impair hepatic glucose production in response to glucagon and epinephrine stimulation. *Diabetes* **32**, 439–44 (1983).

64. Koch-Weser J. Potentiation by glucagon of the hypoprothrombinemic action of warfarin. *Ann Intern Med* **72**, 331–35 (1970).

65. Afonso S, Hansing CE, Ansfield TJ, et al. Enhancement of cardiovascular effects of glucagon by aminophylline. *Cardiovasc Res* **6**, 235–39 (1972).

CHAPTER 15

DRUG–DRUG INTERACTIONS FOR NUCLEIC ACID-BASED DERIVATIVES

JIAN WANG and HONG ZHAO

15.1 INTRODUCTION

In the last decade, oligonucleotide-based therapeutics have been under active development for treatment of diseases such as cancers, genetic disorders, and or infections. Fomivirsen was the first antisense oligonucleotide drug approved by the FDA in 1998 for the treatment of cytomegalovirus retinitis in AIDS patients who have failed therapy with other anticytomegalovirus drugs or who are intolerant or have a contraindication to other therapies. Six years later, pegaptanib (pegylated aptamer), a selective vascular endothelial growth factor (VEGF) antagonist was approved for the treatment of neovascular (wet) age-related macular degeneration (AMD). Both drugs are administered via intravitreal injection.

Antisense oligonucleotides (ASOs) are perhaps the most direct therapeutic strategy to target RNA. Currently ASOs make up the major investigational new drug (IND) submissions for oligonucleotide drug development. These ASOs are designed to bind to the target RNA by well-characterized Watson-Crick base pairing, and once bound to the target RNA, modulate its function through a variety of postbinding events. In this chapter, the primary information used for summarizing drug–drug interactions (DDIs) of oligonucleotides comes from the first- and second-generation ASOs because the currently published DDI studies are mostly focused on these classes of oligonucleotides. When possible, information on other classes of oligonucleotides is presented in the context of these base classes. Caution should be taken in making generalized conclusions based on the limited number of reported studies due to the fact that most of the oligonucleotides are still in the development stage. More information needs to be accumulated to better define the DDI potential between oligonucleotides and other drugs.

Drug–Drug Interactions for Therapeutic Biologics, First Edition.
Edited by Honghui Zhou and Bernd Meibohm.

This chapter will review the results from both preclinical and clinical DDI studies for the first- and second-generation ASOs. Phosphorothioate ASOs involve substituting sulfur for one of the two nonbridging oxygens in each internucleotide linkage, which are the most extensively studied oligonucleotide drugs and are referred to as first-generation ASOs. MOE-modified ASOs are the products with 2″-*O*-(2-methoxyethyl) (2″-*O*-MOE) ribose sugar modifications for the 4–5 nucleotides at each terminus of ASOs. This modification increases the resistance of ASOs to exonucleases and increases the binding affinity of ASOs to mRNA. The chimeric structure (oligonucleotide core surrounded by MOE-modified nucleotides) as second-generation ASOs preserves their capacity to serve as ribonuclease (RNase) H substrates. The clinical pharmacokinetics (PK) of the first-generation ASOs (single-stranded phosphorothioate modified DNA) and the second-generation ASOs (single-stranded phosphorothioate modified chimeric with partial 2″-*O*-MOE ribose sugar modifications) have been well characterized.

There are very limited reports available for alternative backbone modified oligonucleotides including the chemical classes of peptide nucleic acids (PNAs), locked nucleic acids (LNAs) and morpholino (neutral backbone) nucleic acids (MNAs). The small interfering RNA (siRNA), a double-stranded RNA motif containing both sense and antisense RNA strands, precludes its practical systemic delivery without chemical modification, conjugation or special formulation.[1]

15.2 CLINICAL PHARMACOKINETICS

The absorption, distribution, metabolism, and excretion of ASOs have been characterized in clinical trials. Their PK properties largely depend on their physical and chemical properties.

After parenteral administration, the first-generation phosphorothioate oligonucleotides are highly bound to plasma proteins through their hydrophilic sites. The protein binding is >90% for ASOs and >60% for siRNA. The binding of oligonucleotides to plasma proteins limits these compounds from glomerular filtration and urinary excretion. As the degree of protein binding decreases, the amount of free or unbound oligonucleotides increases, leading to an increase in urinary excretion. Oligonucleotides distribute rapidly from blood to tissues with a distribution half-life less than 2 h. They highly accumulate in tissues such as kidney, liver, spleen, lymph nodes, adipocytes, and bone marrow. Oligonucleotides do not cross the blood–brain barrier and poorly distribute to muscle, heart, and lung. Oligonucleotides are metabolized by nucleases ubiquitously expressed in tissue cells. The elimination half-life of oligonucleotides reflects the metabolism in tissues, the equilibration of drugs and metabolites between tissues and blood, and excretion by the kidneys.[2]

Chemical modifications of oligonucleotides result in marked difference in their PK profiles.[3] The first-generation ASOs modify the phosphodiester backbone of DNA to a phosphorothioate backbone, which enhances their hydrophobicity and protein binding and increases their nuclease resistance. The 2″-*O*-MOE modification at the 5″ and 3″ termini for the second-generation ASOs blocks the exonuclease

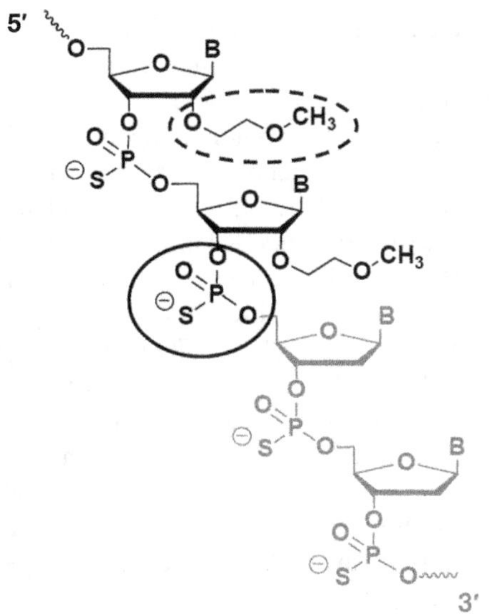

FIGURE 15-1 Chemical modifications. Solid circle: phosphorothioate modification; dotted circle: 2′-methoxyethyl modification

cleavage and therefore they are primarily cleared by slow endonuclease digestion and ultimate excretion of parent drug and metabolites in urine. Figure 15-1 illustrates modification sites of oligonucleotides to produce the first- and second-generation ASOs.

While the terminal elimination half-lives for the first-generation ASOs are <40 h, the second-generation ASOs have longer half-lives up to a month, which allow for a less frequent (once weekly or longer interval) dosing regimen as compared to the first-generation ASOs. Many of the advantages associated with moving from the first-generation to second-generation ASOs are directly related to their improved PK profiles.[1]

The oligonucleotides that lack charge and/or have low binding to plasma proteins are prone to rapid clearance from plasma, with substantially high excretion in urine in the first day after administration, resulting in less overall tissue accumulation and ultimate low target bioavailability. This phenomenon is seen in the classes of PNAs, morpholinos, and unmodified siRNA. The PK properties of unmodified RNAi oligonucleotides preclude systemic delivery. Therefore chemical conjugation or formation approaches, such as pegylation or lipoidal delivery systems are often needed to improve the RNAi stability and distribution into target tissues.

15.3 DRUG–DRUG INTERACTIONS

15.3.1 *In Vitro* Studies

15.3.1.1 DDI via Protein Binding Replacement Oligonucleotides are highly bound to plasma proteins, which protects them from urinary excretion and subsequently increases their plasma exposure. Albumin has been identified as one of the major plasma proteins with high binding capacity to oligonucleotides. The binding of oligonucleotides to plasma proteins is relatively weak with low dissociation constant of 5–20 μmol/L, and the binding sites for this hydrophilic drug class differ from the binding sites for hydrophobic small molecular drugs. Therefore, it is speculated that the potential for PK interactions mediated by protein binding is expected to be low at clinically relevant concentrations. Several *in vitro* and clinical studies have been reported in the literature regarding the protein binding displacement for oligonucleotide-based drugs. Three examples are provided.

ISIS 113715 is a 20-mer first-generation ASO that is complementary to the protein tyrosine phosphatase 1B (PTP-1B) messenger RNA and subsequently reduces translation of the PTP-1B protein, a negative regulator of insulin receptor. ISIS 113715 is currently being studied in early Phase II clinical trials to determine its ability to improve or restore insulin receptor sensitivity in patients with type 2 diabetes mellitus. *In vitro* protein binding displacement studies in human plasma samples showed minor effects on rosiglitazone and no effect on glipizide when co-incubated with ISIS 113715.[4]

ISIS 301012 (mipomersen) is a 20-nucleotide second-generation ASO targeting human ApoB mRNA. Its protein binding is >95% at clinically relevant concentrations. An *in vitro* protein binding interaction study with atorvastatin and simvastatin (both with >95% protein binding) showed no displacement between mipomersen sodium and atorvastatin or simvastatin at clinically relevant plasma concentrations.[5,6]

ISIS 2302, a first-generation ASO that targets human intercellular adhesion molecule-1 (ICAM-1) mRNA, is currently in clinical development for treatment of ulcerative colitis.[7] *In vitro* drug displacement studies with ISIS 2302 and other highly protein-bound drugs, including warfarin, salicylic acid, valproic acid, tolbutamide, acetylsalicylic acid, ibuprofen, and prednisone, were conducted. The results suggest that coadministration of ISIS 2302 with these highly protein-bound drugs did not displace one another at supraclinical plasma concentrations. Based on these findings, it is not anticipated to see PK interactions in the clinical settings as a result of the displacement of plasma protein binding.

In summary, *in vitro* data from protein binding displacement studies suggest that the potential for PK interactions caused by displacement of plasma protein binding is low when ASOs are administrated concomitantly with other highly protein bound drugs. The analyses by Benet et al. have shown that changes in plasma protein binding usually do not influence the clinical exposure of a patient to a drug.[8] As a consequence, no adjustments in dosing regimens will be necessary except in rare cases of a drug with a high extraction ratio and narrow therapeutic index that is given parenterally or, even rarer, a drug with a narrow therapeutic index that is given orally and has a very rapid pharmacokinetic-pharmacodynamic equilibration time.[8]

15.3.1.2 DDI via Metabolic Enzymes Oligonucleotides are mainly metabolized by nucleases and are unlikely substrates for the cytochrome P450 (CYP) mediated oxidative metabolism. Therefore, the PK of oligonucleotides are not expected to be affected by concomitant use of drugs that are CYP inducers or inhibitors. However, the potential for oligonucleotides to act as inhibitors or inducers of CYP enzymes cannot be precluded. *In vitro* studies using human microsomes have shown that several oligonucleotide drugs inhibited multiple CYP enzymes, including CYP3A4, CYP1A2, CYP2D6, CYP2E1, CYP2B6, CYP2A6, and CYP2C8 (unpublished data).

Several studies with negative results are reported in literature. For example, the ability of mipomersen to inhibit the major CYP enzymes—namely CYP1A2, CYP2C9, CYP2C19, CYP2D6, and CYP3A4—in cryo-preserved human hepatocytes was evaluated.[6] The results of this *in vitro* study revealed $<10\%$ inhibition in the activity of all evaluated enzymes at concentrations up to 800 μg/mL (~100 μmol/L), and dose-dependent inhibition was not observed. This study suggested mipomersen is unlikely to be an inhibitor of CYP1A2, CYP2C9, CYP2C19, CYP2D6, and CYP3A4.

15.3.2 DDI Studies in Animals

G3139 is a first-generation ASO targeting the open reading frame of oncogene *Bcl-2* mRNA. The PK interaction between G3139 and anthracycline anticancer drug doxorubixin was studied in severely compromised immunodeficient (SCID) mice model bearing human breast cancer xenografts.[9,10] When doxorubixin was coadministrated intravenously with intraperitoneal (IP) G3139, elevated G3139 peak concentration (C_{max}) was observed, with a fourfold increase in the absorption rate of G3139 from the IP cavity into plasma in the presence of doxorubixin. The systemic exposure (AUC) was only slightly increased.[10]

It is interesting that examination of the exposure of doxorubixin after coadministration with G3139 revealed marked elevation of tumor doxorubixin levels as compared to those after doxorubixin administered alone. Although the plasma exposure of doxorubixin in mice treated with G3139 was minimally reduced compared with no G3139 treatment, tumor exposure to doxorubixin increased twofold to threefold, indicating possible oligonucleotide-drug interactions at the tumor site. The mechanism behind the observed elevation of tumor tissue distribution is not fully understood. Findings from histology slides suggest that it could be through enhancing drug permeability of tumor tissue treated with G3139, thereby increased tumor uptake of doxorubixin-G3139 complexes that could form in the circulation, and increased binding of doxorubixin to tumor cells arising from G3139-induced DNA lesions caused by apoptosis activation.[10]

Similar findings have been reported by Wang et al. for an antisense anti-MDM2 mixed-backbone oligonucleotide, which is a second-generation ASO.[11] There were increased tissue levels of irinotecan and its active metabolite SN-38 in nude xenograft mice models when the two drugs were given in combination.[11]

15.3.3 DDI Studies in Humans

Two ASOs, fomivirsen and pegaptanib have been approved in the United States in 1998 and 2004, respectively. Two early clinical studies conducted for pegaptanib alone and in combination with photodynamic therapy (PDT) or verteporfin showed no apparent difference in the plasma PK of pegaptanib.[12] However, no dedicated DDI studies have been conducted in humans between either of these two ASOs and other drugs.

A number of oligonucleotide-based drugs are currently on clinical trials. For the drugs that are in the development stage, a number of clinical DDI studies have been reported in the literature. An overview of the effects of oligonucleotides on the PK of coadministrated drugs is presented in Table 15-1. The effects of coadministrated small molecular drugs on the PK of oligonucleotides are summarized in Table 15-2. Some of these DDI studies and their study designs tailored based on the PK characteristics of individual drugs are further discussed in the following section. Among them, there are a few formal or dedicated DDI studies. The approach of a cross-study comparison was also used in several studies. A literature search found that the published DDI studies conducted in humans were to assess the potential interactions between first-generation or second-generation ASOs and other chemical drugs that are either in combination therapies with these ASOs, or are likely to be coadministrated to patients.

15.3.3.1 Mipomersen (ISIS 301012) Mipomersen is being developed as an adjunct to maximally tolerated lipid-lowering therapies in patients with familial hypercholesteromia who have severely elevated low-density lipoprotein cholesterol (LDL-C) and are at high risk for coronary heart disease. A dedicated DDI study was conducted to characterize the PK interactions between mipomersen and lipid-lowering agents, including simvastatin and ezetimibe.[6] Both drugs and their metabolites are highly protein bound ($> 90\%$). In addition, simvastatin is a substrate of CYP3A4 and CYP2D6, and ezetimibe is primarily metabolized via glucuronide conjugation with subsequent biliary and renal excretion. In this study, 10 healthy subjects per cohort received an oral dose of 40 mg simvastatin or 10 mg ezetimibe before and after four 2-h intravenous doses of 200 mg mipomersen sodium over an 8-day period. For cohort A, mipomersen sodium was given on days 4, 6, 8, and 11. Mipomersen PK profiles were assessed after the dose on day 8 (mipomersen alone) and on day 11 (in combination with simvastatin). For cohort B, mipomersen sodium was given on days 8, 10, 12, and 15. Mipomersen PK profiles were assessed after the dose on day 12 (mipomersen alone) and on day 15 (in combination with ezetimibe). The dosing regimen for mipomersen sodium was designed to achieve near steady-state tissue accumulation by the end of dosing.

The values for area under the plasma concentration-time curve (AUC_{24h}), maximum plasma concentration (C_{max}), and apparent elimination half-life ($t_{½}$) of each drug were calculated, and their 90% confidence intervals (CIs) of the geometric least-squares means (GLSM) ratios are shown in Table 15-3, which indicates that the systemic exposure of mipomersen was not altered when coadministered with simvastatin or ezetimibe. However, mipomersen coadministration resulted in approximately 50% lower C_{max} and 28% higher $AUC_{24\,h}$ for simvastatin; and a 42% lower C_{max} with no appreciable change in $AUC_{72\,h}$ for ezetimibe. The underlying mechanisms for the observed results are to be elucidated.

TABLE 15-1 Effects of Oligonucleotides on the Pharmacokinetics of Other Drugs

Interacting ASOs	Class	Substrates	Study Design	PK Parameters	Results	References
Mipomersen	Gen 2	Ezetimibe, simvastatin	Healthy subjects; parallel design	$AUC_{72\,h}$, $AUC_{24\,h}$, C_{max}, T_{max}, $t_{1/2}$	Ezetimibe C_{max} ↓ 42%, AUC_{72h} no change; simvastatin C_{max} ↓ 49%, AUC_{24h}, ↑ 28%	5,6
ISIS 113715	Gen 1	Glipizide, metformin, rosiglitazone	Cancer patients; sequential crossover design	$AUC_{24\,h}$, C_{max}, $t_{1/2}$	Rosiglitazone C_{max} ↑ 46%, AUC_{24} ↑ 15%; glipizide and metforminno change	4
ISIS 2503	Gen 1	Gemcitabine	Cancer patients; sequential crossover design	C_{ss}	Gemcitabine C_{max} ↓, AUC ↓; not statistically significant	13
ISIS 3521	Gen 1	5-fluorouracil	Cancer patients; sequential crossover design	C_{ss}	No changes	14
		Gemcitabine, cisplatin	Cancer patients; sequential crossover design	AUC, C_{max}	No changes	15
		Paclitaxel, carboplatin	Cancer patients; sequential crossover design	AUC, C_{max}	No changes	16,17

(continued)

TABLE 15-1 *(Continued)*

Interacting ASOs	Class	Substrates	Study Design	PK Parameters	Results	References
G3139	Gen 1	Carboplatin, paclitaxel	Cancer patients; cross-study PK comparison	AUC, CL, V_{ss}, $t_{1/2}$	No changes	18
GEM231	Gen 2	Paclitaxel, docetaxel	Cancer patients; parallel design	C_{ss}	Ongoing	19
G3139	Gen 1	Doxorubucin	Mice; parallel study	AUC, V_d, $t_{1/2}$	Tumor exposure to DOX ↑ twofold to threefold	9,10
Oligo AS	Gen 2	Irinotecan	Mice; parallel study	Tissue concentration	Tissue irinotecan and SN-38 ↑	20

ASOs: antisense oligonucleotides; AUC: area under the concentration-time curve; CL: clearance; C_{max}: maximum concentration; C_{ss}: steady-state concentrations; DOX: Doxorubucin; Gen 1: first-generation phosphorothioate oligonucleotide; Gen 2: second-generation antisense chimeric 2′-*O*-(2-methoxyethyl) phosphorothioate oligonucleotides; PK: pharmacokinetics; T_{max}: time to maximum concentration; V_d: volume of distribution; V_{ss}: steady-state volume of distribution; $t_{1/2}$: elimination half-life.

TABLE 15-2 Effects of Other Drugs on the Pharmacokinetics of Oligonucleotides

Interacting Drugs	Substrates	Class	Study Design	PK	Results	References
Ezetimibe; simvastatin	Mipomersen	Gen 2	Healthy subjects; parallel design	$AUC_{24\,h}$, C_{max}, T_{max}, $t_{1/2}$	No changes	5,6
Glipizide; metformin; rosiglitazone	ISIS 113715	Gen 1	Cancer patients; sequential crossover design	$AUC_{24\,h}$, C_{max}, $t_{1/2}$	No changes	4
5-Fluorouracil; leucovorin	ISIS 3521	Gen 1	Cancer patients; sequential crossover design	AUC, CL, V_{ss}, $t_{1/2}$	No changes	14
Gemcitabine; cisplatin	ISIS 3521	Gen 1	Cancer patients; sequential crossover design	C_{ss}	C_{ss} ↑ 60% with cisplatin	15
Gemcitabine	ISIS 2503	Gen 1	Cancer patients; sequential crossover design	AUC, C_{max}, CL, V_{ss}, $t_{1/2}$	No changes	13
Doxorubucin	G3139	Gen 1	Mice, parallel study	AUC, C_{max}, CL, $t_{1/2}$	G3139 C_{max} ↑; absorption rate↑	10

AUC: area under the concentration-time curve; CL: clearance; C_{max}: maximum concentration; C_{ss}: steady-state concentrations; Gen 1: first generation phosphorothioate oligonucleotide; Gen 2: second-generation antisense chimeric 2′-*O*-(2-methoxyethyl) phosphorothioate oligonucleotides; PK: pharmacokinetics; V_{ss}: steady-state volume of distribution; $t_{1/2}$: elimination half-life.

TABLE 15-3 Geometric Least-Squares Means (GLSM) Ratios and 90% Confidence Intervals (CIs) of Pharmacokinetic Parameters After Coadministration of Substrates and Interacting Drugs[a]

Substrates	Interacting Drugs	PK Parameters	GLSM Ratio	90% CI
Simvastatin	Mipomersen	C_{max} (ng/mL)	0.51	(0.34, 0.77)
		$AUC_{24\,h}$ (h·ng/mL)	1.28	(0.90, 1.82)
Ezetimibe	Mipomersen	C_{max} (ng/mL)	0.58	(0.40, 0.82)
		$AUC_{72\,h}$ (h·ng/mL)	1.00	(0.78, 1.27)
Mipomersen	Simvastatin	C_{max} (ng/mL)	0.98	(0.93, 1.03)
		$AUC_{24\,h}$ (h·ng/mL)	1.00	(0.94, 1.07)
Mipomersen	Ezetimibe	C_{max} (ng/mL)	1.05	(0.86, 1.28)
		$AUC_{24\,h}$ (h·ng/mL)	1.01	(0.92, 1.11)

[a] $n = 10$.
CI: confidence interval; PK: pharmacokinetics.

The clinical trial design used in this study took into consideration the long elimination half-life of mipomersen and the relatively large interindividual variability in the PK of the oral lipid-lowering drugs. Thus the design of the study included a loading dose approach for mipomersen sodium that ensured achievement of near steady-state concentrations in the target tissues (the liver and kidneys) before testing of potential interactions with concomitantly administered oral lipid-lowering agents. The sequential design allowed each individual in the study to be his or her own control to reduce intraindividual variability. The limitation of the study is that a 24-h PK sampling duration was too short relative to the 30-day elimination half-life of mipomersen; therefore the $AUC_{0-24\,h}$ may not be an appropriate parameter for assessing the effect of coadministered drugs on the PK of mipomersen.

ISIS 2503 is an ASO targeting H-ras mRNA. The potential PK interaction between ISIS 2503 and gemcitabine was evaluated in a Phase I trial.[13,21] A total of 19 patients were treated with a fixed dose of gemcitabine 1000 mg/m^2 on days 1 and 8 and two escalating doses of ISIS 2503 (4 and 6 mg/kg/day) as a 14-day continuous infusion starting on day 1. An additional 8 patients received a flat dose of ISIS 2503 based on their ideal body weight. Blood samples for gemcitabine PK were collected on day 1 (gemcitabine alone) and day 8 (with ISIS 2503). Only steady-state concentration (C_{ss}) data were collected for ISIS 2503, at pregemcitabine and postgemcitabine infusion on day 8.

At three dose levels studied, the C_{ss} of ISIS 2503 achieved during infusion were not altered, and the fraction of total oligonucleotide composed of parent oligonucleotide (intact 20-mer ISIS 2503) was also unchanged when ISIS 2503 was coadministered with gemcitabine. There was a consistent decrease in the gemcitabine plasma concentration observed at the end of the 30-min gemcitabine infusion in all ISIS 2503 three dose cohorts. Alteration of gemcitabine concentration at this single time point resulted in a 31–48% decrease in mean AUC and a 49–57% decrease in mean C_{max} in patients receiving a flat dose or 6 mg/kg/day of ISIS 2503.

ISIS 113715 is a 20-mer first-generation phosphorothioate ASO that is complementary to the protein tyrosine phosphatase 1B (PTP-1B) mRNA. In a clinical DDI study, 23 healthy subjects received a single oral dose of an antidiabetic drug (either metformin 500 mg, glipizide 5 mg or rosiglitazone 2 mg) alone and together with 200 mg of ISIS 113715 subcutaneously in a sequential-crossover design.[4] The results of this study indicated that there were no large deviations outside of the boundaries of 80–125% for PK parameters (C_{max} and AUC) of glipizide and metformin, but some differences were observed for rosiglitazone with the mean ratios and 90% CIs for C_{max} and $AUC_{24\,h}$ of 1.46 (1.04–2.05) and 1.15 (0.80–1.65), respectively when coadministered with ISIS 113715. Conversely, there were no observed significant alterations in the PK of ISIS 113715 when administered in combination with these oral antidiabetic drugs.

15.3.3.2 ISIS 3521 (LY900003)

In Combination with 5-Fluorouracil and Leucovorin ISIS 3521 (LY900003) is a sodium salt form of a 20-mer phosphorothioate oligonucleotide that hybridizes to the 3′-untranslated region of human protein kinase C (PKC) mRNA and inhibits its expression through RNase-mediated cleavage of hybridized PKC mRNA. In a clinical study, patients with refractory solid tumors received ISIS 3521 as a 21-day continuous infusion simultaneously with 5-fluorouracil (5-FU) and leucovorin given daily for 5 days and repeated every 4–5 weeks (one cycle).[14] A PK analysis for 5-FU and ISIS 3521 was performed on samples taken during the first cycle in all patients. Fifteen patients received ISIS 3521 at one of three dose levels: 1.0 mg/kg/day ($n=3$), 1.5 mg/kg/day ($n=3$), and 2.0 mg/kg/day ($n=9$). All patients concomitantly received 5-FU (425 mg/m^2/day) and leucovorin (20 mg/m^2/day) for 5 consecutive days. The washout time period was not specified in the paper. The steady-state plasma concentration (C_{ss}) of ISIS 3521 without 5-FU was determined by averaging the measured plasma concentrations at time 24 h to time 21 days during the continuous infusion. After 5-FU administration on day 4, ISIS 3521 concentrations measured at multiple time points within 90 min of 5-FU administration were averaged to obtain C_{ss}.

The plasma concentration of ISIS 3521 increased in a dose-dependent manner over the range of doses investigated. There was no significant difference in the ISIS 3521 steady-state plasma concentrations when ISIS 3521 was given with or without 5-FU. Furthermore, there was no significant change in the percentage of intact ISIS 3521 in the presence or absence of 5-FU. Because there were only two to three patient observations recorded in each dose treatment cohort, a statistical test cannot be performed with confidence; however, on visual inspection of the data, there was considerable overlap of the ISIS 3521 C_{ss} values in the presence and absence of 5-FU. The PK parameters [AUC, $t_{1/2}$, CL, and volume of distribution at steady-state (V_{ss})] of 5-FU in the presence or absence of ISIS 3521 also showed no evidence for PK interactions.[14]

In Combination with Cisplatin and Gemcitabine A clinical trial of ISIS 3521 in combination with cisplatin/gemcitabin was conducted in patients with advanced non–small cell lung cancer.[15] The PK interaction study design was similar to the previously mentioned study reported by Mani et al. For gemcitabine, the respective GLSM ratios on study days 1 and 8 for each parameter of interest ($AUC_{0-\infty}$ and C_{max}) were ranged from 0.69 to 1.27. Similar findings were observed for cisplatin, and the GLSM ratios for each parameter of interest in cycles 1 and 2 were close to 1.

This study compared the C_{ss} of ISIS 3521 and total oligonucleotide in the absence and presence of gemcitabine, cisplatin, or both drugs. In the presence of gemcitabine, the GLSM ratio for C_{ss} of ISIS 3521 and total oligonucleotide are 0.81 and 0.70, respectively. In the presence of cisplatin, the GLSM ratio for C_{ss} of ISIS 3521 and total oligonucleotide are 1.60 and 1.61, respectively. The clinical significance of this magnitude of change in ISIS 3521 C_{ss} when coadministered with gemcitabine and cisplatin is unknown.

In Combination with Carboplatin and Paclitaxel A clinical trial was conducted with ISIS 3521 in combination with carboplatin and paclitaxel for the treatment of patients with non-small cell lung cancer. Patients initially received carboplatin (target AUC: 5 or 6 mg/mL/min) and paclitaxel (175 mg/m^2) alone in cycle 1. In cycle 2 and beyond, patients received escalating doses of ISIS 3521 (1.0–2.0 mg/kg/day) by continuous intravenous infusion on days 1–14, with carboplatin and paclitaxel administered on day 4. In the expanded phase, all patients received carboplatin (target AUC: 6 mg/mL/min) and paclitaxel (175 mg/m^2) with ISIS 3521 (2.0 mg/kg/day) starting in cycle 1. Treatment was repeated every 21 days until maximum benefit. There was no evidence of a PK interaction between ISIS 3521 and either chemotherapy agent.[16]

Overall, these literature reports conclude that ISIS 3521 has no clinically relevant effect on the PK of those drugs administered concomitantly with ISIS 3521. However, a 60% increase in ISIS 3521 C_{ss} in the presence of cisplatin was observed and the underlying mechanisms are not fully explained. In addition, there are several limitations for these reported DDI studies. Firstly, some of the analyses did not include other PK parameters such as C_{max} which was limited by the design of DDI studies. Secondly, the studies suffered from a small numbers of patients (e.g., only two to three patient observations recorded in each dose cohort in the study with 5-FU and leucovorin). Lastly, the dose and regimen in these studies may not be the maximal clinical doses to detect a significant PK interaction.

Drug Transporters Despite over a decade of research, the mechanism of cellular membrane transport of oligonucleotide-based drugs are still unclear, possibly involving multiple pathways.[22] Oligonucleotides usually have molecular weight greater than 4 kDa and contain more than 12 negative charges per molecule. The general belief is that only lipophilic or small hydrophilic molecules can traverse biological membranes. The mechanism of cellular uptake of naked DNA/RNA is still poorly defined, but one of these pathways may require a cell membrane transporter in mammalian cells.[23]

There are no published reports on the role of P-glycoprotein (P-gp) in oligonucleotide absorption or intracellular transport. A study suggested that at a high dose of 10 μM, an ASO can be transported by P-gp *in vitro* (unpublished data). Until the mechanism of oligonucleotide transport is clearly delineated, the potential for oligonucleotides to be as substrates of transporters such as P-gp should be considered during the drug development.

15.4 OTHER CONSIDERATIONS

15.4.1 Indirect Effects on CYP Enzymes

Some oligonucleotides contain unmethylated CpG motifs within the oligonucleotide sequence that have been shown to be potent immune stimulators.[24] The immunostimulatory oligonucleotides can activate signaling cascades to regulate the expression of a number of host-defense proteins such as interferons and interferon-inducible genes. Cytokines, such as interleukin 6 (IL-6), may affect the oxidative metabolic process by modulating the hepatic microsomal CYP isozymes, and thereby change the metabolism of concomitantly administrated drugs that are substrates of these enzymes.

The delivery systems and formulations of oligonucleotides, which include conjugation with other molecules, such as pegylation, liposome, or lipid nanoparticles (LNPs) are often used to extend elimination half-lives of the siRNA or ASOs or to have a better delivery of them to the target cells. Some delivery systems or formulations may have the ability to modulate cytokine release and increase the complexities of drug interaction potential.

15.4.2 Immunogenicity and Its Effects on PK

Experimental induction of antibodies in mice suggests the potential for DNA to be immunogenic. In normal subjects, anti-DNA B cell precursors make up as much as 1–5% of the total B cell population. The introduced antisense DNA or RNA may induce autoantibody or anti-ASO antibody production (immunogenicity) that would not normally occur for the endogenous sense sequence. There are several mechanisms that may be responsible for the incidence of ASO immunogenicity: (1) the use of modified analogs designed to increase nuclease resistance may cause immunogenicity, (2) protection from digestion and enhanced cellular uptake achieved by linking oligonucleotides to protein carriers causes additional problems, and (3) some oligonucleotide drugs produce potent immune stimulation *in vivo*, thereby enhancing the anti-PEG antibody response when these immunostimulatory agents are encapsulated in PEGylated liposomes.[25,26]

Altered PK profiles associated with antiproduct antibody formation have been seen in therapeutic biological products. The potential for immunogenicity of oligonucleotide-based drugs may also alter their PK. Currently most of the oligonucleotide-based drugs are still in the preclinical or early clinical development stages; therefore, the immunogenicity data are not available for most compounds under development.

15.4.3 Considerations of DDI Studies

The first-generation ASOs often need to be administrated chronically to attain persistent modulation of target gene expression, especially in oncology. When both the substrate and the interacting drug are likely to be given over an extended period of time, the often used DDI study design is to administer the substrate to steady state with collection of blood samples over one or more dosing intervals followed by multiple dose administration of an interacting drug.[27] The second-generation ASOs have long elimination half-life of 10–30 days. Design of the DDI studies for such long half-life drugs may consider inclusion of a loading dose to ensure faster achievement of the steady-state concentrations of the study drug. In addition, a cross-over design may not be practical because of the long elimination half-life.

15.4.4 Regulatory Perspective

Currently, there is no specific regulatory guidance on clinical pharmacology studies for oligonucleotide-based drugs. The current regulatory recommendations for evaluation of oligonucleotide drug–drug interactions are similar to the approaches for small molecule drugs.[27] In general, the metabolism and transport of an investigational new drug and its interactions with other drugs as a substrate, inhibitor, or inducer needs to be assessed during drug development as part of an adequate assessment of the drug's safety and effectiveness. At the early stage of clinical development, *in vitro* screens need to be conducted to determine if the parent or active metabolite(s) is a substrate, inhibitor or inducer of CYP isozymes or P-gp transporter. These studies will help determine the need for *in vivo* drug–drug interactions.

In oncology, many oligonucleotides are developed in combination with the current standard therapeutics. If oligonucleotides are to be used in combination with other drugs such as chemotherapy, the potential for a PK drug–drug interaction between oligonucleotides and the drugs in the combination will need to be evaluated during the development of such combination therapy. The evaluation can be conducted in a dedicated drug–drug interaction trial or incorporated in a clinical efficacy and safety trial. This evaluation is important because many oncology drugs have narrow therapeutic indices and altered drug systemic exposure could increase drug-related toxicities or compromise its therapeutic effects.

15.5 SUMMARY

The reported clinical data suggest that there are limited PK drug–drug interactions between oligonucleotide-based drugs and small molecular drugs. A few animal studies demonstrated that tumor exposures of the anticancer drugs are affected by the oligonucleotides coadministration without affecting their plasma pharmacokinetics. *In vitro* data from protein binding displacement studies suggest that the potential for pharmacokinetic interactions caused by displacement of plasma protein binding is low when oligonucleotides are administrated concomitantly with other highly protein-bound drugs.

Oligonucleotides are mainly metabolized by nucleases in tissues and are unlikely to be substrates of CYP metabolic enzymes. However, the current data regarding the CYP enzyme involvement are very limited and cannot exclude the possibility that the exposure of oligonucleotides can be affected by CYP enzyme inducers or inhibitors. Given that *in vitro* studies using human microsomes have shown that several oligonucleotide drugs inhibited multiple CYP enzymes, the inhibition and induction potential of oligonucleotides for CYP enzymes should be addressed during the oligonucleotide-based drug development.

REFERENCES

1. Crooke ST, ed. *Antisense drug technology: principles, strategies, and applications*. Boca Raton, CRC Press, 2007.
2. Geary RS, Antisense oligonucleotide pharmacokinetics and metabolism. *Expert Opin Drug Metab Toxicol* **5**, 381–91 (2009).
3. Geary RS, Watanabe TA, Truong L, et al. Pharmacokinetic properties of 2′-*O*-(2-methoxyethyl)-modified oligonucleotide analogs in rats. *J Pharmacol Exp Ther* **296**, 890–97 (2001).
4. Geary RS, Bradley JD, Watanabe T, et al. Lack of pharmacokinetic interaction for ISIS 113715, a 2′-*O*-methoxyethyl modified antisense oligonucleotide targeting protein tyrosine phosphatase 1B messenger RNA, with oral antidiabetic compounds metformin, glipizide or rosiglitazone. *Clin Pharmacokinet* **45**, 789–801 (2006).
5. Ito MK, ISIS 301012 gene therapy for hypercholesterolemia: sense, antisense, or nonsense? *Ann Pharmacother* **41**, 1669–78 (2007).
6. Yu RZ, Geary RS, Flaim JD, et al. Lack of pharmacokinetic interaction of mipomersen sodium (ISIS 301012), a 2′-*O*-methoxyethyl modified antisense oligonucleotide targeting apolipoprotein B-100 messenger RNA, with simvastatin and ezetimibe. *Clin Pharmacokinet* **48**, 39–50 (2009).
7. Watanabe TA, Geary RS, Levin AA, Plasma protein binding of an antisense oligonucleotide targeting human ICAM-1 (ISIS 2302). *Oligonucleotides* **16**, 169–80 (2006).
8. Benet LZ, Hoener BA, Changes in plasma protein binding have little clinical relevance. *Clin Pharmacol Ther* **71**, 115–21 (2002).
9. Lopes de Menezes DE, Hudon N, McIntosh N, Mayer LD, Molecular and pharmacokinetic properties associated with the therapeutics of bcl-2 antisense oligonucleotide G3139 combined with free and liposomal doxorubicin. *Clin Cancer Res* **6**, 2891–902 (2000).
10. Lopes de Menezes DE, Mayer LD, Pharmacokinetics of Bcl-2 antisense oligonucleotide (G3139) combined with doxorubicin in SCID mice bearing human breast cancer solid tumor xenografts. *Cancer Chemother Pharmacol* **49**, 57–68 (2002).

11. Wang H, Nan L, Yu D, et al. Antisense anti-MDM2 oligonucleotides as a novel therapeutic approach to human breast cancer: in vitro and in vivo activities and mechanisms. *Clin Cancer Res* **7**, 3613–24 (2001).
12. Gilead Sciences. Package insert for Macugen (pegaptanib). San Dimas, CA, 2011. Available at http://www.accessdata.fda.gov/drugsatfda_docs/label/2011/021756s018lbl.pdf.
13. Adjei AA, Dy GK, Erlichman C, et al. A phase I trial of ISIS 2503, an antisense inhibitor of H-ras, in combination with gemcitabine in patients with advanced cancer. *Clin Cancer Res* **9**, 115–23 (2003).
14. Adjei AA, Rowinsky EK, Novel anticancer agents in clinical development. *Cancer Biol Ther* **2**, S5–15 (2003).
15. Mani S, Rudin CM, Kunkel K, et al. Phase I clinical and pharmacokinetic study of protein kinase C-alpha antisense oligonucleotide ISIS 3521 administered in combination with 5-fluorouracil and leucovorin in patients with advanced cancer. *Clin Cancer Res* **8**, 1042–48 (2002).
16. Villalona-Calero MA, Ritch P, Figueroa JA, et al. A phase I/II study of LY900003, an antisense inhibitor of protein kinase C-alpha, in combination with cisplatin and gemcitabine in patients with advanced non-small cell lung cancer. *Clin Cancer Res* **10**, 6086–93 (2004).
17. Yuen AR, Halsey J, Fisher GA, et al. Phase I study of an antisense oligonucleotide to protein kinase C-alpha (ISIS 3521/CGP 64128A) in patients with cancer. *Clin Cancer Res* **5**, 3357–63 (1999).
18. Motl SE, LY900003: a novel compound for the treatment of non-small cell lung cancer. *Cancer Ther* **1**, 237–44 (2003).
19. Liu G, Kolesar J, McNeel DG, et al. A phase I pharmacokinetic and pharmacodynamic correlative study of the antisense Bcl-2 oligonucleotide g3139, in combination with carboplatin and paclitaxel, in patients with advanced solid tumors. *Clin Cancer Res* **14**, 2732–39 (2008).
20. Mani S, Goel S, Nesterova M, et al. Clinical studies in patients with solid tumors using a second-generation antisense oligonucleotide (GEM 231) targeted against protein kinase A type I. *Ann N Y Acad Sci* **1002**, 252–62 (2003).
21. Wang H, Wang S, Nan L, et al. Antisense anti-MDM2 mixed-backbone oligonucleotides enhance therapeutic efficacy of topoisomerase I inhibitor irinotecan in nude mice bearing human cancer xenografts: in vivo activity and mechanisms. *Int J Oncol* **20**, 745–52 (2002).
22. Wu-Pong S, Alternative interpretations of the oligonucleotide transport literature: insights from nature. *Adv Drug Deliv Rev* **44**, 59–70 (2000).
23. Bennett RM, As nature intended? The uptake of DNA and oligonucleotides by eukaryotic cells. *Antisense Res Dev* **3**, 235–41 (1993).
24. Klinman DM, Klaschik S, Tomaru K, et al. Immunostimulatory CpG oligonucleotides: effect on gene expression and utility as vaccine adjuvants. *Vaccine* **28**, 1919–23 (2010).
25. Dawson KH, Bell DA, Production and pathogenic effects of anti-DNA antibodies: relevance to antisense research. *Antisense Res Dev* **1**, 351–60 (1991).
26. Judge AD, Bola G, Lee AC, MacLachlan I, Design of noninflammatory synthetic siRNA mediating potent gene silencing in vivo. *Mol Ther* **13**, 494–505 (2006).

27. US Food and Drug Administration, US Department of Health and Human Services, Center for Drug Evaluation and Research, Center for Biologics Evaluation and Research (CBER). Draft guidance for industry: drug interaction studies—study design, data analysis, implications for dosing, and labeling. Silver Spring, MD, 2012. Available at http://www.fda.gov/downloads/Drugs/GuidanceComplianceRegulatoryInformation/Guidances/UCM292362.pdf.

APPENDIX

MONOGRAPHS FOR DRUG-DRUG INTERACTIONS OF THERAPEUTICS BIOLOGICS

JOCELYN LEU, SIMONE KASEK, CHRISTINE LI, WARARAT LIMOTHAI, DORA BABU MADHURA, CHETAN RATHI, SUMIT RAWAL, JOSIAH RYMAN, MARGARET THOMSON, ASHIT TRIVEDI, HONGHUI ZHOU, and BERND MEIBOHM

Over the last decade, the unprecedented increase in the clinical use of therapeutic biologics for a variety of indications including immune-mediated inflammatory disease indications such as rheumatoid arthritis and inflammatory bowel diseases, and oncology indications has resulted in a new class of drug-drug interactions, therapeutic biologic-drug interactions. However, this kind of drug-drug interactions is generally understudied or underreported due to a perception that drug-drug interactions involving therapeutic biologics are scarce and of limited relevance. Moreover, the underlying mechanisms of many of those drug-drug interactions are still poorly understood.

In order to better assess the landscape of drug-drug interactions for therapeutic biologics, we have compiled drug-drug interaction monographs involving therapeutic biologics either as victims or perpetrators based on current US labeling information and scientific literature. To broaden our knowledgebase in this area, our monographs also include some important 'theoretical' drug-drug interactions with well established plausible mechanisms. In order to make those monographs more useful to the readers, we have also included other relevant references if available from which the drug-drug interaction contents was obtained. We feel these monographs might be useful for readers as they might guide expectations in what might be encountered as potential therapeutic biologic-drug interaction in clinical development programs and applied pharmacotherapy.

Drug–Drug Interactions for Therapeutic Biologics, First Edition.
Edited by Honghui Zhou and Bernd Meibohm.

Drug-Drug Interaction Monograph for ACCRETROPIN (Somatropin) #1

Perpetrator drug:

INN name/US brand name:	Somatropin/Accretropin, Bio-Tropin,
Class/Molecular weight:	Hormone/~22 kDa

Victim drug:

INN name/US brand name:	Cortisone, prednisone
Class/Molecular weight:	Corticosteroids

Description of the effect of the DDI:
Inhibits metabolism of corticosteroids to the biologically active metabolites (e.g., cortisone to cortisol).

Description of the (proven, putative or hypothesized) mechanism of the DDI:
Inhibits 11 β-hydroxysteroid dehydrogenase type 1 (11βHSD-1) in adipose/hepatic tissue since conversion of these agents to the biologically active metabolites is dependent on the activity of the 11βHSD-1.

Clinical consequences:
Patients treated with glucocorticoid replacement therapy for previously diagnosed hypoadrenalism may require an increase in their maintenance or stress doses.

Reference: Accretropin prescribing information

Drug-Drug Interaction Monograph for ACCRETROPIN (Somatropin) #2

Perpetrator drug:

INN name/US brand name:	Somatropin/Accretropin, Bio-Tropin
Class/Molecular weight:	Hormone/~22 kDa

Victim drug:

INN name/US brand name:	CYP450 3A4 substrates
Class/Molecular weight:	Corticosteroids, sex steroids, anticonvulsants, cyclosporine

Description of the effect of the DDI:
Potentially increases clearance of CYP450 3A4 substrates.

Description of the (proven, putative or hypothesized) mechanism of the DDI:
Increases CYP450-mediated antipyrine clearance. These data suggest that somatropin administration may alter the clearance of compounds known to be metabolized by CYP450 liver enzymes (e.g., corticosteroids, sex steroids, anticonvulsants, cyclosporine). No formal DDI studies have been conducted.

Clinical consequences:
Careful monitoring is advisable.

Reference:

Accretropin US prescribing information

Lee JI, Zhang L, Men AY, Kenna LA, Huang SM. CYP-mediated therapeutic protein-drug interactions: clinical findings, proposed mechanisms and regulatory implications. Clin Pharmacokinet. 2010 May;49(5):295–310.

Drug-Drug Interaction Monograph for ACCRETROPIN (Somatropin) #3

Perpetrator drug:

INN name/US brand name:	Somatropin/Accretropin
Class/Molecular weight:	Hormone/∼22 kDa

Victim drug:

INN name/US brand name:	Insulin and oral antidiabetic agents
Class/Molecular weight:	Hormone/5.8 kDa

Description of the effect of the DDI:
Pharmacodynamic effect inducing insulin resistance.

Description of the (proven, putative or hypothesized) mechanism of the DDI:
Decreases insulin sensitivity at high doses. Growth hormone induces insulin resistance by the stimulation of lipolysis and a concomitant switch from oxidation of glucose to oxidation of lipids, during both acute and chronic treatment.

Clinical consequences:
This could lead to glucose intolerance and possibly diminish the effects of insulin. The dose of insulin and/or oral agents may require adjustment when somatropin therapy is initiated.

Reference:

Accretropin prescribing information

Attallah H, Friedlander AL, Hoffman AR: Visceral obesity, impaired glucose tolerance, metabolic syndrome, and growth hormone therapy. Growth Horm IGF Res (2006) 16 (Suppl A/Jul): S62–7.

Groop L, Segerlantz M, Bramnert M: Insulin sensitivity in adults with growth hormone deficiency and effect of growth hormone treatment. Horm Res (2005) 64 (Suppl 3): 45–50.

Drug-Drug Interaction Monograph for ACCRETROPIN (Somatropin) #4

Perpetrator drug:

INN name/US brand name:	Glucocorticoids
Class/Molecular weight:	Small Molecule

Victim drug:

INN name/US brand name:	Somatropin/Accretropin
Class/Molecular weight:	Hormone/∼22 kDa

Description of the effect of the DDI:
Pharmacodynamic effect on growth promoting effects of somatropin.

Description of the (proven, putative or hypothesized) mechanism of the DDI:
Excessive glucocorticoid therapy may attenuate the growth promoting effects of somatropin in children.

Clinical consequences:
Carefully adjust glucocorticoid replacement therapy with concomitant growth hormone.

(Continued)

Reference: Accretropin prescribing information

Mehls I, Tonshoff B, Kovacs G, Mayer C, Scerek J, Oh J: Interaction between glucocorticoids and growth hormone. Acta Paediatrica (2008) 82 (Suppl s389): 77–82.

Drug-Drug Interaction Monograph for ACTEMRA (Tocilizumab) #1

Perpetrator drug:
INN name/US brand name: Methotrexate
Class/Molecular weight: Small Molecule/476 Da

Victim drug:
INN name/US brand name: Tocilizumab/Actemra
Class/Molecular weight: Monoclonal Antibody/~148 kDa

Description of the effect of the DDI:
No effect on tocilizumab clearance by coadministered methotrexate evaluated by population PK analysis.

Description of the (proven, putative or hypothesized) mechanism of the DDI:
None.

Clinical consequences:
None.

Reference: Actemra US prescribing information

Frey N, Grange S, Woodworth T. Population pharmacokinetic analysis of tocilizumab in patients with rheumatoid arthritis. J Clin Pharmacol. 2010;50(7):754–66.

Drug-Drug Interaction Monograph for ACTEMRA (Tocilizumab) #2

Perpetrator drug:
INN name/US brand name: NSAIDS
Class/Molecular weight: Small Molecule

Victim drug:
INN name/US brand name: Tocilizumab/Actemra
Class/Molecular weight: Monoclonal Antibody/~148 kDa

Description of the effect of the DDI:
No effect on tocilizumab clearance by coadministered NSAIDS evaluated by population PK analysis.

Description of the (proven, putative or hypothesized) mechanism of the DDI:
None.

Clinical consequences:
None.

Reference: Actemra US prescribing information

Frey N, Grange S, Woodworth T. Population pharmacokinetic analysis of tocilizumab in patients with rheumatoid arthritis. J Clin Pharmacol. 2010 Jul;50(7):754–66.

Drug-Drug Interaction Monograph for ACTEMRA (Tocilizumab) #3

Perpetrator drug:

INN name/US brand name:	Corticosteroids
Class/Molecular weight:	Small Molecule

Victim drug:

INN name/US brand name:	Tocilizumab/Actemra
Class/Molecular weight:	Monoclonal Antibody/~148 kDa

Description of the effect of the DDI:
No effect on tocilizumab clearance by coadministered corticosteroids evaluated by population PK analysis.

Description of the (proven, putative or hypothesized) mechanism of the DDI:
None.

Clinical consequences:
None.

Reference:

Actemra US prescribing information

Frey N, Grange S, Woodworth T. Population pharmacokinetic analysis of tocilizumab in patients with rheumatoid arthritis. J Clin Pharmacol. 2010; 50(7):754–66.

Drug-Drug Interaction Monograph for ACTEMRA (Tocilizumab) #4

Perpetrator drug:

INN name/US brand name:	Tocilizumab/Actemra
Class/Molecular weight:	Monoclonal Antibody/~148 kDa

Victim drug:

INN name/US brand name:	Simvastatin/Zocor
Class/Molecular weight:	Small Molecule/419 Da

Description of the effect of the DDI:
In 12 RA patients not treated with Actemra, receiving 40 mg simvastatin, exposures of simvastatin and its metabolite, simvastatin acid, was 4- to 10-fold and 2-fold higher, respectively, than the exposures observed in healthy subjects. One week following administration of a single infusion of Actemra (10 mg/kg), exposure of simvastatin and simvastatin acid decreased by 57% and 39%, respectively, to exposures that were similar or slightly higher than those observed in healthy subjects. Exposures of simvastatin and simvastatin acid increased upon withdrawal of Acemtra in RA patients.

Description of the (proven, putative or hypothesized) mechanism of the DDI:
Simvastatin is a CYP3A4 and OATP1B1 substrate. *In vitro* data suggested that IL-6 reduced mRNA expression for CYP3A4, and this reduced expression was reversed by co-incubation with tocilizumab at clinically relevant concentrations. Accordingly, inhibition of IL-6 signaling in RA patients treated with tocilizumab may restore CYP3A4 activity to higher levels than those in the absence of tocilizumab leading to increased metabolism of drugs that are CYP3A4 substrate.

(Continued)

Clinical consequences:
Selection of a particular dose of simvastatin in RA patients should take into account the potentially lower exposures that may result after initiation of Actemra (due to normalization of CYP3A4) or higher exposures after discontinuation of Actemra.

Reference:

Actemra US prescribing information

Kim S. Rheumatol Int (2012) 32:2601 – 2604.
Schmitt C, et al. Disease-drug-drug interaction involving tocilizumab and simvastatin in patients with rheumatoid arthritis. Clin Pharmacol Ther 2011; 89(5): 735–740.

Zhang X, Schmitt C, Grange S, Terao K, Miya K. 2009. Disease-drug interaction studies of tocilizumab with cytochrome p450 substrate in vitro and in vivo. Clin. Pharmacol. Ther. 85:S59 (Abstr.)

Drug-Drug Interaction Monograph for ACTEMRA (Tocilizumab) #5

Perpetrator drug:

INN name/US brand name:	Tocilizumab/Actemra
Class/Molecular weight:	Monoclonal Antibody/~148 kDa

Victim drug:

INN name/US brand name:	Omeprazole/Prilosec
Class/Molecular weight:	Small Molecule/345 Da

Description of the effect of the DDI:
In RA patients receiving 10 mg omeprazole, exposure to omeprazole was approximately 2 fold higher than that observed in healthy subjects. In RA patients receiving 10 mg omeprazole, before and one week after Actemra infusion (8 mg/kg), the omeprazole AUC_{inf} decreased by 12% for poor (N = 5) and intermediate metabolizers (N = 5) and by 28% for extensive metabolizers (N = 8) and were slightly higher than those observed in healthy subjects.

Description of the (proven, putative or hypothesized) mechanism of the DDI:
Omeprazole is a CYP2C19 and CYP3A4 substrate. *In vitro* data suggested that IL-6 reduced mRNA expression for several CYP450 isoenzymes including CYP1A2, CYP2B6, CYP2C9, CYP2C19, CYP2D6 and CYP3A4, and this reduced expression was reversed by co-incubation with tocilizumab at clinically relevant concentrations. Accordingly, inhibition of IL-6 signaling in RA patients treated with tocilizumab may restore CYP450 activities to higher levels than those in the absence of tocilizumab leading to increased metabolism of drugs that are CYP450 substrates.

Clinical consequences:
None specifically mentioned for omeprazole.

Reference:

Actemra US prescribing information

Zhou H and Mascelli M, Annu. Mechanisms of Monoclonal Antibody-Drug Interactions. Rev. Pharmacol. Toxicol. 2011. 51:359–72.

Drug-Drug Interaction Monograph for ACTEMRA (Tocilizumab) #6

Perpetrator drug:

INN name/US brand name:	Tocilizumab/Actemra
Class/Molecular weight:	Monoclonal Antibody/~148 kDa

Victim drug:

INN name/US brand name:	Dextromethorphan/Robitussin, Delsym
Class/Molecular weight:	Small Molecule/271 Da

Description of the effect of the DDI:
In 13 RA patients receiving 30 mg dextromethorphan, exposure to dextromethorphan was comparable to that in healthy subjects. However, exposure to its metabolite, dextrorphan (a CYP3A4 substrate), was a fraction of that observed in healthy subjects. One week following administration of a single infusion of Actemra (8 mg/kg), dextromethorphan exposure was decreased by approximately 5%. However, a larger decrease (29%) in dextrorphan levels was noted after Actemra infusion.

Description of the (proven, putative or hypothesized) mechanism of the DDI:
Dextromethorphan is a CYP2D6 and CYP3A4 substrate. *In vitro* data suggested that IL-6 reduced mRNA expression for several CYP450 isoenzymes including CYP1A2, CYP2B6, CYP2C9, CYP2C19, CYP2D6 and CYP3A4, and this reduced expression was reversed by co-incubation with tocilizumab at clinically relevant concentrations. Accordingly, inhibition of IL-6 signaling in RA patients treated with tocilizumab may restore CYP450 activities to higher levels than those in the absence of tocilizumab leading to increased metabolism of drugs that are CYP450 substrates.

Clinical consequences:
None specifically mentioned for dextromethorphan.

Reference:

Actemra US prescribing information

Zhang X, Schmitt C, Grange S, Terao K, Miya K. 2009. Disease-drug interaction studies of tocilizumab with cytochrome p450 substrate in vitro and in vivo. Clin. Pharmacol. Ther. 85:S59 (Abstr.)

Drug-Drug Interaction Monograph for ACTIMMUNE (Interferon Gamma-1B)

Perpetrator drug:

INN name/US brand name:	Interferon Gamma-1B/Actimmune
Class/Molecular weight:	Polypeptide/~17 kDa

Victim drug:

INN name/US brand name:	CYP450 substrates
Class/Molecular weight:	Small Molecule

Description of the effect of the DDI:
Depression of hepatic metabolism.

Description of the (proven, putative or hypothesized) mechanism of the DDI:
Preclinical studies in rodents demonstrated a decrease in hepatic microsomal CYP450 concentrations.

(Continued)

Clinical consequences:
None mentioned.

Reference:

Actimmune prescribing information

Mahmood I, Green MD. Drug interaction studies of therapeutic proteins or monoclonal antibodies. J Clin Pharmacol. 2007 Dec;47(12):1540–54.

Drug-Drug Interaction Monograph for ADCETRIS (Brentuximab vedotin) #1

Perpetrator drug:

INN name/US brand name:	Brentuximab vedotin/Adcetris
Class/Molecular weight:	Antibody-drug conjugate/~153 kDa

Victim drug:

INN name/US brand name:	Midazolam or other CYP3A4 substrates
Class/Molecular weight:	Small molecule/326 Da

Description of the effect of the DDI:
Co-administration of Adcetris did not affect exposure to midazolam, a CYP3A4 substrate. Monomethyl auristatin E (MMAE) does not inhibit other CYP enzymes at relevant clinical concentrations. Adcetris is not expected to alter the exposure to drugs that are metabolized by CYP3A4 enzymes.

Description of the (proven, putative or hypothesized) mechanism of the DDI:
None.

Clinical consequences:
None mentioned.

Reference:
Adcetris prescribing information

Drug-Drug Interaction Monograph for ADCETRIS (Brentuximab vedotin) #2

Perpetrator drug:

INN name/US brand name:	Ketoconazole/Nizoral
Class/Molecular weight:	Small molecule/531 Da

Victim drug:

INN name/US brand name:	Brentuximab vedotin/Adcetris
Class/Molecular weight:	Antibody-drug conjugate/~153 kDa

Description of the effect of the DDI:
Co-administration of Adcetris with ketoconazole, a potent CYP3A4 inhibitor, increased exposure to MMAE by approximately 34%.

Description of the (proven, putative or hypothesized) mechanism of the DDI:
Adcetris is a CD30-directed antibody-drug conjugate (ADC) consisting of three components: 1) the chimeric IgG1 antibody cAC10, specific for human CD30, 2) the microtubule

disrupting agent monomethyl auristatin E (MMAE), and 3) a protease-cleavable linker that covalently attaches MMAE to cAC10. MMAE is primarily metabolized by CYP3A.

Clinical consequences:
Patients who are receiving strong CYP3A4 inhibitors concomitantly with Adcetris should be closely monitored for adverse reactions.

Reference:
Adcetris prescribing information

Drug-Drug Interaction Monograph for ADCETRIS (Brentuximab vedotin) #3

Perpetrator drug:

INN name/US brand name:	Rifampin/Rifadin, Rimactane
Class/Molecular weight:	Small molecule/823 Da

Victim drug:

INN name/US brand name:	Brentuximab vedotin/Adcetris
Class/Molecular weight:	antibody-drug conjugate/~153 kDa

Description of the effect of the DDI:
Co-administration of Adcetris with rifampin, a potent CYP3A4 inducer, reduced exposure to MMAE by approximately 46%.

Description of the (proven, putative or hypothesized) mechanism of the DDI:
Adcetris is a CD30-directed antibody-drug conjugate (ADC) consisting of three components: 1) the chimeric IgG1 antibody cAC10, specific for human CD30, 2) the microtubule disrupting agent monomethyl auristatin E (MMAE), and 3) a protease-cleavable linker that covalently attaches MMAE to cAC10. MMAE is primarily metabolized by CYP3A.

Clinical consequences:
None mentioned.

Reference:
Adcetris prescribing information

Drug-Drug Interaction Monograph for AMEVIVE (Alefacept)

Perpetrator drug:

INN name/US brand name:	Alefacept/Amevive
Class/Molecular weight:	Fc fusion protein/91.4 kDa

Victim drug:

INN name/US brand name:	CYP450 substrates
Class/Molecular weight:	Small molecule

Description of the effect of the DDI:
No drug interaction studies have been conducted.

(Continued)

Description of the (proven, putative or hypothesized) mechanism of the DDI:
The formation of CYP450 enzymes may be suppressed by increased levels of cytokines (e.g., TNFα, IL-1, IL-6, IL-10, IFN) during chronic inflammation. Therefore, a molecule that exerts its pharmacological effect by inhibiting the release of cytokines, such as Amevive, could normalize the formation of CYP450 enzymes.

Clinical consequences:
Upon initiation or discontinuation of Amevive in patients being treated with CYP450 substrates with a narrow therapeutic index, monitoring of the effect (e.g., warfarin) or drug concentration (e.g., cyclosporine or theophylline) is recommended and the individual dose of the drug product may be adjusted as needed.

Reference:
Amevive prescribing information

Lee JI, Zhang L, Men AY, Kenna LA, Huang SM. CYP-mediated therapeutic protein-drug interactions: clinical findings, proposed mechanisms and regulatory implications. Clin Pharmacokinet. 2010 May;49(5):295–310.

Drug-Drug Interaction Monograph for ANGIOMAX (Bivalirudin)

Perpetrator drug:

INN name/US brand name:	Bivalirudin/Angiomax
Class/Molecular weight:	Peptide/2.2 kDa

Victim drug:

INN name/US brand name:	Anti-coagulants (e.g., heparin, warfarin, thrombolytics, glycoprotein IIb/IIIa-receptor inhibitors (GPI))
Class/Molecular weight:	Small Molecule

Description of the effect of the DDI:
In clinical trials in patients undergoing Percutaneous Coronary Intervention (PCI) or Percutaneous Transluminal Coronary Angioplasty (PTCA), co-administration of Angiomax with heparin, warfarin, thrombolytics, or GPIs was associated with increased risks of major bleeding events compared to patients not receiving these concomitant medications.

Description of the (proven, putative or hypothesized) mechanism of the DDI:
Bivalirudin is a direct thrombin inhibitor used as an anticoagulant. Thus use with other anticoagulants can increase the risk of bleeding events.

Clinical consequences:
None mentioned in label.

Reference:
Angiomax prescribing information

Kastrati A, Neumann FJ, Mehilli J et al: Bivalirudin versus unfractionated heparin during percutaneous coronary intervention. N Engl J Med (2008) 359 (7/Aug 14): 688–96.

Drug-Drug Interaction Monograph for APIDRA, HUMULIN, HUMALOG, NOVOLOG, LANTUS (Insulin) #1

Perpetrator drug:

INN name/US brand name:	Oral antidiabetic products
Class/Molecular weight:	Small molecule

Victim drug:

INN name/US brand name:	Insulin/Apidra, Humulin, Humalog, Novolog, Lantus, etc
Class/Molecular weight:	Hormone/∼5.8 kDa

Description of the effect of the DDI:
Pharmacodynamic interaction: Drugs that may increase the blood glucose-lowering effect of insulins can increase the risk of hypoglycemia. Insulin may be used as adjunctive therapy with oral antihyperglycemic agents when glycemic control deteriorates on oral antidiabetic agents in patients with type 2 diabetes mellitus. However lower daily doses of insulin are required to maintain desirable glycemic control.

Description of the (proven, putative or hypothesized) mechanism of the DDI:
Both drugs reduce blood glucose.

Clinical consequences:
May require insulin dose adjustment and particularly close monitoring

Reference:
Apidra prescribing information, Humulin prescribing information, Humalog prescribing information, Lantus prescribing information, Novolog prescribing information

Kabadi MU, Kabadi UM. Efficacy of sulfonylureas with insulin in type 2 diabetes mellitus. Ann Pharmacother. 2003 Nov;37(11):1572–6.

Drug-Drug Interaction Monograph for APIDRA, HUMULIN, HUMALOG, NOVOLOG, LANTUS (Insulin) #2

Perpetrator drug:

INN name/US brand name:	Pramlintide/Symlin
Class/Molecular weight:	Peptide/4 kDa

Victim drug:

INN name/US brand name:	Insulin/Apidra, Humulin, Humalog, Novolog, Lantus, etc
Class/Molecular weight:	Hormone/∼5.8 kDa

Description of the effect of the DDI:
Pharmacodynamic interaction: Drugs that may increase the blood glucose-lowering effect of insulins can increase the risk of hypoglycemia. Pramlintide is used as an adjunct to insulin and has been associated with an increased risk of insulin-induced severe hypoglycemia, particularly in patients with type 1 diabetes.

Description of the (proven, putative or hypothesized) mechanism of the DDI:
Both drugs reduce blood glucose.

(Continued)

Clinical consequences:
May require insulin dose adjustment and particularly close monitoring. To minimize the risk of insulin-induced hypoglycemia, all patients should reduce premeal insulin by 50% upon initiation of pramlintide therapy, monitor blood glucose frequently, and contact their healthcare provider if symptoms of nausea and/or hypoglycemia are persistent. Careful patient selection, patient instruction, and insulin dose adjustments are critical for reducing the risk of hypoglycemia.

Reference:
Pullman J, Darsow T, Frias J. Pramlintide in the Management of Insulin-Using Patients with Type 2 and Type 1 Diabetes. Vasc Health Risk Manag. 2006 September; 2(3): 203–212.

Apidra prescribing information, Humulin prescribing information, Humalog prescribing information, Lantus prescribing information, Novolog prescribing information
Symlin prescribing information.

Drug-Drug Interaction Monograph for APIDRA, HUMULIN, HUMALOG, NOVOLOG, LANTUS (Insulin) #3

Perpetrator drug:

INN name/US brand name:	ACE Inhibitors (e.g. lisinopril, enalapril, captopril)
Class/Molecular weight:	Small molecule/peptides

Victim drug:

INN name/US brand name:	Insulin/Apidra, Humulin, Humalog, Novolog, Lantus, etc
Class/Molecular weight:	Hormone/~5.8 kDa

Description of the effect of the DDI:
Pharmacodynamic interaction: Drugs that may increase the blood glucose-lowering effect of insulins can increase the risk of hypoglycemia. A nested case-control study was performed with diabetic patients treated with insulin or with oral antidiabetic drugs, who were admitted to hospital with hypoglycemia. Hypoglycemia was significantly associated with current use of ACE inhibitors (odds ratio 2.8 [95% CI 1.4–5.7]). In another nested case-control study, hypoglycemia was associated with the use of ACE inhibitors (odds ratio [OR] 3.2, 95% CI 1.2–8.3, $P = 0.023$),

Description of the (proven, putative or hypothesized) mechanism of the DDI:
Insulin resistance appears to be a syndrome that is associated with a clustering of metabolic disorders, including non-insulin-dependent diabetes mellitus, obesity, hypertension, lipid abnormalities, and atherosclerotic cardiovascular disease. Thus if a facet of metabolic disorders is normalized, this may perhaps increase insulin sensitivity

Clinical consequences:
May require insulin dose adjustment and particularly close monitoring

Reference:
Herings RM, de Boer A, Stricker BH, Leufkens HG, Porsius A. c. Hypoglycemia associated with use of inhibitors of angiotensin converting enzyme. Lancet. 1995;345(8959):1195–8.

DeFronzo R and Ferrannini E. Insulin resistance: a multifaceted syndrome responsible for NIDDM, obesity, hypertension, dyslipidemia, and atherosclerotic cardiovascular disease. Diabetes Care 1991;14:173–194.

Morris AD, Boyle DI, McMahon AD, Pearce H, Evans JM, Newton RW, Jung RT, MacDonald TM. ACE inhibitor use is associated with hospitalization for severe hypoglycemia in patients with diabetes. DARTS/MEMO Collaboration. Diabetes Audit and Research in Tayside, Scotland. Medicines Monitoring Unit. Diabetes Care. 1997 Sep; 20(9):1363–7.

Apidra prescribing information, Humulin prescribing information, Humalog prescribing information, Lantus prescribing information, Novolog prescribing information.

Drug-Drug Interaction Monograph for APIDRA, HUMULIN, HUMALOG, NOVOLOG, LANTUS (Insulin) #4

Perpetrator drug:

INN name/US brand name:	Disopyramide/Norpace, Rythmodan
Class/Molecular weight:	Small molecule/339 Da

Victim drug:

INN name/US brand name:	Insulin/Apidra, Humulin, Humalog, Novolog, Lantus, etc
Class/Molecular weight:	Hormone/∼5.8 kDa

Description of the effect of the DDI:
Pharmacodynamic interaction: Drugs that may increase the blood glucose-lowering effect of insulins can increase the risk of hypoglycemia. Case reports describe disopyramide-induced hypoglycemia as well as hypoglycemia on disopyramide treatment with concomitant insulin.

Description of the (proven, putative or hypothesized) mechanism of the DDI:
Both drugs have the potential to cause hypoglycemia.

Clinical consequences:
May require insulin dose adjustment and particularly close monitoring

Reference:

Cacoub P, Deray G, Baumelou A, Grimaldi A, Soubrie C, Jacobs C. Disopyramide-induced hypoglycemia: case report and review of the literature. Fundam Clin Pharmacol. 1989; 3(5):527–35.

Taketa K, Yamamoto Y. Hypoglycemic effect of disopyramide in a case of diabetes mellitus under insulin treatment. Acta Med Okayama. 1980 Sep;34(4):289–92.Croxson MS, Shaw DW, Henley PG, Gabriel HD. Disopyramide-induced hypoglycaemia and increased serum insulin.

Apidra prescribing information, Humulin prescribing information, Humalog prescribing information, Lantus prescribing information, Novolog prescribing information.

Drug-Drug Interaction Monograph for APIDRA, HUMULIN, HUMALOG, NOVOLOG, LANTUS (Insulin) #5

Perpetrator drug:

INN name/US brand name:	Fibrates (e.g. clofibrate, gemfibrozil, fenofibrate)
Class/Molecular weight:	Small molecule

Victim drug:

INN name/US brand name:	Insulin/Apidra, Humulin, Humalog, Novolog, Lantus, etc
Class/Molecular weight:	Hormone/~5.8 kDa

Description of the effect of the DDI:
Pharmacodynamic interaction: Drugs that may increase the blood glucose-lowering effect of insulins can increase the risk of hypoglycemia. In a case report where the patient added gemfibrozil to her medications, the patient experienced recurrent hypoglycemic episodes concurrant with a massive reduction in insulin requirements.

Description of the (proven, putative or hypothesized) mechanism of the DDI:
Insulin resistance appears to be a syndrome that is associated with a clustering of metabolic disorders, including non-insulin-dependent diabetes mellitus, obesity, hypertension, lipid abnormalities, and atherosclerotic cardiovascular disease. Thus if a facet of metabolic disorders is normalized, this may perhaps increase insulin sensitivity.

Clinical consequences:
May require insulin dose adjustment and particularly close monitoring.

Reference:
Klein J, Ott V, Schütt M, Klein HH. Recurrent hypoglycaemic episodes in a patient with Type 2 diabetes under fibrate therapy. J Diabetes Complications. 2002;16(3):246–8.

DeFronzo R and Ferrannini E. Insulin resistance: a multifaceted syndrome responsible for NIDDM, obesity, hypertension, dyslipidemia, and atherosclerotic cardiovascular disease. Diabetes Care 1991;14:173–194.

Apidra prescribing information, Humulin prescribing information, Humalog prescribing information, Lantus prescribing information, Novolog prescribing information.

Drug-Drug Interaction Monograph for APIDRA, HUMULIN, HUMALOG, NOVOLOG, LANTUS (Insulin) #6

Perpetrator drug:

INN name/US brand name:	Fluoxetine/Prozac
Class/Molecular weight:	Small molecule/309 Da

Victim drug:

INN name/US brand name:	Insulin/Apidra, Humulin, Humalog, Novolog, Lantus, etc
Class/Molecular weight:	Hormone/~5.8 kDa

Description of the effect of the DDI:
Pharmacodynamic interaction: Drugs that may increase the blood glucose-lowering effect of insulins can increase the risk of hypoglycemia. Fluoxetine treatment was associated with a large decline in HbA1c and insulin levels in obese diabetics compared to placebo. Fasting glucose levels also declined significantly.

Description of the (proven, putative or hypothesized) mechanism of the DDI:
Serotonergic antidepressants appear to enhance insulin sensitivity, reduce hyperglycemia and facilitate the storage of glucose as glycogen in muscle tissue.

Clinical consequences:
May require insulin dose adjustment and particularly close monitoring.

Reference:
MA Deeg and EW Lipkin. Hypoglycemia associated with the use of fluoxetine. West J Med. 1996 March; 164(3): 262–263.

McIntyre RS, Soczynska JK, Konarski JZ, Kennedy SH. The effect of antidepressants on glucose homeostasis and insulin sensitivity: synthesis and mechanisms. Expert Opin Drug Saf. 2006 Jan;5(1):157–68.

Gray DS, Fujioka K, Devine W et al.: A randomized double-blind clinical trial of fluoxetine in obese diabetics.Int. J. Obes. Relat. Metab. Disord. (1992)16(Suppl. 4): S67–S72.

Maheux P, Ducros F, Bourque J et al.: Fluoxetine improves insulin sensitivity in obese patients with non-insulindependent diabetes mellitus independently of weight loss. Int. J. Obes. Relat. Metab. Disord. (1997) 21(2):97–102.

Apidra prescribing information, Humulin prescribing information, Humalog prescribing information, Lantus prescribing information, Novolog prescribing information.

Drug-Drug Interaction Monograph for APIDRA, HUMULIN, HUMALOG, NOVOLOG, LANTUS (Insulin) #7

Perpetrator drug:

INN name/US brand name:	Monoamine oxidase inhibitors (e.g. hydralazine, phenelzine)
Class/Molecular weight:	Small molecule

Victim drug:

INN name/US brand name:	Insulin/Apidra, Humulin, Humalog, Novolog, Lantus, etc
Class/Molecular weight:	Hormone/~5.8 kDa

Description of the effect of the DDI:
Pharmacodynamic interaction: Drugs that may increase the blood glucose-lowering effect of insulins can increase the risk of hypoglycemia. Depressed subjects responding to monoamine oxidase inhibitor (MAOI) therapy displayed a decrease in peak glucose concentration, a lower fasting blood glucose level and shorter duration of the hyperglycemic period after the administration of glucose ($p < 0.05$).

Description of the (proven, putative or hypothesized) mechanism of the DDI:
MAOIs are associated with hypoglycemia.

Clinical consequences:
May require insulin dose adjustment and particularly close monitoring.

Reference:
McIntyre RS, Soczynska JK, Konarski JZ, Kennedy SH. The effect of antidepressants on glucose homeostasis and insulin sensitivity: synthesis and mechanisms. Expert Opin Drug Saf. 2006;5(1):157–68.

(Continued)

Adnitt PI. Hypoglycemic action of monoamineoxidase inhibitors (MAOI'S). Diabetes. 1968;17(10):628–33.

Apidra prescribing information, Humulin prescribing information, Humalog prescribing information, Lantus prescribing information, Novolog prescribing information.

Drug-Drug Interaction Monograph for APIDRA, HUMULIN, HUMALOG, NOVOLOG, LANTUS (Insulin) #8

Perpetrator drug:

INN name/US brand name:	Propoxyphene/Darvon
Class/Molecular weight:	Small molecule/339 Da

Victim drug:

INN name/US brand name:	Insulin/Apidra, Humulin, Humalog, Novolog, Lantus, etc
Class/Molecular weight:	Hormone/~5.8 kDa

Description of the effect of the DDI:
Pharmacodynamic interaction: Drugs that may increase the blood glucose-lowering effect of insulins can increase the risk of hypoglycemia. Case reports of this interaction were reported in patients with chronic renal failure.

Description of the (proven, putative or hypothesized) mechanism of the DDI:
Both drugs reduce blood glucose.

Clinical consequences:
May require insulin dose adjustment and particularly close monitoring.

Reference:

Almirall J, Montoliu J, Torras A, Revert L Propoxyphene-Induced Hypoglycemia in a Patient with Chronic Renal Failure. Nephron 1989;53:273–275.

Shah P, Aniszweski J, Service FJ. Propoxyphene-induced hypoglycemia in renal failure. Endocrine Practice 2006:12: 170–173.

Apidra prescribing information, Humulin prescribing information, Humalog prescribing information, Lantus prescribing information, Novolog prescribing information.

Drug-Drug Interaction Monograph for APIDRA, HUMULIN, HUMALOG, NOVOLOG, LANTUS (Insulin) #9

Perpetrator drug:

INN name/US brand name:	Pentoxifylline/Trental
Class/Molecular weight:	Small molecule/278 Da

Victim drug:

INN name/US brand name:	Insulin/Apidra, Humulin, Humalog, Novolog, Lantus, etc
Class/Molecular weight:	Hormone/~5.8 kDa

Description of the effect of the DDI:
Pharmacodynamic interaction: Drugs that may increase the blood glucose-lowering effect of insulins can increase the risk of hypoglycemia.

Description of the (proven, putative or hypothesized) mechanism of the DDI:
Both drugs reduce blood glucose.

Clinical consequences:
May require insulin dose adjustment and particularly close monitoring.

Reference:
Apidra prescribing information, Humulin prescribing information, Humalog prescribing information, Lantus prescribing information, Novolog prescribing information.

Drug-Drug Interaction Monograph for APIDRA, HUMULIN, HUMALOG, NOVOLOG, LANTUS (Insulin) #10

Perpetrator drug:

INN name/US brand name:	Salicylates (e.g. aspirin)
Class/Molecular weight:	Small molecule

Victim drug:

INN name/US brand name:	Insulin/Apidra, Humulin, Humalog, Novolog, Lantus, etc
Class/Molecular weight:	Hormone/~5.8 kDa

Description of the effect of the DDI:
Pharmacodynamic interaction: Drugs that may increase the blood glucose-lowering effect of insulins can increase the risk of hypoglycemia.

Description of the (proven, putative or hypothesized) mechanism of the DDI:
Salicylates lower blood sugar and enhance glucose-stimulated insulin secretion in normal and diabetic subjects. Although the hypoglycemic effect appears to be mediated by enhanced insulin secretion, extrapancreatic mechanisms cannot be excluded. The mechanism of the enhanced insulin secretion appears to be mediated by prostaglandin synthesis inhibition.

Clinical consequences:
May require insulin dose adjustment and particularly close monitoring.

Reference:
Limbeck G, Ruvalcaba R, Samols E, Kelley V. Salicylates and Hypoglycemia. Archives of Pediatrics & Adolescent Medicine. 1965:109:165–167.

Baron S. Salicylates as Hypoglycemic Agents. Diabetes Care. 1982:5:64–71.

Apidra prescribing information, Humulin prescribing information, Humalog prescribing information, Lantus prescribing information, Novolog prescribing information.

Drug-Drug Interaction Monograph for APIDRA, HUMULIN, HUMALOG, NOVOLOG, LANTUS (Insulin) #11

Perpetrator drug:

INN name/US brand name:	Somatostatin analogs/(see Accretropin)
Class/Molecular weight:	Hormone

Victim drug:

INN name/US brand name:	Insulin/Apidra, Humulin, Humalog, Novolog, Lantus, etc
Class/Molecular weight:	Hormone/∼5.8 kDa

Description of the effect of the DDI:

Pharmacodynamic interaction: Drugs that may increase the blood glucose-lowering effect of insulins can increase the risk of hypoglycemia. Infusion of somatostatin (combined deficiency of glucagon and growth hormone) accentuated insulin-induced hypoglycemia (plasma glucose nadir: 36+/−2 ng/dl during infusion of somatostatin vs. 47+/−2 mg/dl in control studies, $P < 0.01$) and impaired restoration of normoglycemia (plasma glucose at min 90: 73+/−3 mg/dl at end of somatostatin infusion vs. 92+/−3 mg/dl in control studies, $P < 0.01$).

Description of the (proven, putative or hypothesized) mechanism of the DDI:

Impaired recovery of plasma glucose was due to blunting of the compensatory increase in glucose appearance since glucose disappearance was not augmented, and was attributable to suppression of glucagon secretion.

Clinical consequences:

May require insulin dose adjustment and particularly close monitoring.

Reference:

Rizza R, Cryer P, and Gerich J. Role of glucagon, catecholamines, and growth hormone in human glucose counterregulation. J Clin Invest. 1979;64: 62–71.

Apidra prescribing information, Humulin prescribing information, Humalog prescribing information, Lantus prescribing information, Novolog prescribing information.

Drug-Drug Interaction Monograph for APIDRA, HUMULIN, HUMALOG, NOVOLOG, LANTUS (Insulin) #12

Perpetrator drug:

INN name/US brand name:	Sulfonamide antibiotics
Class/Molecular weight:	Small molecule

Victim drug:

INN name/US brand name:	Insulin/Apidra, Humulin, Humalog, Novolog, Lantus, etc
Class/Molecular weight:	Hormone/∼5.8 kDa

Description of the effect of the DDI:

Pharmacodynamic interaction: Drugs that may increase the blood glucose-lowering effect of insulins can increase the risk of hypoglycemia.

Description of the (proven, putative or hypothesized) mechanism of the DDI:

Because of their structural similarities to sulfonylureas, sulfonamides are liable to facilitate hypoglycemia by increasing insulin release in susceptible individuals.

Clinical consequences:
May require insulin dose adjustment and particularly close monitoring.

Reference:
Arem R, Garber A, Field J. Sulfonamide-induced hypoglycemia in chronic renal failure. Archives of Int Med. 1983;143:827–829.

Hekimsoy Z, Biberoğlu S, Cömlekçi A, Tarhan O, Mermut C, Biberoğlu K. Trimethoprim/sulfamethoxazole-induced hypoglycemia in a malnourished patient with severe infection. Eur J Endocrinol. 1997;136(3):304–6.

Apidra prescribing information, Humulin prescribing information, Humalog prescribing information, Lantus prescribing information, Novolog prescribing information.

Drug-Drug Interaction Monograph for APIDRA, HUMULIN, HUMALOG, NOVOLOG, LANTUS (Insulin) #13

Perpetrator drug:

INN name/US brand name:	Corticosteroids
Class/Molecular weight:	Small molecule

Victim drug:

INN name/US brand name:	Insulin/Apidra, Humulin, Humalog, Novolog, Lantus, etc
Class/Molecular weight:	Hormone/~5.8 kDa

Description of the effect of the DDI:
Pharmacodynamic interaction: Drugs that may reduce the blood glucose-lowering effect of insulins can increase the risk of hypoglycemia. Case report of hypoglycemic seizure occurring after high-dose inhaled corticosteroids.

Description of the (proven, putative or hypothesized) mechanism of the DDI:
None.

Clinical consequences:
May require insulin dose adjustment and particularly close monitoring.

Reference:
Dunlop KA, Carson DJ, Shields MD. Hypoglycemia due to adrenal suppression secondary to high-dose nebulized corticosteroid. Pediatr Pulmonol. 2002;34(1):85–6.

Apidra prescribing information, Humulin prescribing information, Humalog prescribing information, Lantus prescribing information, Novolog prescribing information.

Drug-Drug Interaction Monograph for APIDRA, HUMULIN, HUMALOG, NOVOLOG, LANTUS (Insulin) #14

Perpetrator drug:

INN name/US brand name:	Niacin
Class/Molecular weight:	Small molecule/123 Da

Victim drug:

INN name/US brand name:	Insulin/Apidra, Humulin, Humalog, Novolog, Lantus, etc
Class/Molecular weight:	Hormone/~5.8 kDa

(Continued)

Description of the effect of the DDI:
Pharmacodynamic interaction: Drugs that may reduce the blood glucose-lowering effect of insulins can increase the risk of hypoglycemia. Case reports of high dose niacin causing hypoglycemia.

Description of the (proven, putative or hypothesized) mechanism of the DDI:
None.

Clinical consequences:
May require insulin dose adjustment and particularly close monitoring.

Reference:
Arcinegas-Rodriguez S et al. Metabolic acidosis, hypoglycemia, and severe myalgias: an attempt to mask urine drug screen results. Pediatr Emerg Care. 2011;27(4): 315–7.

Apidra prescribing information, Humulin prescribing information, Humalog prescribing information, Lantus prescribing information, Novolog prescribing information.

Drug-Drug Interaction Monograph for APIDRA, HUMULIN, HUMALOG, NOVOLOG, LANTUS (Insulin) #15

Perpetrator drug:

INN name/US brand name:	Danazol/Danocrine
Class/Molecular weight:	Small molecule/338 Da

Victim drug:

INN name/US brand name:	Insulin/Apidra, Humulin, Humalog, Novolog, Lantus, etc
Class/Molecular weight:	Hormone/~5.8 kDa

Description of the effect of the DDI:
Pharmacodynamic interaction: Drugs that may reduce the blood glucose-lowering effect of insulins can increase the risk of hypoglycemia. In subjects treated with danazol for lupus thrombocytopenia, side effects including hypoglycemia were reported.

Description of the (proven, putative or hypothesized) mechanism of the DDI:
None.

Clinical consequences:
May require insulin dose adjustment and particularly close monitoring.

Reference:
Wong KL. Danazol in treatment of lupus thrombocytopenia. Asian Pac J Allergy Immunol. 1991;9(2):125–9.

Apidra prescribing information, Humulin prescribing information, Humalog prescribing information, Lantus prescribing information, Novolog prescribing information.

Drug-Drug Interaction Monograph for APIDRA, HUMULIN, HUMALOG, NOVOLOG, LANTUS (Insulin) #16

Perpetrator drug:

INN name/US brand name:	Diuretics (e.g. furosemide, hydrochlorothiazide)
Class/Molecular weight:	Small molecule

Victim drug:

INN name/US brand name:	Insulin/Apidra, Humulin, Humalog, Novolog, Lantus, etc
Class/Molecular weight:	Hormone/~5.8 kDa

Description of the effect of the DDI:
Pharmacodynamic interaction: Drugs that may reduce the blood glucose-lowering effect of insulins can increase the risk of hypoglycemia.

Description of the (proven, putative or hypothesized) mechanism of the DDI:
Insulin resistance appears to be a syndrome that is associated with a clustering of metabolic disorders, including non-insulin-dependent diabetes mellitus, obesity, hypertension, lipid abnormalities, and atherosclerotic cardiovascular disease. Thus if a facet of metabolic disorders is normalized, this may perhaps increase insulin sensitivity

Clinical consequences:
May require insulin dose adjustment and particularly close monitoring.

Reference:
DeFronzo R and Ferrannini E. Insulin resistance: a multifaceted syndrome responsible for NIDDM, obesity, hypertension, dyslipidemia, and atherosclerotic cardiovascular disease. Diabetes Care 1991;14:173–194.

Apidra prescribing information, Humulin prescribing information, Humalog prescribing information, Lantus prescribing information, Novolog prescribing information.

Drug-Drug Interaction Monograph for APIDRA, HUMULIN, HUMALOG, NOVOLOG, LANTUS (Insulin) #17

Perpetrator drug:

INN name/US brand name:	Sympathomimetic agents (eg. epinephrine, albuterol, terbutaline)
Class/Molecular weight:	Small molecule

Victim drug:

INN name/US brand name:	Insulin/Apidra, Humulin, Humalog, Novolog, Lantus, etc
Class/Molecular weight:	Hormone/~5.8 kDa

Description of the effect of the DDI:
Pharmacodynamic interaction: Drugs that may reduce the blood glucose-lowering effect of insulins can increase the risk of hypoglycemia.

(Continued)

Description of the (proven, putative or hypothesized) mechanism of the DDI:
None.

Clinical consequences:
May require insulin dose adjustment and particularly close monitoring.

Reference:
Apidra prescribing information, Humulin prescribing information, Humalog prescribing information, Lantus prescribing information, Novolog prescribing information.

Drug-Drug Interaction Monograph for APIDRA, HUMULIN, HUMALOG, NOVOLOG, LANTUS (Insulin) #18

Perpetrator drug:

INN name/US brand name:	Glucagon
Class/Molecular weight:	Peptide hormone

Victim drug:

INN name/US brand name:	Insulin/Apidra, Humulin, Humalog, Novolog, Lantus, etc
Class/Molecular weight:	Hormone/∼5.8 kDa

Description of the effect of the DDI:
Pharmacodynamic interaction: Drugs that may reduce the blood glucose-lowering effect of insulins can increase the risk of hypoglycemia.

Description of the (proven, putative or hypothesized) mechanism of the DDI:
None.

Clinical consequences:
May require insulin dose adjustment and particularly close monitoring.

Reference:
Apidra prescribing information, Humulin prescribing information, Humalog prescribing information, Lantus prescribing information, Novolog prescribing information.

Drug-Drug Interaction Monograph for APIDRA, HUMULIN, HUMALOG, NOVOLOG, LANTUS (Insulin) #19

Perpetrator drug:

INN name/US brand name:	Isoniazid/Laniazid, Nydrazid
Class/Molecular weight:	Small molecule/137 Da

Victim drug:

INN name/US brand name:	Insulin/Apidra, Humulin, Humalog, Novolog, Lantus, etc
Class/Molecular weight:	Hormone/∼5.8 kDa

Description of the effect of the DDI:
Pharmacodynamic interaction: Drugs that may reduce the blood glucose-lowering effect of insulins can increase the risk of hypoglycemia.

Description of the (proven, putative or hypothesized) mechanism of the DDI:
None.

Clinical consequences:
May require insulin dose adjustment and particularly close monitoring.

Reference:
Apidra prescribing information, Humulin prescribing information, Humalog prescribing information, Lantus prescribing information, Novolog prescribing information.

Drug-Drug Interaction Monograph for APIDRA, HUMULIN, HUMALOG, NOVOLOG, LANTUS (Insulin) #20

Perpetrator drug:

INN name/US brand name:	Phenothiazine derivatives (e.g. chlorpromazine, thioridazine)
Class/Molecular weight:	Small molecule

Victim drug:

INN name/US brand name:	Insulin/Apidra, Humulin, Humalog, Novolog, Lantus, etc
Class/Molecular weight:	Hormone/~5.8 kDa

Description of the effect of the DDI:
Pharmacodynamic interaction: Drugs that may reduce the blood glucose-lowering effect of insulins can increase the risk of hypoglycemia.

Description of the (proven, putative or hypothesized) mechanism of the DDI:
None.

Clinical consequences:
May require insulin dose adjustment and particularly close monitoring.

Reference:
Apidra prescribing information, Humulin prescribing information, Humalog prescribing information, Lantus prescribing information, Novolog prescribing information.

Drug-Drug Interaction Monograph for APIDRA, HUMULIN, HUMALOG, NOVOLOG, LANTUS (Insulin) #21

Perpetrator drug:

INN name/US brand name:	Thyroid hormones
Class/Molecular weight:	Small molecule

Victim drug:

INN name/US brand name:	Insulin/Apidra, Humulin, Humalog, Novolog, Lantus, etc
Class/Molecular weight:	Hormone/~5.8 kDa

Description of the effect of the DDI:
Pharmacodynamic interaction: Drugs that may reduce the blood glucose-lowering effect of insulins can increase the risk of hypoglycemia.

(Continued)

Description of the (proven, putative or hypothesized) mechanism of the DDI:
None.

Clinical consequences:
May require insulin dose adjustment and particularly close monitoring.

Reference:
Apidra prescribing information, Humulin prescribing information, Humalog prescribing information, Lantus prescribing information, Novolog prescribing information.

Drug-Drug Interaction Monograph for APIDRA, HUMULIN, HUMALOG, NOVOLOG, LANTUS (Insulin) #22

Perpetrator drug:

INN name/US brand name: Estrogens
Class/Molecular weight: Small molecule

Victim drug:

INN name/US brand name: Insulin/Apidra, Humulin, Humalog, Novolog, Lantus, etc
Class/Molecular weight: Hormone/~5.8 kDa

Description of the effect of the DDI:
Pharmacodynamic interaction: Drugs that may reduce the blood glucose-lowering effect of insulins can increase the risk of hypoglycemia.

Description of the (proven, putative or hypothesized) mechanism of the DDI:
none.

Clinical consequences:
May require insulin dose adjustment and particularly close monitoring.

Reference:
Apidra prescribing information, Humulin prescribing information, Humalog prescribing information, Lantus prescribing information, Novolog prescribing information.

Drug-Drug Interaction Monograph for APIDRA, HUMULIN, HUMALOG, NOVOLOG, LANTUS (Insulin) #23

Perpetrator drug:

INN name/US brand name: Progestins (e.g., in oral contraceptics)
Class/Molecular weight: Small molecule

Victim drug:

INN name/US brand name: Insulin/Apidra, Humulin, Humalog, Novolog, Lantus, etc
Class/Molecular weight: Hormone/~5.8 kDa

Description of the effect of the DDI:
Pharmacodynamic interaction: Drugs that may reduce the blood glucose-lowering effect of insulins can increase the risk of hypoglycemia.

Description of the (proven, putative or hypothesized) mechanism of the DDI:
None.

Clinical consequences:
May require insulin dose adjustment and particularly close monitoring.

Reference:
Apidra prescribing information, Humulin prescribing information, Humalog prescribing information, Lantus prescribing information, Novolog prescribing information.

Drug-Drug Interaction Monograph for APIDRA, HUMULIN, HUMALOG, NOVOLOG, LANTUS (Insulin) #24

Perpetrator drug:

INN name/US brand name:	Protease inhibitors
Class/Molecular weight:	Small molecule

Victim drug:

INN name/US brand name:	Insulin/Apidra, Humulin, Humalog, Novolog, Lantus, etc
Class/Molecular weight:	Hormone/~5.8 kDa

Description of the effect of the DDI:
Pharmacodynamic interaction: Drugs that may reduce the blood glucose-lowering effect of insulins can increase the risk of hypoglycemia.

Description of the (proven, putative or hypothesized) mechanism of the DDI:
None.

Clinical consequences:
May require insulin dose adjustment and particularly close monitoring.

Reference:
Apidra prescribing information, Humulin prescribing information, Humalog prescribing information, Lantus prescribing information, Novolog prescribing information.

Drug-Drug Interaction Monograph for APIDRA, HUMULIN, HUMALOG, NOVOLOG, LANTUS (Insulin) #25

Perpetrator drug:

INN name/US brand name:	Atypical antipsychotics
Class/Molecular weight:	Small molecule

Victim drug:

INN name/US brand name:	Insulin/Apidra, Humulin, Humalog, Novolog, Lantus, etc
Class/Molecular weight:	Hormone/~5.8 kDa

Description of the effect of the DDI:
Pharmacodynamic interaction: Drugs that may reduce the blood glucose-lowering effect of insulins can increase the risk of hypoglycemia.

(Continued)

Description of the (proven, putative or hypothesized) mechanism of the DDI:
none.

Clinical consequences:
May require insulin dose adjustment and particularly close monitoring.

Reference:
Apidra prescribing information, Humulin prescribing information, Humalog prescribing information, Lantus prescribing information, Novolog prescribing information.

Drug-Drug Interaction Monograph for APIDRA, HUMULIN, HUMALOG, NOVOLOG, LANTUS (Insulin) #26

Perpetrator drug:

INN name/US brand name: Beta-blockers
Class/Molecular weight: Small molecule

Victim drug:

INN name/US brand name: Insulin/Apidra, Humulin, Humalog, Novolog, Lantus, etc
Class/Molecular weight: Hormone/∼5.8 kDa

Description of the effect of the DDI:
Pharmacodynamic interaction: May either increase or decrease the blood-glucose-lovwering effect of insulin.

Description of the (proven, putative or hypothesized) mechanism of the DDI:
None.

Clinical consequences:
May require insulin dose adjustment and particularly close monitoring.

Reference:
Apidra prescribing information, Humulin prescribing information, Humalog prescribing information, Lantus prescribing information, Novolog prescribing information.

Drug-Drug Interaction Monograph for APIDRA, HUMULIN, HUMALOG, NOVOLOG, LANTUS (Insulin) #27

Perpetrator drug:

INN name/US brand name: Clonidine/Catapres
Class/Molecular weight: Small molecule/230 Da

Victim drug:

INN name/US brand name: Insulin/Apidra, Humulin, Humalog, Novolog, Lantus, etc
Class/Molecular weight: Hormone/∼5.8 kDa

Description of the effect of the DDI:
Pharmacodynamic interaction: May either increase or decrease the blood-glucose-lovwering effect of insulin.

Description of the (proven, putative or hypothesized) mechanism of the DDI:
None.

Clinical consequences:
May require insulin dose adjustment and particularly close monitoring.

Reference:
Apidra USPI, Feb 2009, Humulin USPI, May 2011, Humalog USPI, Oct 2012, Lantus USPI, Mar Apidra prescribing information, Humulin prescribing information, Humalog prescribing information, Lantus prescribing information, Novolog prescribing information.

Drug-Drug Interaction Monograph for APIDRA, HUMULIN, HUMALOG, NOVOLOG, LANTUS (Insulin) #28

Perpetrator drug:

INN name/US brand name:	Lithium salts
Class/Molecular weight:	Small molecule

Victim drug:

INN name/US brand name:	Insulin/Apidra, Humulin, Humalog, Novolog, Lantus, etc
Class/Molecular weight:	Hormone/~5.8 kDa

Description of the effect of the DDI:
Pharmacodynamic interaction: May either increase or decrease the blood-glucose-lovwering effect of insulin.

Description of the (proven, putative or hypothesized) mechanism of the DDI:
None.

Clinical consequences:
May require insulin dose adjustment and particularly close monitoring.

Reference:
Apidra prescribing information, Humulin prescribing information, Humalog prescribing information, Lantus prescribing information, Novolog prescribing information.

Drug-Drug Interaction Monograph for APIDRA, HUMULIN, HUMALOG, NOVOLOG, LANTUS (Insulin) #29

Perpetrator drug:

INN name/US brand name:	Alcohol
Class/Molecular weight:	Small molecule

Victim drug:

INN name/US brand name:	Insulin/Apidra, Humulin, Humalog, Novolog, Lantus, etc
Class/Molecular weight:	Hormone/~5.8 kDa

Description of the effect of the DDI:
Pharmacodynamic interaction: May either increase or decrease the blood-glucose-lowering effect of insulin.

(Continued)

Description of the (proven, putative or hypothesized) mechanism of the DDI:
None.

Clinical consequences:
May require insulin dose adjustment and particularly close monitoring.

Reference:
Apidra prescribing information, Humulin prescribing information, Humalog prescribing information, Lantus prescribing information, Novolog prescribing information.

Drug-Drug Interaction Monograph for ARCALYST (Rilonacept) #1

Perpetrator drug:

INN name/US brand name:	TNF-blocking agents
Class/Molecular weight:	

Victim drug:

INN name/US brand name:	Rilonacept/Arcalyst
Class/Molecular weight:	Fusion protein/251 kDa

Description of the effect of the DDI:
Pharmacodynamic interaction: Concomitant administration of another drug that blocks IL-1 with a TNF-blocking agent in another patient population has been associated with an increased risk of serious infections and an increased risk of neutropenia. Concomitant administration of rilonacept with TNF-blocking agents may also result in similar toxicities and is not recommended.

Description of the (proven, putative or hypothesized) mechanism of the DDI:
Both are immunomodulators which carry the risk of increased infections.

Clinical consequences:
Increased risk of serious infections and neutropenia.

Reference:
Arcalyst prescribing information.

Drug-Drug Interaction Monograph for ARCALYST (Rilonacept) #2

Perpetrator drug:

INN name/US brand name:	Rilonacept/Arcalyst
Class/Molecular weight:	Fusion protein/251 kDa

Victim drug:

INN name/US brand name:	CYP450 substrates
Class/Molecular weight:	

Description of the effect of the DDI:
Normalization of CYP450 enzymes.

Description of the (proven, putative or hypothesized) mechanism of the DDI:
The formation of CYP450 enzymes is suppressed by increased levels of cytokines (e.g., IL-1) during chronic inflammation. Thus it is expected that for a molecule that binds to IL-1, such as rilonacept, the formation of CYP450 enzymes could be normalized.

Clinical consequences:
Clinically relevant for CYP450 substrates with narrow therapeutic index. Upon initiation of rilonacept, in patients being treated with these types of medicinal products, therapeutic monitoring of the effect or drug concentration should be performed and the individual dose of the medicinal product may need to be adjusted as needed.

Reference:
Arcalyst prescribing information

Lee JI, Zhang L, Men AY, Kenna LA, Huang SM. CYP-mediated therapeutic protein-drug interactions: clinical findings, proposed mechanisms and regulatory implications. Clin Pharmacokinet. 2010;49(5):295–310.

Drug-Drug Interaction Monograph for AVASTIN (Bevacizumab) #1

Perpetrator drug:

INN name/US brand name:	Bevacizumab/Avastin
Class/Molecular weight:	Monoclonal antibody/149 kDa

Victim drug:

INN name/US brand name:	Irinotecan/Camptosar, Campto
Class/Molecular weight:	Small molecule/587–677 Da

Description of the effect of the DDI:
No significant effect of bevacizumab on the pharmacokinetics of irinotecan or its active metabolite SN38.

Description of the (proven, putative or hypothesized) mechanism of the DDI:
None.

Clinical consequences:
None.

Reference:
Horita Y, Yamada Y, Hirashima Y, Kato K, Nakajima T, Hamaguchi T, Shimada Y. Effects of bevacizumab on plasma concentration of irinotecan and its metabolites in advanced colorectal cancer patients receiving FOLFIRI with bevacizumab as second-line chemotherapy. Cancer Chemother Pharmacol. 2010;65(3):467–71. Epub 2009 Jun 25.

Avastin prescribing information.

Drug-Drug Interaction Monograph for AVASTIN (Bevacizumab) #2

Perpetrator drug:

INN name/US brand name:	Bevacizumab/Avastin
Class/Molecular weight:	Monoclonal antibody/149 kDa

Victim drug:

INN name/US brand name:	Carboplatin/Paraplatin
Class/Molecular weight:	Small molecule/371 Da

(Continued)

Description of the effect of the DDI:
Based on limited data, there did not appear to be a difference in the mean exposure of carboplatin when each was administered alone or in combination with bevacizumab.

Description of the (proven, putative or hypothesized) mechanism of the DDI:
None.

Clinical consequences:
None.

Reference:
Avastin prescribing information.

Drug-Drug Interaction Monograph for AVASTIN (Bevacizumab) #3

Perpetrator drug:

INN name/US brand name:	Bevacizumab/Avastin
Class/Molecular weight:	Monoclonal antibody/149 kDa

Victim drug:

INN name/US brand name:	Paclitaxel/Taxol
Class/Molecular weight:	Small Molecule/854 Da

Description of the effect of the DDI:
Based on limited data, there did not appear to be a difference in the mean exposure of paclitaxel when each was administered alone or in combination with bevacizumab. However, 3 of the 8 patients receiving bevacizumab plus paclitaxel/carboplatin had substantially lower paclitaxel exposure after 4 cycles of treatment (Day 63) than those at Day 0. Patients receiving paclitaxel/carbopatin without Avastin had a greater paclitaxel exposure at Day 63 than at Day 0.

Description of the (proven, putative or hypothesized) mechanism of the DDI:
None.

Clinical consequences:
None.

Reference:
Avastin prescribing information.

Drug-Drug Interaction Monograph for AVASTIN (Bevacizumab) #4

Perpetrator drug:

INN name/US brand name:	Bevacizumab/Avastin
Class/Molecular weight:	Monoclonal antibody/149 kDa

Victim drug:

INN name/US brand name:	Interferon alfa/Multiferon
Class/Molecular weight:	Interferon

Description of the effect of the DDI:
No difference in the mean exposure of interferon alfa when administered alone or in combination with bevacizumab.

Description of the (proven, putative or hypothesized) mechanism of the DDI:
None.

Clinical consequences:
None.

Reference:
Avastin prescribing information.

Drug-Drug Interaction Monograph for BENLYSTA (Belimumab) #1

Perpetrator drug:

INN name/US brand name:	Corticosteroids
Class/Molecular weight:	Small molecule

Victim drug:

INN name/US brand name:	Belimumab/Benlysta
Class/Molecular weight:	Monoclonal antibody/147 kDa

Description of the effect of the DDI:
Formal drug interaction studies have not been performed. In clinical trials of patients with systemic lupus erythematosus (SLE), there was no evidence of a clinically meaningful effect of concomitant medications on belimumab pharmacokinetics.

Description of the (proven, putative or hypothesized) mechanism of the DDI:
None.

Clinical consequences:
None.

Reference:
Benlysta prescribing information.

Drug-Drug Interaction Monograph for BENLYSTA (Belimumab) #2

Perpetrator drug:

INN name/US brand name:	Antimalarials
Class/Molecular weight:	Small molecule

Victim drug:

INN name/US brand name:	Belimumab/Benlysta
Class/Molecular weight:	Monoclonal antibody/147 kDa

Description of the effect of the DDI:
Formal drug interaction studies have not been performed. In clinical trials of patients with systemic lupus erythematosus (SLE), there was no evidence of a clinically meaningful effect of concomitant antimalarials on belimumab pharmacokinetics.

Description of the (proven, putative or hypothesized) mechanism of the DDI:
None.

Clinical consequences:
None.

Reference:
Benlysta prescribing information.

Drug-Drug Interaction Monograph for BENLYSTA (Belimumab) #3

Perpetrator drug:

INN name/US brand name:	Immunomodulatory and immunosuppressive agents (incl azathioprine, methotrexate, and mycophenolate)
Class/Molecular weight:	

Victim drug:

INN name/US brand name:	Belimumab/Benlysta
Class/Molecular weight:	Monoclonal antibody/147 kDa

Description of the effect of the DDI:
Formal drug interaction studies have not been performed. In clinical trials of patients with systemic lupus erythematosus (SLE), there was no evidence of a clinically meaningful effect of concomitant immunomodulatory and immunosuppressive agents on belimumab pharmacokinetics.

Description of the (proven, putative or hypothesized) mechanism of the DDI:
None.

Clinical consequences:
None.

Reference:
Benlysta prescribing information.

Drug-Drug Interaction Monograph for BENLYSTA (Belimumab) #4

Perpetrator drug:

INN name/US brand name:	Angiotensin pathway antihypertensives
Class/Molecular weight:	Small molecules

Victim drug:

INN name/US brand name:	Belimumab/Benlysta
Class/Molecular weight:	Monoclonal antibody/147 kDa

Description of the effect of the DDI:
Formal drug interaction studies have not been performed. In clinical trials of patients with systemic lupus erythematosus (SLE), there was no evidence of a clinically meaningful effect of concomitant angiotensin pathway antihypertensives on belimumab pharmacokinetics.

Description of the (proven, putative or hypothesized) mechanism of the DDI:
None.

Clinical consequences:
None.

Reference:
Benlysta prescribing information.

Drug-Drug Interaction Monograph for BENLYSTA (Belimumab) #5

Perpetrator drug:

INN name/US brand name:	HMG-CoA reductase inhibitors (statins)
Class/Molecular weight:	Small molecules

Victim drug:

INN name/US brand name:	Belimumab/Benlysta
Class/Molecular weight:	Monoclonal antibody/147 kDa

Description of the effect of the DDI:
Formal drug interaction studies have not been performed. In clinical trials of patients with systemic lupus erythematosus (SLE), there was no evidence of a clinically meaningful effect of concomitant HMG-CoA reductase inhibitors on belimumab pharmacokinetics.

Description of the (proven, putative or hypothesized) mechanism of the DDI:
None.

Clinical consequences:
None.

Reference:
Benlysta prescribing information.

Drug-Drug Interaction Monograph for BENLYSTA (Belimumab) #6

Perpetrator drug:

INN name/US brand name:	NSAIDS (e.g. ibuprofen)
Class/Molecular weight:	Small molecules

Victim drug:

INN name/US brand name:	Belimumab/Benlysta
Class/Molecular weight:	Monoclonal antibody/147 kDa

Description of the effect of the DDI:
Formal drug interaction studies have not been performed. In clinical trials of patients with systemic lupus erythematosus (SLE), there was no evidence of a clinically meaningful effect of concomitant NSAIDS on belimumab pharmacokinetics.

Description of the (proven, putative or hypothesized) mechanism of the DDI:
None.

Clinical consequences:
None.

Reference:
Benlysta prescribing information.

Drug-Drug Interaction Monograph for BIO-TROPIN/(Somatropin)

See ACCRETROPIN (Somatropin) drug-drug interaction monographs #1– #4

Reference:	Bio-Tropin US prescribing information

Drug-Drug Interaction Monograph for BOTOX, DYSPORT (Onabotulinumtoxin A) #1

Perpetrator drug:

INN name/US brand name:	Aminoglycosides or other agents interfering with neuromuscular transmission (curare-like compounds)
Class/Molecular weight:	Small molecules

Victim drug:

INN name/US brand name:	Onabotulinumtoxin A/Botox, Dysport
Class/Molecular weight:	Protein Toxin/149.3 kDa

Description of the effect of the DDI:

<Note: Besides cosmetic uses, Botox is also used for the treatment of destusor overactivity associated with a neurologic condition, chronic migraine, upper limb spasticity, cervical dystonia, primary axillary hyperhidrosis, blepharospasm and strabismus>

Effect of the onabotulinumtoxin A may be potentiated with coadministration of aminoglycosides or other agents interfering with neuromuscular transmission (curare-like compounds).

Description of the (proven, putative or hypothesized) mechanism of the DDI:

None.

Clinical consequences:

None.

Reference:

Botox prescribing information.

Dysport prescribing information.

Drug-Drug Interaction Monograph for BOTOX, DYSPORT (Onabotulinumtoxin) #2

Perpetrator drug:

INN name/US brand name:	Onabotulinumtoxin A/Botox, Dysport
Class/Molecular weight:	Protein Toxin/149.3 kDa

Victim drug:

INN name/US brand name:	Anticholinergic drugs
Class/Molecular weight:	

Description of the effect of the DDI:

<Note: Besides cosmetic uses, Onabotulinumtoxin A is also used for the treatment of destusor overactivity associated with a neurologic condition, chronic migraine, upper limb spasticity, cervical dystonia, primary axillary hyperhidrosis, blepharospasm and strabismus>

Use of anticholinergic drugs after administration of Botox may potentiate systemic anticholinergic effects.

Description of the (proven, putative or hypothesized) mechanism of the DDI:

None.

Clinical consequences:

None.

Reference:

Botox prescribing information.

Dysport prescribing information.

Drug-Drug Interaction Monograph for BOTOX, DYSPORT (Onabotulinumtoxin) #3

Perpetrator drug:
INN name/US brand name: Muscle relaxants
Class/Molecular weight:

Victim drug:
INN name/US brand name: Onabotulinumtoxin A/Botox, Dysport
Class/Molecular weight: Protein Toxin/149.3 kDa

Description of the effect of the DDI:
<Note: Besides cosmetic uses, Botox is also used for the treatment of destusor overactivity associated with a neurologic condition, chronic migraine, upper limb spasticity, cervical dystonia, primary axillary hyperhidrosis, blepharospasm and strabismus>

Excessive weakness may be exaggerated by administration of a muscle relaxant before or after administration of onabotulinumtoxin A.

Description of the (proven, putative or hypothesized) mechanism of the DDI:
None.

Clinical consequences:
None.

Reference:
Botox prescribing information.

Dysport prescribing information.

Drug-Drug Interaction Monograph for BYDUREON (Exenatide) #1

Perpetrator drug:
INN name/US brand name: Orally administered drugs
Class/Molecular weight:

Victim drug:
INN name/US brand name: Exenatide/Bydureon
Class/Molecular weight: Peptide/4.2 kDa

Description of the effect of the DDI:
Exenatide has the potential to reduce the extent and rate of absorption of orally administered drugs.

Description of the (proven, putative or hypothesized) mechanism of the DDI:
Exenatide slows gastric emptying.

Clinical consequences:
Use caution when administering oral medications with exenatide. Exenatide should be used with caution in patients receiving oral medications that have narrow therapeutic index or require rapid gastrointestinal absorption. For oral medications that are dependent on threshold concentrations for efficacy, such as contraceptives and antibiotics, patients should be advised to take those drugs at least 1 hour before exenatide injection. If such drugs are to

(Continued)

be administered with food, patients should be advised to take them with a meal or snack when exenatide is not administered

Reference:
Bydureon prescribing information.

Byetta prescribing information.

Drug-Drug Interaction Monograph for BYDUREON (Exenatide) #2

Perpetrator drug:

INN name/US brand name:	Acetaminophen/Tylenol
Class/Molecular weight:	Small molecule/151 Da

Victim drug:

INN name/US brand name:	Exenatide/Bydureon
Class/Molecular weight:	Peptide/4.2 kDa

Description of the effect of the DDI:
In patients with type 2 diabetes, exenatide did not affect the absorption of orally administered acetaminophen to any clinically relevant degree.

Description of the (proven, putative or hypothesized) mechanism of the DDI:
No DDI.

Clinical consequences:
None.

Reference:
Bydureon prescribing information.

(not in Byetta labeling)

Drug-Drug Interaction Monograph for BYDUREON (Exenatide) #3

Perpetrator drug:

INN name/US brand name:	Warfarin/Coumadin
Class/Molecular weight:	Small molecule/308 Da

Victim drug:

INN name/US brand name:	Exenatide/Bydureon, Byetta
Class/Molecular weight:	Peptide/4.2 kDa

Description of the effect of the DDI:
There are postmarketing reports of increased INR with concomitant use of warfarin, sometimes associated with bleeding. In a drug interaction study, exenatide did not have a significant effect on INR.

Description of the (proven, putative or hypothesized) mechanism of the DDI:
None.

Clinical consequences:
In patients taking warfarin, the INR should be monitored more frequently after initiating BYDUREON. Once a stable INR has been documented, the INR can be monitored at the intervals usually recommended for patients on warfarin.

Reference:
Bydureon prescribing information.

Byetta prescribing information.

Drug-Drug Interaction Monograph for BYETTA/(Exenatide)

See BYDUREON (Exenatide) drug-drug interaction monographs #1 and #3
Reference: Bio-Tropin US prescribing information

Drug-Drug Interaction Monograph for CIMZIA (Certolizumab)

Perpetrator drug:

INN name/US brand name:	TNF-blocking agents and other immune modulators (e.g. anakinra, abatacept, rituximab, and natalizumab)
Class/Molecular weight:	

Victim drug:

INN name/US brand name:	Certolizumab pegol /Cimzia
Class/Molecular weight:	Antibody Fab' fragment/91 kDa

Description of the effect of the DDI:
An increased risk of serious infections has been seen in clinical studies of other TNF-blocking agents used in combination with anakinra or abatacept, with no added benefit. Formal drug interaction studies have not been performed with rituximab or natalizumab. Because of the nature of the adverse events seen with these combinations with TNF blocker therapy, similar toxicities may also result from the use of Cimzia in these combinations.

Description of the (proven, putative or hypothesized) mechanism of the DDI:
Immunomodulatory effects of both agents.

Clinical consequences:
There is not enough information to assess the safety and efficacy of such combination therapy. Therefore, the use of certolizumab pegol in combination with anakinra, abatacept, rituximab, or natalizumab is not recommended

Reference:
Cimzia prescribing information.

Johnston BL, Conly JM. Tumour necrosis factor inhibitors and infection: What is there to know for infectious diseases physicians? Can J Infect Dis Med Microbiol. 2006; 17(4): 209–212.

Katikireddi VS, Whittle SL, Hill CL. Tumour necrosis factor inhibitors and risk of serious infection in rheumatoid arthritis. Int J Rheum Dis. 2010;13(1):12–26.

Drug-Drug Interaction Monograph for COPAXONE (Glatiramer acetate)

Perpetrator drug:

INN name/US brand name:	Glatiramer acetate/Copaxone
Class/Molecular weight:	Peptide/5 kDa–9 kDa

Victim drug:

INN name/US brand name:	Corticosteroids
Class/Molecular weight:	

Description of the effect of the DDI:
Results from existing clinical trials do not suggest any significant interactions of glatiramer acetate with therapies commonly used in multiple sclerosis (MS) patients, including the concurrent use of corticosteroids for up to 28 days.

Description of the (proven, putative or hypothesized) mechanism of the DDI:
None.

Clinical consequences:
None.

Reference:

Magraner MJ, Coret F, Navarré A, Boscá I, Simó M, Escutia M, Bernad A, Navarro L, Casanova B. Pulsed steroids followed by glatiramer acetate to prevent inflammatory activity after cessation of natalizumab therapy: a prospective, 6-month observational study. J Neurol. 2011;258(10):1805–11.

De Stefano N, Filippi M, Hawkins C. Short-term combination of glatiramer acetate with i.v. steroid treatment preceding treatment with GA alone assessed by MRI-disease activity in patients with relapsing-remitting multiple sclerosis. J Neurol Sci. 2008;266(1–2):44–50.

Copaxone prescribing information.

Drug-Drug Interaction Monograph for CUBICIN (Daptomycin)

Perpetrator drug:

INN name/US brand name:	Daptomycin/Cubicin
Class/Molecular weight:	Cyclic lipopeptide/1.6 kDa

Victim drug:

INN name/US brand name:	Simvastatin/Zocor
Class/Molecular weight:	Small molecule/419 Da

Description of the effect of the DDI:
In healthy subjects, concomitant administration of daptomycin and simvastatin had no effect on plasma trough concentrations of simvastatin, and there were no reports of skeletal myopathy.

Description of the (proven, putative or hypothesized) mechanism of the DDI:
None.

Clinical consequences:
Experience with the coadministration of HMG-CoA reductase inhibitors and daptomycin in patients is limited; therefore, consideration should be given to suspending use of HMG-CoA reductase inhibitors temporarily in patients receiving daptomycin.

Reference:

Cubicin prescribing information.

Drug-Drug Interaction Monograph for DDAVP (Desmopressin) #1

Perpetrator drug:

INN name/US brand name:	Pressor Agents
Class/Molecular weight:	

Victim drug:

INN name/US brand name:	Desmopressin/DDAVP
Class/Molecular weight:	Peptide/1.2 kDa

Description of the effect of the DDI:
Desmopressin generally has very low pressor activity compared to the antidiuretic activity. However large doses of desmopressin with other pressor agents could potentiate the pressor effect.

Description of the (proven, putative or hypothesized) mechanism of the DDI:
Pharmacodynamic synergy.

Clinical consequences:
Although the pressor activity of desmopressin is very low compared to the antidiuretic activity, use of large doses of intranasal mesmopressin with other pressor agents should only be done with careful patient monitoring.

Reference:
DDAVP prescribing information.

Drug-Drug Interaction Monograph for DDAVP (Desmopressin) #2

Perpetrator drug:

INN name/US brand name:	Drugs that may cause hyponatremia (e.g. tricyclic antidepressants, selective serotonin re-uptake inhibitors, chlorpromazine, opiate analgesics, NSAIDs, lamotrigine and carbamazepine)
Class/Molecular weight:	

Victim drug:

INN name/US brand name:	Desmopressin/DDAVP
Class/Molecular weight:	Peptide/1.2 kDa

Description of the effect of the DDI:
Increased risk of hyponatremia. Several cases of mild to moderate hyponatremia have been reported as well as a severe case of hyponatremia after administration of desmopressin.

Description of the (proven, putative or hypothesized) mechanism of the DDI:
Pharmacodynamic synergy. Desmopressin is an synthetic analog of vasopressin and stimulates collecting duct aquaporin expression and subsequent resorption of water to the renal parenchyma.

(Continued)

Clinical consequences:
The concomitant administration of drugs that may increase the risk of water intoxication with hyponatremia, (e.g. tricyclic antidepressants, selective serotonin re-uptake inhibitors, chlorpromazine, opiate analgesics, NSAIDs, lamotrigine and carbamazepine) should be performed with caution.

Reference:
Kelleher HB, Henderson SO. Severe hyponatremia due to desmopressin. J Emerg Med. 2006;30(1):45–7.

DDAVP prescribing information.

Drug-Drug Interaction Monograph for ERBITUX (Cetuximab)

Perpetrator drug:

INN name/US brand name:	Cetuximab/Erbitux
Class/Molecular weight:	Monoclonal Antibody/152 kDa

Victim drug:

INN name/US brand name:	Irinotecan/Camptosar
Class/Molecular weight:	Small Molecule/677.2 Da

Description of the effect of the DDI:
There was no evidence of any pharmacokinetic interactions between cetuximab and irinotecan.

Description of the (proven, putative or hypothesized) mechanism of the DDI:
Not applicable.

Clinical consequences:
No dose adjustment is needed.

Reference:
Delbaldo C et al (2005). Pharmacokinetic profile of cetuximab (Erbitux) alone and in combination with irinotecan in patients with advanced EGFR-positive adenocarcinoma. Eur J Cancer 41(12):1739–1745.

Drug-Drug Interaction Monograph for ENBREL (Etanercept) #1

Perpetrator drug:

INN name/US brand name:	Abatacept/Orencia
Class/Molecular weight:	Fusion Protein/92 kDa

Victim drug:

INN name/US brand name:	Etanercept/Enbrel
Class/Molecular weight:	Fusion Protein/150 kDa

Description of the effect of the DDI:
The incidence of total serious adverse events in the combination therapy of abatacept and background biologic treatment (64.1% etanercept) was much higher (23%) than that (8%) of background biologic treatment (65.6% etanercept).

Description of the (proven, putative or hypothesized) mechanism of the DDI:
Mechanism is not known.

Clinical consequences:
Use of etanercept with abatacept is not recommended.

Reference:
Enbrel prescribing information

Weinblatt M et al (2006). Safety of the selective costimulation modulator abatacept in rheumatoid arthritis patients receiving background biologic and nonbiologic disease-modifying antirheumatic drugs: A one-year randomized, placebo-controlled study. Arthritis Rheumatism 54(9): 2807–2816.

Drug-Drug Interaction Monograph for ENBREL (Etanercept) #2

Perpetrator drug:

INN name/US brand name:	Anakinra/Kineret
Class/Molecular weight:	IL-1 Receptor Antagonist/17.3 kDa

Victim drug:

INN name/US brand name:	Etanercept/Enbrel
Class/Molecular weight:	Fusion Protein/150 kDa

Description of the effect of the DDI:
In a study in which patients with active rheumatoid arthritis were treated for up to 24 weeks with concurrent etanercept and anakinra therapy, a 7% rate of serious infections was observed, which was higher than that observed with etanercept alone (0%). Two percent of patients treated concurrently with etanercept and anakinra developed neutropenia (ANC $< 1 \times 109$/L).

Description of the (proven, putative or hypothesized) mechanism of the DDI:
Mechanism is not known.

Clinical consequences:
Use of etanercept with anakinra is not recommended.

Reference:
Enbrel prescribing information.

Genovese MC et al (2004). Combination therapy with etanercept and anakinra in the treatment of patients with rheumatoid arthritis who have been treated unsuccessfully with methotrexate. Arthritis Rheumatism 50(5): 1412–1419.

Drug-Drug Interaction Monograph for ENBREL (Etanercept) #3

Perpetrator drug:

INN name/US brand name:	Etanercept/Enbrel
Class/Molecular weight:	Fusion Protein/150 kDa

Victim drug:

INN name/US brand name:	Digoxin/Lanoxin
Class/Molecular weight:	Small Molecule/780.95 Da

(Continued)

Description of the effect of the DDI:
Digoxin maximum plasma concentration (C_{max}) and area under the plasma concentration versus time curve(AUC) were not altered by the coadministration of etanercept. The mean Cmax and AUC of etanercept were 4.2% and 12.5% lower, respectively, when etanercept was given with digoxin than when administered alone. There were no clinically relevant changes in the electrocardiogram (ECG) parameters when both drugs were combined.

Description of the (proven, putative or hypothesized) mechanism of the DDI:
Not applicable.

Clinical consequences:
There is no clinically relevant interaction between etanercept and digoxin, and both drugs can be safely coadministered without the need for a dosage adjustment.

Reference:
Zhou H et al (2004). Absence of a clinically relevant interaction between etanercept and digoxin. J Clin Pharmacol 44: 1244–1251.

Drug-Drug Interaction Monograph for ENBREL (Etanercept) #4

Perpetrator drug:

INN name/US brand name:	Etanercept/Enbrel
Class/Molecular weight:	Fusion Protein/150 kDa

Victim drug:

INN name/US brand name:	Pneumococcal Polysaccharide Vaccine/Pneumovax
Class/Molecular weight:	Vaccine/NA

Description of the effect of the DDI:
Most patients with psoriatic arthritis receiving etanercept were able to mount effective B-cell immune responses to pneumococcal polysaccharide vaccine, but titers in aggregate were moderately lower and fewer patients had 2-fold rises in titers compared to patients not receiving Enbrel.

Description of the (proven, putative or hypothesized) mechanism of the DDI:
Mechanism is not known.

Clinical consequences:
The clinical relevance is unknown.

Reference:
Enbrel prescribing information.

Drug-Drug Interaction Monograph for ENBREL (Etanercept) #5

Perpetrator drug:

INN name/US brand name:	Etanercept/Enbrel
Class/Molecular weight:	Fusion Protein/150 kDa

Victim drug:

INN name/US brand name:	Warfarin/Coumadin, Jantoven, Marevan, Lawarin, etc.
Class/Molecular weight:	Small Molecule/330.3 Da

Description of the effect of the DDI:
Etanercept did not affect the pharmacokinetics and pharmacodynamics of warfarin. All ratios of maximum serum concentration (C_{max}) and area under the serum concentration versus time curve (AUC) for pharmacokinetics (R- and S-enantiomers of warfarin) and International Normalized Ratio (INR) fell within the confidence interval of 0.8 to 1.25.

Warfarin also did not cause a clinically significant alteration in the pharmacokinetics of etanercept.

Description of the (proven, putative or hypothesized) mechanism of the DDI:
Not applicable.

Clinical consequences:
Coadministration of etanercept and warfarin would not be expected to change the pharmacokinetics of either medication; therefore, no dosage adjustment is needed in cases in which warfarin and etanercept are coadministered.

Reference:	Zhou H et al (2004). Absence of a pharmacokinetic interaction between etanercept and warfarin. J Clin Pharmacol 44: 543–550.

Drug-Drug Interaction Monograph for ENBREL (Etanercept) #6

Perpetrator drug:

INN name/US brand name:	Methotrexate/Mexate, Folex, Rheumatrex
Class/Molecular weight:	Small Molecule/454.5 Da

Victim drug:

INN name/US brand name:	Etanercept/Enbrel
Class/Molecular weight:	Fusion Protein/150 kDa

Description of the effect of the DDI:
Based on the bioequivalence criteria, the Bayesian-estimated clearance for patients receiving etanercept alone (mean: 0.070 L/h) was comparable to that for patients receiving a combination of etanercept and methotrexate (mean = 0.066 L/h). The pharmacokinetics of etanercept were not altered by the concurrent administration of methotrexate in patients with rheumatoid arthritis.

Description of the (proven, putative or hypothesized) mechanism of the DDI:
Not applicable.

Clinical consequences:
No etanercept dose adjustment is needed for patients taking concurrent methotrexate.

Reference:
Zhou H et al (2004). Unaltered etanercept pharmacokinetics with concurrent methotrexate in patients with rheumatoid arthritis. J Clin Pharmacol 44: 1235–1243.

Drug-Drug Interaction Monograph for EXUBERA (Insulin human)

Perpetrator drug:

INN name/US brand name:	Albuterol/Ventolin, Proventil, Proventil-HFA, AccuNeb, Vospire, ProAir
Class/Molecular weight:	Small Molecule/576.7 Da

Victim drug:

INN name/US brand name:	Insulin Human/Exubera
Class/Molecular weight:	Peptide Hormone/5.8 kDa

Description of the effect of the DDI:
Administration of albuterol 30 minutes prior to administration of insulin in non-diabetic subjects with both mild asthma ($n = 36$) and moderate asthma ($n = 31$) resulted in a mean increase in insulin AUC and Cmax of between 25 and 50% compared to when insulin was administered alone.

Description of the (proven, putative or hypothesized) mechanism of the DDI:
The mechanism is not known. Albuterol, a short-acting beta-agonist that works as a bronchodilator, may enhance insulin's inhalation/absorption through its bronchodilatory effect.

Clinical consequences:
Consistent timing of dosing of bronchodilators relative to insulin administration, close monitoring of blood glucose concentrations and dose titration as appropriate are recommended.

Reference:

http://www.accessdata.fda.gov/drugsatfda_docs/nda/2006/021868s000_ClinPharmR.pdf.

http://www.accessdata.fda.gov/drugsatfda_docs/label/2008/021868s016s017lbl.pdf.

Drug-Drug Interaction Monograph for HUMIRA (Adalimumab) #1

Perpetrator drug:

INN name/US brand name:	Adalimumab/Humira
Class/Molecular weight:	Monoclonal Antibody/148 kDa

Victim drug:

INN name/US brand name:	Methotrexate/Rheumatrex, Trexall
Class/Molecular weight:	Small Molecule/454.5 Da

Description of the effect of the DDI:
Repeated administration of adalimumab had no statistically significant effect on the pharmacokinetics of methotrexate.

Description of the (proven, putative or hypothesized) mechanism of the DDI:
None.

Clinical consequences:
No dose adjustment is needed.

Reference: Weisman M et al. Efficacy, pharmacokinetic, and safety assessment of adalimumab, a fully human anti-tumor necrosis factor-alpha monoclonal antibody, in adults with rheumatoid arthritis receiving concomitant methotrexate: a pilot study. Clin Therapeutics 2003;25(6):1700–1721.

Drug-Drug Interaction Monograph for HUMIRA (Adalimumab) #2

Perpetrator drug:

INN name/US brand name:	Methotrexate/Rheumatrex, Trexall
Class/Molecular weight:	Small Molecule/454.5 Da

Victim drug:

INN name/US brand name:	Adalimumab/Humira
Class/Molecular weight:	Monoclonal Antibody/148 kDa

Description of the effect of the DDI:
Methotrexate reduced adalimumab apparent clearance after single and multiple dosing by 29% and 44%, respectively.

Description of the (proven, putative or hypothesized) mechanism of the DDI:
Methotrexate lowers adalimumab's clearance by decreasing the incidence of immunogenicity and/or operating via interacting Fcγ receptor.

Clinical consequences:
No dose adjustment is needed.

Reference: Humira prescribing information.

Drug-Drug Interaction Monograph for INTRON A (Interferon alfa-2B)

Perpetrator drug:

INN name/US brand name:	Interferon Alfa-2B/Intron A
Class/Molecular weight:	Cytokine/19.3 kDa

Victim drug:

INN name/US brand name:	Theophylline/Elixophyllin, Theo-24, Theolair, etc.
Class/Molecular weight:	Small Molecule/180.2 Da

Description of the effect of the DDI:
Concomitant use of alpha interferon and theophylline decreases theophylline clearance, resulting in a significant increase in serum theophylline levels.

Description of the (proven, putative or hypothesized) mechanism of the DDI:
Intron A inhibits metabolism of theophylline by CYP1A2.

Clinical consequences:
No dose adjustment is needed.

Reference:
Intron A prescribing information.

Willimas SJ et al. Inhibition of theophylline metabolism by interferon. Lancet 1987;24; 2(8565):939–41.

Drug-Drug Interaction Monograph for IPRIVASK (Desirudin) #1

Perpetrator drug:

INN name/US brand name:	Desirudin Recombinant/Iprivask
Class/Molecular weight:	Polypeptide/19.2 kDa

Victim drug:

INN name/US brand name:	Warfarin/Coumadin, Jantoven, Marevan, Lawarin, etc.
Class/Molecular weight:	Small Molecule/330.3 Da

Description of the effect of the DDI:
Warfarin (2×5 mg daily oral administration) did not significantly affect the pharmacokinetics of desirudin.

There was an additive effect of desirudin on activated partial thromboplastin time (aPTT), prothrombin time (PT), and international normalized ratio (INR) when co-administered with warfarin.

Description of the (proven, putative or hypothesized) mechanism of the DDI:
The exact mechanism is not yet understood.

Clinical consequences:
If a patient is switched from oral anticoagulants to desirudin therapy or from desirudin to oral anticoagulants, the anticoagulant activity should continue to be closely monitored with appropriate methods. That activity should be taken into account in the evaluation of the overall coagulation status of the patient during the switch.

Reference:
Iprivask prescribing information.

http://www.accessdata.fda.gov/drugsatfda_docs/nda/2003/21-271_Iprivask_BioPharmr.pdf.

Drug-Drug Interaction Monograph for IPRIVASK (Desirudin) #2

Perpetrator drug:

INN name/US brand name:	Acetylsalicylic acid/Aspirin
Class/Molecular weight:	Small Molecule/180.1 Da

Victim drug:

INN name/US brand name:	Desirudin Recombinant/Iprivask
Class/Molecular weight:	Polypeptide/19.2 kDa

Description of the effect of the DDI:
Acetylsalicylic acid (ASA) had no effect on the pharmacokinetics of desirudin. Administration of ASA had no effect on aPTT and PT produced by intravenous infusion of desirudin. Prolonged bleeding times observed in some patients.

Description of the (proven, putative or hypothesized) mechanism of the DDI:
The exact mechanism is not yet understood.

Clinical consequences:
Desirudin should be used with caution in conjunction with drugs which affect platelet function. These medications include systemic salicylates, NSAIDS including ketorolac,

acetylsalicylic acid, ticlopidine, dipyridamole, sulfinpyrazone, clopidogrel, abciximab and other glycoprotein IIb/IIIa antagonists.

Reference:
Iprivask prescribing information.

http://www.accessdata.fda.gov/drugsatfda_docs/nda/2003/21-271_Iprivask_BioPharmr.pdf.

Drug-Drug Interaction Monograph for IPRIVASK (Desirudin) #3

Perpetrator drug:

INN name/US brand name:	Piroxicam/Feldene
Class/Molecular weight:	Small Molecule/331.35 Da

Victim drug:

INN name/US brand name:	Desirudin Recombinant/Iprivask
Class/Molecular weight:	Polypeptide/19.2 kDa

Description of the effect of the DDI:
When desirudin was coadministered with piroxicam (half of the usual daily dose of 20 mg), no significant pharmacokinetic interaction were observed. Neither was there any obvious alteration of the anticoagulant effect of desirudin.

Description of the (proven, putative or hypothesized) mechanism of the DDI:
The exact mechanism is not yet understood.

Clinical consequences:
As with other anticoagulants, desirudin should be used with caution in conjunction with drugs which affect platelet function. These medications include systemic salicylates, NSAIDS including ketorolac, acetylsalicylic acid, ticlopidine, dipyridamole, sulfinpyrazone, clopidogrel, abciximab and other glycoprotein IIb/IIIa antagonists

Reference:
Iprivask prescribing information.

http://www.accessdata.fda.gov/drugsatfda_docs/nda/2003/21-271_Iprivask_BioPharmr.pdf.

Drug-Drug Interaction Monograph for IPRIVASK (Desirudin) #4

Perpetrator drug:

INN name/US brand name:	Heparin/Calciparine and Hep-Lock
Class/Molecular weight:	Sulfated Glycosaminoglycan/3–30 kDa

Victim drug:

INN name/US brand name:	Desirudin Recombinant/Iprivask
Class/Molecular weight:	Polypeptide/19.2 kDa

Description of the effect of the DDI:
The pharmacokinetics of desirudin was not altered by concomitant infusion of heparin (bolus of 5000 units followed by infusion of 1000 units/hour).

(Continued)

Description of the (proven, putative or hypothesized) mechanism of the DDI:
The exact mechanism is not yet understood.

Clinical consequences:
During prophylaxis of venous thromboembolism, concomitant treatment with heparins (unfractionated and low-molecular weight heparins) or dextrans is not recommended. The effects of desirudin and unfractionated heparins on prolongation of aPTT are additive.

Reference:
Iprivask prescribing information.

http://www.accessdata.fda.gov/drugsatfda_docs/nda/2003/21-271_Iprivask_BioPharmr.pdf.

Drug-Drug Interaction Monograph for IPRIVASK (Desirudin) #5

Perpetrator drug:

INN name/US brand name: Desmopressin/DDAVP, DesmoMelt, Stimate, Minirin
Class/Molecular weight: Peptide/1069.2 Da

Victim drug:

INN name/US brand name: Desirudin Recombinant/Iprivask
Class/Molecular weight: Polypeptide/19.2 kDa

Description of the effect of the DDI:
There was no effect of desmopressin on the pharmacokinetics of desirudin. However, it was shown that an infusion of desmopressin (0.3 μg/kg) can partially reverse the desirudin induced increase in aPTT and can dramatically increase factor VIII:C levels depending on the rate of ongoing desirudin infusion.

Description of the (proven, putative or hypothesized) mechanism of the DDI:
The exact mechanism is not yet understood.

Clinical consequences:
The clinical effectiveness of desmopressin in treating bleeding due to desirudin overdose has not been studied.

Reference:
Iprivask prescribing information.

http://www.accessdata.fda.gov/drugsatfda_docs/nda/2003/21-271_Iprivask_BioPharmr.pdf.

Drug-Drug Interaction Monograph for KINERET (Anakinra)

Perpetrator drug:

INN name/US brand name: Methotrexate/Mexate, Folex, Rheumatrex
Class/Molecular weight: Small Molecule/454.5 Da

Victim drug:

INN name/US brand name: Anakinra/Kineret
Class/Molecular weight: IL-1 Receptor Antagonist/17.3 kDa

Description of the effect of the DDI:
Toxicologic and toxicokinetic studies in rats did not demonstrate any alterations in the clearance or toxicologic profile of either methotrexate or anakinra when the two agents were administered together.

Description of the (proven, putative or hypothesized) mechanism of the DDI:
The exact mechanism is not yet understood.

Clinical consequences:
No dose adjustment is needed.

Reference:
Kineret prescribing information.

Drug-Drug Interaction Monograph for LUVERIS (Lutropin alfa)

Perpetrator drug:

INN name/US brand name:	Follitropin Alfa/Gonal-f
Class/Molecular weight:	Human hormone/22.7 kDa

Victim drug:

INN name/US brand name:	Lutropin Alfa/Luveris
Class/Molecular weight:	Recombinant human hormone/23.4 kDa

Description of the effect of the DDI:
There are no pharmacokinetic interactions between lutropin alfa (150 IU) with follitropin alfa (150 IU) when administered simultaneously.

Description of the (proven, putative or hypothesized) mechanism of the DDI:
The exact mechanism is not yet understood.

Clinical consequences:
No dose adjustment is needed.

Reference:
Luveris prescribing information.

http://www.accessdata.fda.gov/drugsatfda_docs/nda/2004/021322s000_Luveris_biopharmr.pdf.

Drug-Drug Interaction Monograph for MIACALCIN (Calcitonin)

Perpetrator drug:

INN name/US brand name:	Calcitonin-salmon/Miacalcin
Class/Molecular weight:	Polypeptide/3455 Da

Victim drug:

INN name/US brand name:	Lithium
Class/Molecular weight:	Small Molecule

Description of the effect of the DDI:
Concurrent use of calcitonin and lithium may result in decreased lithium concentrations and loss of lithium efficacy. The co-administration of calcitonin and lithium causes a significant decrease in lithium plasma levels, with the average decrease reaching a maximum of 30%.

Description of the (proven, putative or hypothesized) mechanism of the DDI:
Increased renal lithium excretion and/or decreased lithium intestinal absorption.

(Continued)

Clinical consequences:
Lithium levels should be closely monitored in patients receiving calcitonin therapy. Lithium dose adjustments may be necessary.

Reference: Miacalcin prescribing information.

Passiu G, et al. Calcitonin decreases lithium plasma levels in man. Preliminary report. Int J Clin Pharm Res 1998; 18:179–181.

Micromedex

Drug-Drug Interaction Monograph for MIRCERA (Methoxy polyethylene glycol-epoetin beta)

Perpetrator drug:
INN name/US brand name: Methoxy polyethylene glycol-epoetin beta/Mircera
Class/Molecular weight: Erythropoietin analog/60 kDa

Victim drug:
INN name/US brand name: None
Class/Molecular weight:

Description of the effect of the DDI:
No drug interactions have been identified.

Description of the (proven, putative or hypothesized) mechanism of the DDI:
None available.

Clinical consequences:
None.

Reference: Mircera prescribing information

Drug-Drug Interaction Monograph for MYOBLOC (Rimabotulinumtoxin B)

Perpetrator drug:
INN name/US brand name: Aminoglycosides/other agents interfering neuromuscular transmission (e.g., curare-like compounds)
Class/Molecular weight: Small molecule

Victim drug:
INN name/US brand name: Rimabotulinumtoxin B/Myobloc
Class/Molecular weight: Two chain polypeptide/150 kDa

Description of the effect of the DDI:
Caution should be exercised when co-administering rimabotulinumtoxin B and aminoglycosides or other agents interfering with neuromuscular transmission (e.g., curare-like compounds) as the effect of the toxin may be potentiated.

Description of the (proven, putative or hypothesized) mechanism of the DDI:
None available

Clinical consequences:
Toxicity of rimabotulinumtoxin B may increase.

Reference: Myobloc prescribing information.

Drug-Drug Interaction Monograph for MYOZYME (Alglucosidase alfa)

Perpetrator drug:
INN name/US brand name: Alglucosidase
Class/Molecular weight: Enzyme/110 kDa

Victim drug:
INN name/US brand name: None
Class/Molecular weight:

Description of the effect of the DDI:
No drug interactions have been identified.

Description of the (proven, putative or hypothesized) mechanism of the DDI:
None available.

Clinical consequences:
None.

Reference: Myozyme prescribing information.

Drug-Drug Interaction Monograph for NEUMEGA (Oprelvekin)

Perpetrator drug:
INN name/US brand name: Oprelvekin/Neumega
Class/Molecular weight: Recombinant interleukin/19 kDa

Victim drug:
Class/Molecular weight: Opioids

Description of the effect of the DDI:
Patient on treatment with opioids and oprelvekin developed respiratory depression and sedation suggested from a sedative medication.

Description of the (proven, putative or hypothesized) mechanism of the DDI:
Oprelvekin alters renal handling of water and sodium. Morphine and its metabolites are known to undergo renal elimination, and oprelvekin may have interfered with this process.

Clinical consequences:
Precaution should be taken while prescribing oprelvekin to patients already on opioid therapy due to possibility of drug interaction.

Reference: Moorehead et al. Suspected interaction of oprelvekin (rhIL-11) with a long-acting opioid. Pediatr Blood Cancer. 2008 Dec;51(6):837

Drug-Drug Interaction Monograph for NEUPOGEN (Filgrastim)

Perpetrator drug:

INN name/US brand name:	Filgrastim/Neupogen
Class/Molecular weight:	Hematopoietic growth factor/18.8 kDa

Victim drug:

INN name/US brand name:	Oprelvekin/Neumega
Class/Molecular weight:	Recombinant interleukin/19 kDa

Description of the effect of the DDI:
No change in pharmacokinetic parameters or toxicity has been observed under co-administration of filgrastim and oprelvekin.

Description of the (proven, putative or hypothesized) mechanism of the DDI:
NA

Clinical consequences:
No clinical consequences have been observed.

Reference: Neupogen prescribing information.

Drug-Drug Interaction Monograph for NULOJIX (Belatacept) #1

Perpetrator drug:

INN name/US brand name:	Belatacept/Nulojix
Class/Molecular weight:	FC-CTLA-4 fusion protein/90.6 kDa

Victim drug:

INN name/US brand name:	Mycophenolic acid (MPA)
Class/Molecular weight:	Small molecule/320.3 Da

Description of the effect of the DDI:
In a pharmacokinetic study, the plasma concentrations of mycophenolic acid (MPA) was measured in 41 patients who received fixed mycophenolate mofetil (MMF) doses of 500 mg to 1500 mg twice daily with either 5 mg per kg of belatacept or cyclosporine. The mean dose-normalized MPA C_{max} and $AUC_{0\text{-}12}$ were approximately 20% and 40% higher, respectively, with belatacept coadministration than with cyclosporine coadministration.

Description of the (proven, putative or hypothesized) mechanism of the DDI:
Belatacept may cause a decrease in CYP450 activity leading to a decrease in systemic clearance of MPA.

Clinical consequences:
Clinicians should be aware that there is a potential change of MPA exposure after crossover from cyclosporine to belatacept or from belatacept to cyclosporine in patients concomitantly receiving MMF.

Reference: Nulojix prescribing information.

Drug-Drug Interaction Monograph for NULOJIX (Belatacept) #2

Perpetrator drug:
INN name/US brand name: Belatacept/Nulojix
Class/Molecular weight: FC-CTLA-4 fusion protein/90.6 kDa

Victim drug:
INN name/US brand name: CYP450 substrates
Class/Molecular weight:

Description of the effect of the DDI:
No formal drug interaction studies have been conducted with belatacept, but biologic therapies that are cytokines or cytokine modulators have been shown to affect the expression and/or functional activities of CYP450 enzymes.

Description of the (proven, putative or hypothesized) mechanism of the DDI:
In vitro studies have shown that belatacept inhibits the production of certain cytokines during an alloimmune response.

Clinical consequences:
The potential for belatacept to alter the systemic concentrations of drugs that are CYP450 substrates has not been studied; however, in the event that kidney transplant patients receiving belatacept exhibit signs and symptoms of altered efficacy or adverse events associated with co-administered drugs which are known to be metabolized by CYP450, the clinician should be aware of potentially altered CYP450 metabolism of these drugs.

Reference: Nulojix prescribing information.

Drug-Drug Interaction Monograph for OMNITROPE (Somatropin recombinant)

See ACCRETROPIN (Somatropin) drug-drug interaction monographs #1–#4

Reference: Omnitrope prescribing information

Drug-Drug Interaction Monograph for ONCASPAR (Pegaspargase)

Perpetrator drug:
INN name/US brand name: N/A
Class/Molecular weight:

Victim drug:
INN name/US brand name: N/A
Class/Molecular weight:

Description of the effect of the DDI:
No formal drug interaction studies between pegaspartase and other drugs have been performed, as per drug label.

Description of the (proven, putative or hypothesized) mechanism of the DDI:
No dose adjustments have been recommended.

Clinical consequences:
No clinical consequences have been observed.

Reference: Oncaspar prescribing information.

Drug-Drug Interaction Monograph for ORENCIA (Abatacept)

Perpetrator drug:

INN name/US brand name:	Abatacept/Orencia
Class/Molecular weight:	CTLA-4/human IgG1 Fc fusion protein Immune Modulator (Antirheumatic)/92 kDa

Victim drug:

Class:	TNF antagonists/inhibitors (etanercept, adalimumab, anakinra, infliximab, golimumab)

Description of the effect of the DDI:
Increased risk of serious infection.

Description of the (proven, putative or hypothesized) mechanism of the DDI:
Potentiating effect on inhibition of immune reactivity.

Clinical consequences:
Concomitant therapy is not recommended.

Reference: Micromedex.

Drug-Drug Interaction Monograph for ORTHOCLONE OKT3 (Muromonab-CD3)

Perpetrator drug:

INN name/US brand name:	Echinacea preparations
Class/Molecular weight:	Herbal preparation

Victim drug:

INN name/US brand name:	Muromonab-CD3/Orthoclone OKT3
Class/Molecular weight:	Murine IgG2α monoclonal antibody/146 kDa

Description of the effect of the DDI:
Concurrent use of Echinacea preparations and muromonab-CD3 may result in decreased effectiveness of muromonab-CD3.

Description of the (proven, putative or hypothesized) mechanism of the DDI:
The probable mechanism of interaction is stimulation of the immune system by Echinacea, resulting in increased formation of neutralizing anti-drug antibodies.

Clinical consequences:
Since reduced muromonab effectiveness may be life threatening in organ transplantation, the concomitant use of Echinacea and muromonab-CD3 should be avoided.

Reference: Micromedex.

Drug-Drug Interaction Monograph for PERJETA (Pertuzumab) #1

Perpetrator drug:

INN name/US brand name:	Pertuzumab/Perjeta
Class/Molecular weight:	Monoclonal Antibody/148 kDa

Victim drug:

INN name/US brand name:	Trastuzumab/Herceptin
Class/Molecular weight:	Monoclonal Antibody/148 kDa

Description of the effect of the DDI:
No change in pharmacokinetic parameters or toxicity has been observed under co-administration of pertuzumab and trastuzumab.

Description of the (proven, putative or hypothesized) mechanism of the DDI:
No clinical consequences have been observed.

Clinical consequences:
No dose adjustments have been recommended.

Reference: Perjeta prescribing information.

Drug-Drug Interaction Monograph for PERJETA (Pertuzumab) #2

Perpetrator drug:

INN name/US brand name:	Pertuzumab/Perjeta
Class/Molecular weight:	Monoclonal Antibody/148 kDa

Victim drug:

INN name/US brand name:	Docetaxel/Taxotere
Class/Molecular weight:	Chemotherapeutic/807.9 Da

Description of the effect of the DDI:
No change in pharmacokinetic parameters or toxicity has been observed under co-administration of pertuzumab and docetaxel.

Description of the (proven, putative or hypothesized) mechanism of the DDI:
No clinical consequences have been observed.

Clinical consequences:
No dose adjustments have been recommended.

Reference: Perjeta prescribing information.

Drug-Drug Interaction Monograph for PRIALT (Ziconotide Acetate)

Perpetrator drug:

INN name/US brand name:	Antidepressants, Anxiolytics, Antiepileptics,
Class/Molecular weight:	Neuroleptics, Sedatives

Victim drug:

INN name/US brand name:	Ziconotide Acetate/Prialt
Class/Molecular weight:	Peptide/2.6 kD

Description of the effect of the DDI:
The use of drugs with CNS-depressant activities in conjunction with ziconotide acetate may be associated with an increased incidence of CNS-depressant effects.

(Continued)

Description of the (proven, putative or hypothesized) mechanism of the DDI:
The patient may experience cognitive impairment when ziconotide acetate is used in conjunction with CNS-depressant drugs due to the additive effects of the two therapies.

Clinical consequences:
It may be necessary to decrease the dose of the perpetrator medication.

Reference: Prialt prescribing information.

Drug-Drug Interaction Monograph for PROLEUKIN (Aldesleukin) #1

Perpetrator drug:
Class: Glucocorticoids

Victim drug:
INN name/US brand name: Aldesleukin/Proleukin
Class/Molecular weight: Cytokine/15.3 kDa

Description of the effect of the DDI:
Reduced antitumor effectiveness of aldesleukin with a concomitantly administered glucocorticoid.

Description of the (proven, putative or hypothesized) mechanism of the DDI:
This is a class related pharmacodynamics interaction of unknown mechanism.

Clinical consequences:
NA

Reference: Proleukin prescribing information.

Drug-Drug Interaction Monograph for PROLEUKIN (Aldesleukin) #2

Perpetrator drug:
Class Antihypertensives

Victim drug:
INN name/US brand name: Aldesleukin/Proleukin
Class/Molecular weight: Cytokine/15.3 kDa

Description of the effect of the DDI:
Antihypertensives potentiate the hypotensive effects of aldesleukin.

Description of the (proven, putative or hypothesized) mechanism of the DDI:
This is a class related pharmacodynamics interaction of unknown mechanism.

Clinical consequences:
Monitor patient's blood pressure.

Reference: Proleukin prescribing information.

Drug-Drug Interaction Monograph for RAPTIVA (Efazulimab)

Perpetrator drug:
INN name/US brand name: Efalizumab/Raptiva
Class/Molecular weight: Monoclonal Antibody/150 kDa

Victim drug:
Class/Molecular weight: Immunosuppressants

Description of the effect of the DDI:
Interactions have not been studied but an increased risk of infection or malignancy is expected.

Description of the (proven, putative or hypothesized) mechanism of the DDI:
Efalizumab and other immunosuppressants have concurrent mechanisms of action and increase the risk of toxicity.

Clinical consequences:
Do not coadminister efalizumab and immunosuppressants.

Reference: Raptiva prescribing information.

Drug-Drug Interaction Monograph for REGRANEX (Becaplermin)

Perpetrator drug:
INN name/US brand name: Becaplermin/Regranex
Class/Molecular weight: Peptide/25 kDa

Victim drug:
INN name/US brand name: NA
Class/Molecular weight:

Description of the effect of the DDI:
No drug-drug interaction studies have been performed.

Description of the (proven, putative or hypothesized) mechanism of the DDI:
NA

Clinical consequences:
None

Reference: Regranex prescribing information.

Drug-Drug Interaction Monograph for REMICADE (Infliximab) #1

Perpetrator drug:
INN name/US brand name: Infliximab/Remicade
Class/Molecular weight: Chimeric Monoclonal Antibody/148 kDa

Victim drug:
INN name/US brand name: Abatacept/Orencia
Class/Molecular weight: Fusion Protein/92 kDa

(Continued)

Description of the effect of the DDI:
Concomitant administration of infliximab and abatacept appeared to increase the risk for severe infection.

Description of the (proven, putative or hypothesized) mechanism of the DDI:
The increased risk of infection is hypothesized to be due to increased immunosuppression by TNF-alpha blockade if both drugs are given together.

Clinical consequences:
Infliximab should not be given concomitantly with abatacept.

Reference: Remicade prescribing information.

Drug-Drug Interaction Monograph for REMICADE (Infliximab) #2

Perpetrator drug:
INN name/US brand name: Infliximab/Remicade
Class/Molecular weight: Chimeric Monoclonal Antibody/148 kDa

Victim drug:
INN name/US brand name: Anakinra/Kineret
Class/Molecular weight: Recombinant Peptide/17 kDa

Description of the effect of the DDI:
Concomitant administration of infliximab and anakinra appeared to increase the risk for severe infection.

Description of the (proven, putative or hypothesized) mechanism of the DDI:
The increased risk of infection is hypothesized to be due to increased immunosuppression by TNF-alpha blockade if both drugs are given together.

Clinical consequences:
Infliximab should not be given concomitantly with anakinra.

Reference: Remicade prescribing information.

Drug-Drug Interaction Monograph for REOPRO (Abciximab)

Perpetrator drug:
INN name/US brand name: Abciximab/Reopro
Class/Molecular weight: chimeric monoclonal antibody fragment/47.6 kDa

Victim drug:
INN name/US brand name: Heparin
Class/Molecular weight: Anticoagulant

Description of the effect of the DDI:
Increased risk of bleeding has been described.

Description of the (proven, putative or hypothesized) mechanism of the DDI:
No change in pharmacokinetic parameters has been described. This appears to be a pharmacodynamic interaction.

Clinical consequences:
Use of heparin and abciximab during PTCA procedures should be monitored. Starting doses of heparin should not be higher than 70 IU/kg.

Reference: Reopro prescribing information.

The CAPTURE Investigators. Randomised placebo-controlled trial of abciximab before and during coronary intervention in refractory unstable angina: the CAPTURE Study. Lancet 1997, 349, 1429–35.

Drug-Drug Interaction Monograph for SANDOSTATIN (Octreotide acetate)

Perpetrator drug:

INN name/US brand name:	Octreotide acetate/Sandostatin
Class/Molecular weight:	Peptide/1.02 kDa

Victim drugs:

INN name/US brand name:	Various small molecule drugs
Class/Molecular weight:	Small molecules

Description of the effect of the DDI:
Increased risk of cardiotoxicity (QT prolongation, torsades de pointes, cardiac arrest)

Description of the (proven, putative or hypothesized) mechanism of the DDI:
Additive effects on QT prolongation was the mechanism involved in drug interactions of Sandostatin with ziprasidone, bepridil, thioridazine, pimozide, terfenadine, mesoridazine, sparfloxacin, dronedarone, cisapride, levomethadyl, gemifloxacin, Phenothiazines, Erythromycin, Iloperidone, moxifloxacin, enflurane, apomorphine, perflutren lipid microsphere, pazopanib, ondansetron, trazodone, crizotinib, astemizole, spiramycin, solifenacin, fingolimod, mifepristone, vandetanib, sodium phosphate, clarithromycin, chloral hydrate, droperidol, telavancin, saquinavir, isoflurane, toremifene, pentamidine, foscarnet, salmeterol, domperidone, levofloxacin, class ia antiarrhythmics, antipsychotics, gatifloxacin, paliperidone, lidoflazine, dasatinib, telithromycin, chloroquine, artemether/lumefantrine, quinine, dolasetron, tricyclic antidepressants, voriconazole, arsenic trioxide, tetrabenazine,. methadone, clozapine, vardenafil, isradipine, norfloxacin, lopinavir/ritonavir, sorafenib, ofloxacin, cotrimoxazole, probucol, vasopressin, nilotinib, clomipramine, sunitinib, azithromycin, ciprofloxacin, class i antiarrhythmic agents, fluoxetine, lapatinib, asenapine, ranolazine, zolmitriptan, mefloquine, class iii antiarrhythmics, halofantrine, citalopram, fluconazole, posaconazole, halothane.

Clinical consequences:
Concomitant use of drugs with other drugs that prolong the QT interval, such as Sandostatin/octreotide is contraindicated due to the potential for additive effects on QT interval prolongation and an increased risk of serious ventricular arrhythmias, serious cardiac events including torsade de pointes.

Reference: Sandostatin prescribing information, Micromedex.

Drug-Drug Interaction Monograph for SEROSTIM/SAIZEN (Somatropin)

See ACCRETROPIN (Somatropin) drug-drug interaction monographs #1– #4

Reference: Serostim prescribing information
Saizen prescribing information

Drug-Drug Interaction Monograph for SIMPONI (Golimumab) #1

Perpetrator drug:
INN name/US brand name: Golimumab/Simponi
Class/Molecular weight: Human IgG1 monoclonal antibody/150 kDa

Victim drugs:
INN name/US brand name: Abatacept/Orencia
Class/Molecular weight: CTLA-4/human IgG1 Fc fusion protein/92 kDa

Description of the effect of the DDI:
Increased risk of serious infection.

Description of the (proven, putative or hypothesized) mechanism of the DDI:
Interference with the immune system at multiple levels.

Clinical consequences:
In controlled trials, the concurrent administration of another TNF-blocker and abatacept was associated with a greater proportion of serious infections than the use of a TNF-locker alone; and the combination therapy, compared to the use of a TNF-blocker alone, has not demonstrated improved clinical benefit in the treatment of psoriatic arthritis. Therefore, the combination of TNF-blockers including golimumab and abatacept is not recommended.

Reference: Simponi prescribing information, Micromedex.

Drug-Drug Interaction Monograph for SIMPONI (Golimumab) #2

Perpetrator drug:
INN name/US brand name: Golimumab/Simponi
Class/Molecular weight: Human IgG1 monoclonal antibody/150 kDa

Victim drugs:
INN name/US brand name: Anakinra/Kineret
Class/Molecular weight: recombinant, nonglycosylated form of the human interleukin-1 receptor antagonist/17.3 kDa
INN name/US brand name: Various
Class/Molecular weight: Live Vaccines

Description of the effect of the DDI:
Increased risk of serious infections and neutropenia.

Description of the (proven, putative or hypothesized) mechanism of the DDI:
Increased risk of serious infections has been seen in clinical rheumatoid arthritis studies of other TNF-blockers used in combination with anakinra, with no added benefit; therefore, use of golimumab with anakinra is not recommended. A higher rate of serious infections has also been observed in rheumatoid arthritis patients treated with rituximab who received subsequent treatment with a TNF-blocker. There is insufficient information to provide recommendations regarding the concomitant use of golimumab and other biologic products approved to treat rheumatoid arthritis or psoriatic arthritis.

Clinical consequences:
Concurrent administration of anakinra (an interleukin-1 antagonist) and another TNF-blocker was associated with a greater portion of serious infections and neutropenia and no additional benefits compared to TNF-blocker alone. Therefore, the combination of anakinra with TNF blockers, including golimumab, is not recommended.

Reference: Simponi prescribing information, Micromedex.

Drug-Drug Interaction Monograph for SIMPONI (Golimumab) #3

Perpetrator drug:
INN name/US brand name: Golimumab/Simponi
Class/Molecular weight: Human IgG1 monoclonal antibody/150 kDa

Victim drugs:
INN name/US brand name: Various
Class/Molecular weight: Live Vaccines

Description of the effect of the DDI:
Live vaccines: Increased risk of secondary transmission of infection by live vaccines.

Description of the (proven, putative or hypothesized) mechanism of the DDI:
IgG antibodies are known to cross the placenta during pregnancy and have been detected in the serum of infants born to patients treated with these antibodies. Since golimumab is an IgG antibody, infants born to women treated with golimumab during their pregnancy may be at increased risk of infection for up to 6 months.

Clinical consequences:
Administration of live vaccines to infants exposed to golimumab *in utero* is not recommended for 6 months following the mother's last golimumab injection during pregnancy.

Reference: Simponi prescribing information, Micromedex

Drug-Drug Interaction Monograph for SIMPONI (Golimumab) #4

Perpetrator drug:
INN name/US brand name: Golimumab/Simponi
Class/Molecular weight: Human IgG1 monoclonal antibody/150 kDa

(Continued)

Victim drugs:

INN name/US brand name: Various
Class/Molecular weight: CYP 450 substrates

Description of the effect of the DDI:
Coadministration of Simponi may affect the clearance of concurrently administered CYP 450 substrates.

Description of the (proven, putative or hypothesized) mechanism of the DDI:
The formation of CYP450 enzymes may be suppressed by increased levels of cytokines (e.g., TNF-α) during chronic inflammation. Therefore, it is expected that for a molecule that antagonizes cytokine activity, such as golimumab, the formation of CYP450 enzymes could be normalized.

Clinical consequences:
Upon initiation or discontinuation of golimumab in patients being treated with CYP450 substrates with a narrow therapeutics index, monitoring of the effect (e.g., warfarin) or drug concentration, cyclosporine or theophylline) is recommended and the individual dose of drug product may be adjusted as needed.

Reference: Simponi prescribing information, Micromedex.

Drug-Drug Interaction Monograph for SIMULECT (Basiliximab) #1

Perpetrator drug:

INN name/US brand name: Basiliximab/Simulect
Class/Molecular weight: Monoclonal antibody/143.8 kDa

Victim drug:

INN name/US brand name: Tacrolimus/Prograf
Class/Molecular weight: Immunosuppressant/804 Da

Description of the effect of the DDI:
Concurrent use of basiliximab and tacrolimus may result in increased plasma tacrolimus trough concentrations and increased risk of tacrolimus toxicity.

Description of the (proven, putative or hypothesized) mechanism of the DDI:
Cytokine-induced alterations in CYP3A4-mediated metabolism of tacrolimus.

Clinical consequences:
Monitor tacrolimus plasma concentrations closely in patients receiving concomitant basiliximab treatment. Tacrolimus blood trough levels should be closely monitored 1 to 2 months post-transplant and dose-adjustments made as necessary.

Reference: Simulect labeling information.
Micromedex.

Drug-Drug Interaction Monograph for SIMULECT (Basiliximab) #2

Perpetrator drug:

INN name/US brand name:	Echinacea
Class/Molecular weight:	Herbal preparation

Victim drug:

INN name/US brand name:	Basiliximab/Simulect
Class/Molecular weight:	Monoclonal antibody/143.8 kDa

Description of the effect of the DDI:
Concurrent use of Echinacea and basiliximab may result in decreased effectiveness of basiliximab. Theoretically, Echinacea may antagonize the immunosuppressive effect of basiliximab. Human studies have shown an increase in granulocytes, monocytes, and lymphocytes in response to treatment with Echinacea, increased phagocytic activity, and increased natural killer cell function. Animals treated with Echinacea have shown increased immune response to antigens, and increased macrophage activation.

Description of the (proven, putative or hypothesized) mechanism of the DDI:
Immune system stimulation by Echinacea.

Clinical consequences:
Concomitant use of Echinacea and basiliximab should be avoided due to the antagonistic effect an immunostimulant such as echinacea may have on basiliximab, an immunosuppressant.

Reference: Simulect labeling information.
Micromedex.

Drug-Drug Interaction Monograph for SOLIRIS (Eculizumab)

Perpetrator drug:

INN name/US brand name:	Eculizumab/Soliris
Class/Molecular weight:	Human monoclonal antibody/148 kDa

Victim drugs:

INN name/US brand name:	Bioflavonoids; zinc glycinate; vitamin E; cholecalciferol, zinc sulfate
Class/Molecular weight:	Vitamins and immunostimulants

Description of the effect of the DDI:
Drug interaction studies have not been performed with eculizumab.

Description of the (proven, putative or hypothesized) mechanism of the DDI:
Theoretically, agents that are thought to have immunostimulant properties such as echinacea, vitamin E and zinc may antagonize the pharmacologic effects of immunosuppressants.

Clinical consequences:
Clinical cases of drug interactions have not been reported.

Reference: Soliris prescribing information.

Drug-Drug Interaction Monograph for SOMAVERT (Pegvisomat)

Perpetrator drug:

INN name/US brand name:	Pegvisomant/Somavert
Class/Molecular weight:	Recombinant protein/52 kDa

Victim drugs:

INN name/US brand name:	Octreotide/Sandostatin Somatuline/Lanreotide
Class/Molecular weight:	Sandostatin: Cyclic octapeptide/1,02 kDa Lanreotide: Cyclical octapeptide analog of the natural hormone, somatostatin/1,1 kDa

Description of the effect of the DDI:
Concurrent use of octreotide or lanreotide and pegvisomant may result in elevated liver enzymes.

Description of the (proven, putative or hypothesized) mechanism of the DDI:
Concomitant use of somatostatin analogs, like octreotide and lanreotide, and pegvisomant may result in increased hepatic enzyme levels. Monitoring of liver function tests and appropriate dosage adjustment of pegvisomant is recommended during coadministration.

Clinical consequences:
Concomitant use of octreotide or lanreotide with pegvisomant may result in increased hepatic enzyme levels.

Reference: Somavert prescribing information, Micromedex.

Drug-Drug Interaction Monograph for STELARA (Ustekinumab) #1

Perpetrator drug:

INN name/US brand name:	Ustekinumab/Stelara
Class/Molecular weight:	Human IgG1 monoclonal antibody/146 kDa

Victim drugs:

INN name/US brand name:	Various
Class/Molecular weight:	Live Vaccines

Description of the effect of the DDI:
Concurrent use of ustekinumab and live vaccines may result in an increased risk of secondary transmission of infection by the live vaccine.

Description of the (proven, putative or hypothesized) mechanism of the DDI:
Altered cellular immune response.

Clinical consequences:
Although data are not available on the effects of concurrent vaccination, the manufacturer does not recommend the administration of live vaccines in patients receiving ustekinumab. Bacillus of Calmette and Guerin live vaccines should not be administered during treatment with ustekinumab or for 1 year prior to initiation or 1 year after discontinuation of ustekinumab.

Reference: Stelara prescribing information, Micromedex.

Drug-Drug Interaction Monograph for STELARA (Ustekinumab) #2

Perpetrator drug:

INN name/US brand name:	Ustekinumab/Stelara
Class/Molecular weight:	Human IgG1 monoclonal antibody/146 kDa

Victim drugs:

INN name/US brand name:	Various
Class/Molecular weight:	CYP 450 substrates

Description of the effect of the DDI:
Coadministration of Stelara may affect the clearance of concurrently administered CYP 450 substrates.

Description of the (proven, putative or hypothesized) mechanism of the DDI:
The formation of CYP450 enzymes can be altered by increased levels of certain cytokines (e.g., IL-1, 289 IL-6, IL-10, TNFα, IFN) during chronic inflammation. Thus, ustekinumab, an antagonist of IL-12 and IL-23, could normalize the formation of CYP450 enzymes.

Clinical consequences:
Upon initiation of ustekinumab in patients who are receiving concomitant CYP450 substrates, particularly those with a narrow therapeutic index, monitoring for therapeutic effect (e.g., for warfarin) or drug concentration (e.g., for cyclosporine) should be considered and the individual dose of the drug adjusted as needed.

Reference: Stelara prescribing information, Micromedex.

Drug-Drug Interaction Monograph for STIMATE (Desmopressin acetate)

See DDAVP (Desmopressin) drug-drug interaction monographs #1-#2

Reference: Stimate prescribing information

Drug-Drug Interaction Monograph for SYLATRON (Peginterferon)

Perpetrator drug:

INN name/US brand name:	Peginterferon alfa-2B/Sylatron
Class/Molecular weight:	Recombinant interferon alfa-2b)/40 kDa

Victim drug:

Class:	Drugs metabolized by CYP2C9 or CYP2D6

Description of the effect of the DDI:
When administering peginterferon alfa with medications metabolized by CYP2C9 or CYP2D6, the therapeutic effect of these drugs may be altered.

The coadministration of interferons and zidovudine may result in severe bone marrow toxicity, most often manifested as granulocytopenia and/or thrombocytopenia.

The risk of seizures may be increased during coadministration of tramadol or bupropion with any substance that can reduce the seizure threshold, including inferons. Interferons also exhibit CNS- and/or respiratory-depressant effects, which may be enhanced during their concomitant use with tramadol.

(Continued)

Description of the (proven, putative or hypothesized) mechanism of the DDI:
In a study in healthy subjects, a measure of CYP2C9 activity increased to 125% (90% CI: 116% to 135%) of baseline, whereas a measure of CYP2D6 activity decreased to 51% (90% CI: 38% to 67%) of baseline when coadministered with peginterferon alfa-2b at 1 mcg/kg/week. A measure of CYP1A2 and CYP3A4 activity did not show clinically meaningful changes.

Interferons and zidovudine agents are individually myelotoxic, with potentially additive or synergistic effects.

Clinical consequences:
Close monitoring for drugs metabolized by CYP2C9 or CYP2D6.

Caution is advised if concomitant therapy is necessary, particularly with the alfa interferons. Complete blood counts, differential and platelet counts should be performed regularly and drug dosages adjusted accordingly. Patients should be advised to promptly report symptoms such as chills, fever, sore throat, pale skin, or unusual fatigue to their physician. Cessation of one or all drugs may be necessary if toxicity develops

Caution is advised if tramadol or bupropion is administered with any substance that can reduce the seizure threshold such as interferons, particularly in the elderly and in patients with epilepsy, a history of seizures, or other risk factors for seizures (e.g., head trauma, brain tumor, metabolic disorders, alcohol and drug withdrawal, CNS infections).

Reference: Sylatron prescribing information.

Drug-Drug Interaction Monograph for SYNAGIS (Palivizumab)

Perpetrator drug:

INN name/US brand name:	Palivizumab/Synagis
Class/Molecular weight:	Humanized IgG1 monoclonal antibody/148 kDa

Victim drug:

Class:	Abciximab/Reopro; Belimumab/Benlysta
Class/Molecular weight:	Monoclonal antibodies

Description of the effect of the DDI:
When abciximab or belimumab is used in combination with palivizumab, drug interactions could occur. As a result, the risk for allergic reactions, infections, or low blood cell counts can increase.

Description of the (proven, putative or hypothesized) mechanism of the DDI:
If patients received abciximab in the past, they may have developed antibodies to the medication. These antibodies can increase the chance of having an allergic reaction to Synagis or other monoclonal antibody medications.

Using palivizumab with belimumab may increase the risk for side effects, such as infection or low blood cell counts.

Clinical consequences:
No specific recommendations.

Reference: Synagis prescribing information.

Drug-Drug Interaction Monograph for TEV-TROPIN (Somatropin)

See ACCRETROPIN (Somatropin) drug-drug interaction monographs #1– #4

Reference: Tev-Tropin prescribing information

Drug-Drug Interaction Monograph for TNKASE (Tenecteplase)

Perpetrator drug:

INN name/US brand name:	Tenecteplase/TNKase
Class/Molecular weight:	Enzyme/58.9 kDa

Victim drug:

Class:	Low molecular-weight heparins (enoxaparin, dalteparin, parnaparin, nadroparin, reviparin, tinzaparin, ardeparin, and certoparin)
Class	Anticoagulants (heparin, drotrecogin alfa, dabigatran, warfarin, phenprocoumon, phenindione, acenocoumarol, danaparoid, bivalirudin, argatroban, lepirudin, desirudin, fondaparinux, rivaroxaban, protein C (Human)).
Class	Platelet aggregation inhibitors (acetylsalicylic acid, dipyridamole, GP IIb/IIIa inhibitors, ticlopidine)
Class	Thrombolytic agents (streptokinase, urokinase, alteplase, anistreplase, reteplase, lepirudin)

Description of the effect of the DDI:
Increasing risk of bleeding might be observed if anticoagulants and drugs that alter platelet function are administered prior to, during, or after TNKase therapy.

Description of the (proven, putative or hypothesized) mechanism of the DDI:
NA

Clinical consequences:
Monitor for signs of bleeding.

Reference: Tnkase prescribing information, Micromedex.

Drug-Drug Interaction Monograph for TRASYLOL (Aprotinin) #1

Perpetrator drug:

INN name/US brand name:	Aprotinin/Trasylol
Class/Molecular weight:	Bovine proteinase inhibitor/6.5 kDa

Victim drug:

Class:	Fibrinolytic agents

(Continued)

Description of the effect of the DDI:
Aprotinin is known to have antifibrinolytic activity and, therefore, may inhibit the effects of fibrinolytic agents.

Description of the (proven, putative or hypothesized) mechanism of the DDI:
NA

Clinical consequences:
No specific recommendations are available.

Reference: Trasylol prescribing information.

Drug-Drug Interaction Monograph for TRASYLOL (Aprotinin) #2

Perpetrator drug:

INN name/US brand name:	Aprotinin/Trasylol
Class/Molecular weight:	Bovine proteinase inhibitor/6.5 kDa

Victim drug:

INN name/US brand name:	Captopril/Capoten
Class/Molecular weight:	Small molecule

Description of the effect of the DDI:
In study of nine patients with untreated hypertension, aprotinin infused intravenously in a dose of 2 million KIU over two hours blocked the acute hypotensive effect of 100 mg of captopril.

Description of the (proven, putative or hypothesized) mechanism of the DDI:
NA

Clinical consequences:
No specific recommendations are available.

Reference: Trasylol prescribing information.

Drug-Drug Interaction Monograph for TRASYLOL (Aprotinin) #3

Perpetrator drug:

INN name/US brand name:	Aprotinin/Trasylol
Class/Molecular weight:	Bovine proteinase inhibitor/6.5 kDa

Victim drug:

INN name/US brand name:	Heparin
Class/Molecular weight:	Carbohydrate/12–15 kDa

Description of the effect of the DDI:
Aprotinin, in the presence of heparin, has been found to prolong the activated clotting time

Description of the (proven, putative or hypothesized) mechanism of the DDI:
Aprotinin prolongs whole blood clotting times by a different mechanism than heparin. In the presence of aprotinin, prolongation is dependent on the type of whole blood clotting test employed. If an activated clotting time (ACT) is used to determine the effectiveness of heparin anticoagulation, the prolongation of the ACT by aprotinin may lead to an overestimation of the degree of anticoagulation, thereby leading to inadequate anticoagulation.

Clinical consequences:
No specific recommendations are available.

Reference: Trasylol prescribing information.

Drug-Drug Interaction Monograph for TYSABRI (Natalizumab) #1

Perpetrator drug:

INN name/US brand name:	Interferon beta-1a/Avonex
Class:/Molecular weight:	Protein/20.0 kDa

Victim drug:

INN name/US brand name:	Natalizumab/Tysabri
Class/Molecular weight:	Monoclonal Antibody/149 kDa

Description of the effect of the DDI:
Natalizumab exposure parameters (i.e., C_{max}, AUC) and terminal elimination half-life appeared to be somewhat higher when administered in combination with interferon beta 1a. C_{max} increased from 151.9 to 174.2 μg/L while the half-life increased from 142.6 to 175.3 h.

Description of the (proven, putative or hypothesized) mechanism of the DDI:
No dose adjustments have been recommended.

Clinical consequences:
No clinical consequences have been observed

Reference: Tysabri prescribing information.

Vollmer et al. An open-label safety and drug interaction study of natalizumab (Antegren™) in combination with interferon-beta (Avonex®) in patients with multiple sclerosis. Mult Scler. 2004;10(5):511–20.

Drug-Drug Interaction Monograph for TYSABRI (Natalizumab) #2

Perpetrator drug:

INN name/US brand name:	Immunosuppressants

Victim drug:

INN name/US brand name:	Natalizumab/Tysabri Change formatting
Class/Molecular weight:	Monoclonal Antibody/149 kDa

Description of the effect of the DDI:
Because of the potential for increased risk of PML(Progressive multifocal leukoencephalopathy) and other infections, Crohn's disease patients receiving natalizumab should not be treated with concomitant immunosuppressants (e.g., 6- mercaptopurine, azathioprine, cyclosporine, or methotrexate) or inhibitors of TNF-α. Corticosteroids should be reduced in patients with Crohn's disease who will begin with natalizumab therapy.

Description of the mechanism of the DDI:
NA

(Continued)

Clinical consequences:
In Crohn's disease clinical studies, opportunistic infections (*Pneumocystis carinii* pneumonia, pulmonary *Mycobacterium avium intracellulare*, bronchopulmonary aspergillosis, and *Burkholderia cepacia*) have been observed in <1% of TYSABRI-treated patients; some of these patients were receiving concurrent immunosuppressants.

Multiple Sclerosis patients receiving chronic immunosuppressant or immunomodulatory therapy should not be treated with natalizumab.

Reference: Tysabri prescribing information.

Drug-Drug Interaction Monograph for ULTRESA/[Pancrelipase(Amylase, lipase, protease)]

Perpetrator drug:
INN name/US brand name: Pancrelipase/Ultresa
Class/Molecular weight: Enzyme mixture/~131 kDa

Victim drug:
INN name/US brand name: L-Methylfolate
Class/Molecular weight: Small Molecule/459 Da

Description of the effect of the DDI:
Concurrent use of L-methylfolate and pancrelipase may result in decreased L-methylfolate serum concentrations.

Description of the (proven, putative or hypothesized) mechanism of the DDI:
Decreased folate absorption.

Clinical consequences:
No clinical consequences have been observed.

Reference: Micromedex.

Drug-Drug Interaction Monograph for VORAXAZE (Glucarpidase)

Perpetrator drug:
INN name/US brand name: Leucovorin/Wellcovorin
Class/Molecular weight: Folate Analog

Victim drug:
INN name/US brand name: Glucarpidase/Voraxaze
Class/Molecular weight: Homodimer Protein/83 kDa

Description of the effect of the DDI:
Coadministration of leucovorin and glucarpidase results in increased plasma concentrations of methotrexate.

Description of the (proven, putative or hypothesized) mechanism of the DDI:
Leucovorin competes with methotrexate for the same binding sites on glucarpidase. This action diminishes the efficacy of glucarpidase in decreasing elevated methotrexate concentrations.

Clinical consequences:
Do not give leucovorin within 2 hours of glucarpidase administration.

Reference: Voraxaze prescribing information.

Drug-Drug Interaction Monograph for WYDASE (Hyaluronidase)

Perpetrator drug:

INN name/US brand name:	Hyaluronidase/Wydase
Class/Molecular weight:	Enzyme/53.9 kDa

Victim drug:

Class:	Anesthetics (Mepivacaine, bupivacaine, benzocaine, etidocaine, cocaine, lidocaine, propxycaine, tetracaine, ropivacaine, bucaine, proparacaine, dibucaine, procaine, chloroprocaine, prilocaine)

Description of the effect of the DDI:
Concurrent use of anesthetics and hyaluronidase may result in an increased incidence of a systemic reaction to the anesthetic.

Description of the (proven, putative or hypothesized) mechanism of the DDI:
Hyaluronidase increases the duration of anesthetics in inferior alveolar nerve blocks.

Clinical consequences:
Monitor patients for local anesthetic toxicity. Hyaluronidase effectively increases topical diffusion of local anesthetics. However, due to its propensity to increase the incidence of systemic reactions to local anesthetics and because there are newer anesthetic agents which penetrate very effectively, its use can no longer be justified.

Reference: Horliana TAC, et al. Hyaluronidase increases the duration of mepivacaine in inferior alveolar nerve blocks. J Oral Maxillofac Surg. 2008;66(2):286–90.
Micromedex.

Drug-Drug Interaction Monograph for XGEVA (Denosumab)

Perpetrator drug:

INN name/US brand name:	Denosumab/Xgeva
Class/Molecular weight:	Human IgG2 Monoclonal Antibody/147 kDa

Victim drug:

Class:	Anticancer chemotherapeutics

Description of the effect of the DDI:
No change in pharmacokinetic parameters or toxicity has been observed under co-administration of denosumab and anticancer agents.

Serum denosumab concentrations at 1 and 3 months were not altered by concomitant chemotherapy.

Description of the (proven, putative or hypothesized) mechanism of the DDI:
No dose adjustments have been recommended.

Clinical consequences:
No clinical consequences have been observed

Reference: Xgeva prescribing information.

Drug-Drug Interaction Monograph for XIAFLEX (Collagenase *Clostridium histolyticum*)

Perpetrator drug:

INN name/US brand name: Collagenase *Clostridium histolyticum*/Xiaflex
Class/Molecular weight: Two polypeptides/114 kDa and 113 kDa

Victim drug:

Class: Anticoagulants

Description of the effect of the DDI:
Possibility of developing an ecchymosis/contusion or an injection site hemorrhage.

Description of the (proven, putative or hypothesized) mechanism of the DDI:
Additive anticoagulation.

Clinical consequences:
Collagenase should be used with caution in patients receiving concomitant anticoagulants (except for low-dose aspirin).

Reference: Xiaflex prescribing information.

Drug-Drug Interaction Monograph for XIGRIS (Drotrecogin alfa)

Perpetrator drug:

INN name/US brand name: Drotrecogin alfa/Xigris
Class/Molecular weight: Glycoprotein/55 kDa

Victim drug:

Class: Anticoagulants

Description of the effect of the DDI
Concurrent use of drotrecogin alfa and anticoagulants may result in increased risk of bleeding.

Description of the (proven, putative or hypothesized) mechanism of the DDI:
Additive anticoagulation.

Clinical consequences:
Use caution with coadministration of drotrecogin alfa and anticoagulants in therapeutic dosages for thromboembolism. No drotrecogin alfa dosage adjustment is necessary when coadministered with heparin using the prophylactic doses for venous thromboembolism.

Reference: Xigris prescribing information.

Drug-Drug Interaction Monograph for ZALTRAP (Ziv-Aflibercept) #1

Perpetrator drug:

INN name/US brand name: Ziv-Aflibercept/Zaltrap
Class/Molecular weight: Recombinant fusion protein/155 kDa

Victim drug:

INN name/US brand name: Irinotecan/Camptosar
Class/Molecular weight: Antineoplastic agent/677.2 Da

Description of the effect of the DDI:
No change in pharmacokinetic parameters or toxicity has been observed under co-administration of ziv-aflibercept and irinotecan.

Description of the (proven, putative or hypothesized) mechanism of the DDI:
No dose adjustments have been recommended.

Clinical consequences:
No clinical consequences have been observed.

Reference: Zaltrap prescribing information.

Drug-Drug Interaction Monograph for ZALTRAP (Ziv-Aflibercept) #2

Perpetrator drug:

INN name/US brand name: Ziv-Aflibercept/Zaltrap
Class/Molecular weight: Recombinant fusion protein/155 kDa

Victim drug:

INN name/US brand name: Fluorouracil/Efudex
Class/Molecular weight: Antineoplastic antimetabolite/130.1 Da

Description of the effect of the DDI:
No change in pharmacokinetic parameters or toxicity has been observed under co-administration of ziv-aflibercept and fluorouracil.

Description of the (proven, putative or hypothesized) mechanism of the DDI:
No dose adjustments have been recommended.

Clinical consequences:
No clinical consequences have been observed.

Reference: Zaltrap prescribing information.

Drug-Drug Interaction Monograph for ZENAPAX (Daclizumab) #1

Perpetrator drug:

INN name/US brand name: Daclizumab/Zenapax
Class/Molecular weight: Humanized IgG1 Monoclonal Antibody/144 kDa

Victim drug:

Class: Immunosuppressive agents (cyclosporine, mycophenolate mofetil, ganciclovir, acyclovir, azathioprine, corticosteroids)

(Continued)

Description of the effect of the DDI:
In cardiac transplant recipients (n = 434) receiving concomitant cyclosporine, mycophenolate mofetil, and corticosteroids, mortality was increased in patients randomized to receive daclizumab compared with those randomized to receive placebo.

The concomitant administration of daclizumab and mycophenolate mofetil had no effect on the pharmacokinetics of mycophenolic acid.

Description of the (proven, putative or hypothesized) mechanism of the DDI:
No dose adjustments have been recommended.

Clinical consequences:
No clinical interventions have been recommended.

Reference: Zenapax prescribing information.

Drug-Drug Interaction Monograph for ZENAPAX (Daclizumab) #2

Perpetrator drug:

Class:	Anti-Lymphocyte Antibody Antibody/144 kDa

Victim drug:

INN name/US brand name:	Daclizumab/Zenapax
Class/Molecular weight:	Immunosuppressive, Humanized IgG1 Monoclonal

Description of the effect of the DDI:
The use of daclizumab was associated with a higher incidence of mortality in patients receiving concomitant anti-lymphocyte antibody therapy, compared to placebo, in a large (n = 434) randomized controlled study of patients receiving cardiac transplants.

Description of the (proven, putative or hypothesized) mechanism of the DDI:
No dose adjustments have been recommended.

Clinical consequences:
No clinical interventions have been recommended.

Reference: Zenapax prescribing information.

Drug-Drug Interaction Monograph for ZEVALIN (Ibritumomab tiuxetan)

Perpetrator drug:

INN name/US brand name:	Ibritumomab tiuxetan/Zevalin
Class/Molecular weight:	Immunoconjugate between monoclonal antibody and linker-chelator/148 kDa

Victim drug:

Class:	Anticoagulants/Antiplatelet agents

Description of the effect of the DDI:
Concurrent use of ibritumomab tiuxetan and anticoagulants/antiplatelet agents may result in an increased risk of bleeding and hemorrhage.

Description of the (proven, putative or hypothesized) mechanism of the DDI:
NA

Clinical consequences:
Monitor for thrombocytopenia more frequently when using ibritumomab and anticoagulants/antiplatelet agents concomitantly. The transfusion practices may need to be modified.

Reference: Zevalin prescribing information.

Drug-Drug Interaction Monograph for ZOLADEX (Goserelin Acetate)

Perpetrator drug:

INN name/US brand name:	Goserelin Acetate/Zoladex
Class/Molecular weight:	Decapeptide/1,269 Da

Victim drug:

INN name/US brand name:	Tamoxifen Citrate/Nolvadex
Class/Molecular weight:	Nonsteroidal Antiestrogen/563.6 Da

Description of the effect of the DDI:
The effect of concurrent goserelin and tamoxifen in treating premenopausal women with early breast cancer is not clear, but may have less benefit than a mono therapy.

Description of the (proven, putative or hypothesized) mechanism of the DDI:
No change in pharmacokinetic parameters or toxicity has been observed under co-administration of goserelin and tamoxifen.

Clinical consequences:
The combination of goserelin and tamoxifen reduces the risk of first event by 24%, compared to 32% with goserelin and 27% with tamoxifen alone.

Reference: Zoladex prescribing information.

Sverrisdottir A, et al. Interaction between goserelin and tamoxifen in a prospective randomised clinical trial of adjuvant endocrine therapy in premenopausal breast cancer. Breast Cancer Res Treat. 2011;128(3):755–63.

Drug-Drug Interaction Monograph for ZORBTIVE (Somatropin)

See ACCRETROPIN (Somatropin) drug-drug interaction monographs #1– #4

Reference: Zorbtive prescribing information

INDEX

Drug–Drug Interactions for Therapeutic Biologics, First Edition.
Edited by Honghui Zhou and Bernd Meibohm.

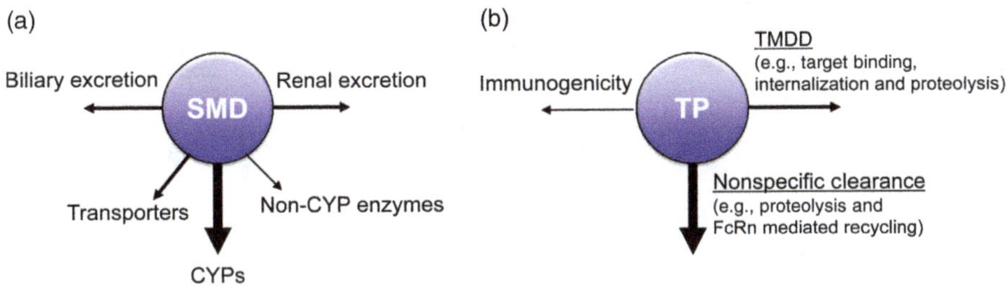

FIGURE 2-1 Comparison of clearance mechanisms of (a) a SMD and (b) a TP. CYP: cytochrome P450; FcRn: neonatal Fc receptor; SMD: small molecule drug; TMDD: target-mediated drug disposition; TP: therapeutic protein.

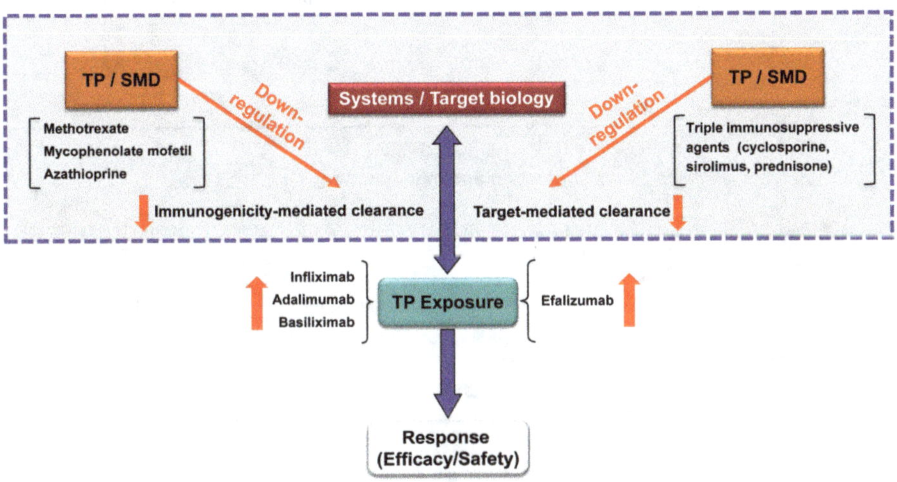

FIGURE 2-2 Theoretical mechanisms of PD-based TP-DIs that change exposure of the TP. SMD: small molecule drug; TP: therapeutic protein.

Drug–Drug Interactions for Therapeutic Biologics, First Edition.
Edited by Honghui Zhou and Bernd Meibohm.

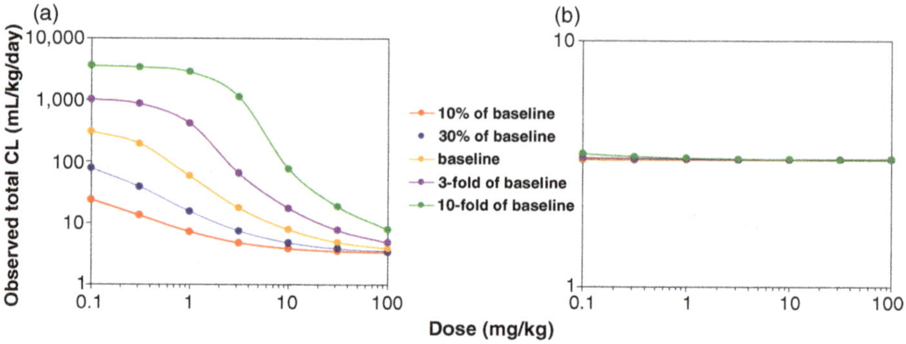

FIGURE 2-3 Effect of receptor expression level on the dose-total CL relationship for a typical mAb. Observed total CL = dose/area under the concentration–time profile. Simulation is performed using a standard TMDD model after a single dose of a typical mAb [46,47]. The baseline free receptor expression levels was altered to 3-fold, 10-fold, 10%, or 30% of baseline, and the observed total CL values were plotted versus dose levels. (a) TMDD accounts for >50% of total clearance in the therapeutically relevant dose range. (b) TMDD accounts for ≤1% of total clearance in the therapeutically relevant dose range. CL: clearance.

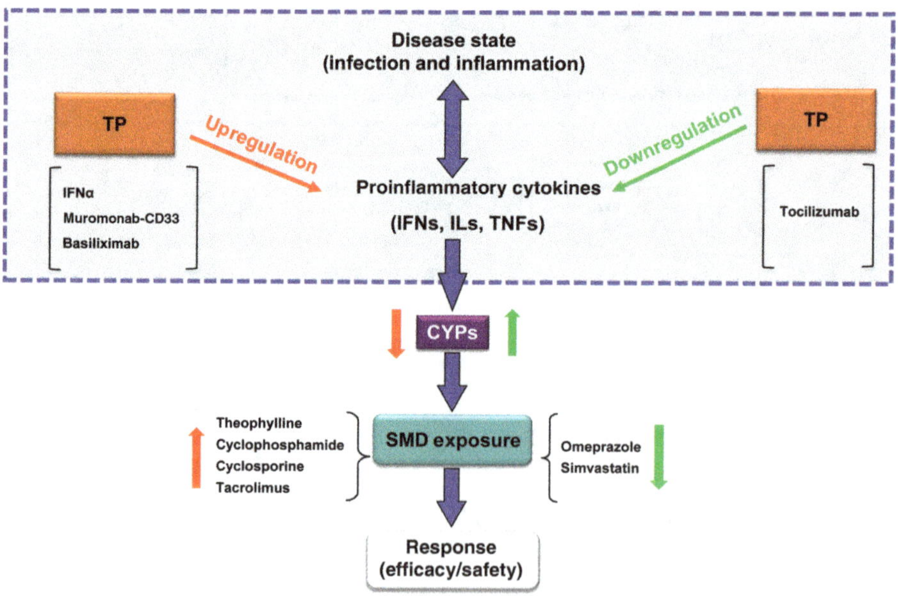

FIGURE 2-4 Theoretical mechanisms of PD-based TP-DIs that have an impact on the PK properties of SMDs through TP-cytokine-CYP modulation effect. CYP: cytochrome P450; IFN: interferon; IL: interleukin; SMD: small molecule drug; TNF: tumor necrosis factor; TP: therapeutic protein.

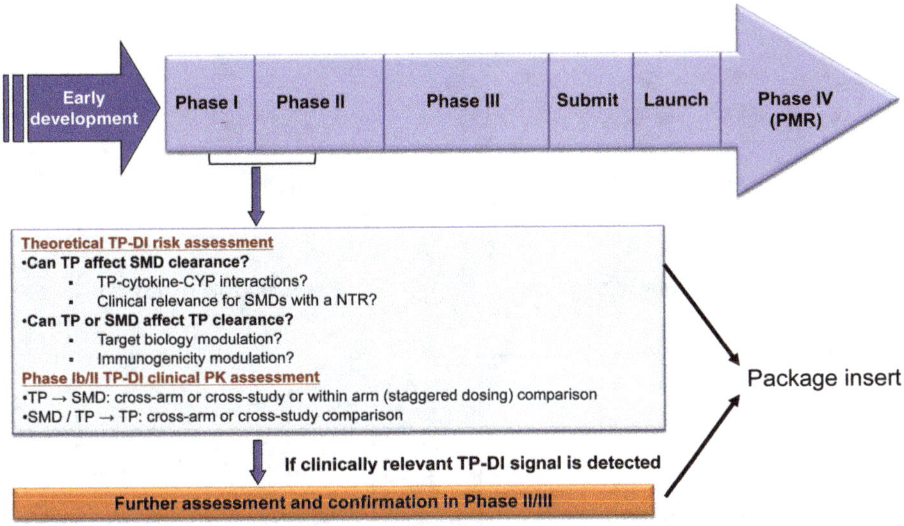

FIGURE 2-5 Question-based TP-DI risk assessment strategy during TP development. CYP: cytochrome P450; NTR: narrow therapeutic range; PK: pharmacokinetic; PMR: post-marketing requirement; SMD: small molecule drug; TP: therapeutic protein; TP-DI: therapeutic protein–drug interaction.

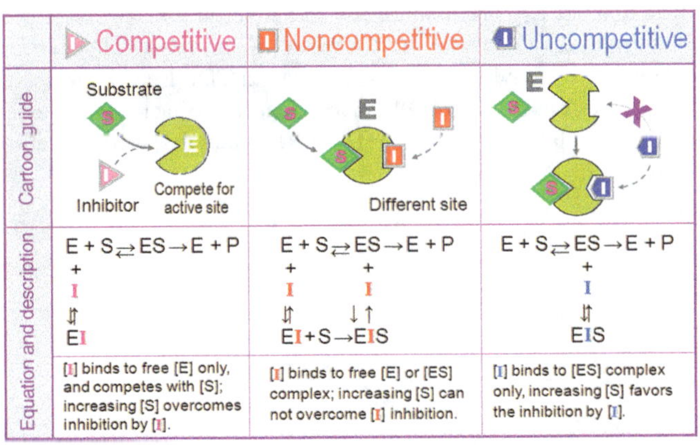

FIGURE 3-1 Enzyme-substrate-inhibitor interactions and kinetic equilibria for the three major models of reversible inhibition. E: enzyme; ES: enzyme–substrate; I: inhibitor; P: product; S: substrate. (Reprinted with permission from Rong-Huay Juang, National Taiwan University.[70])

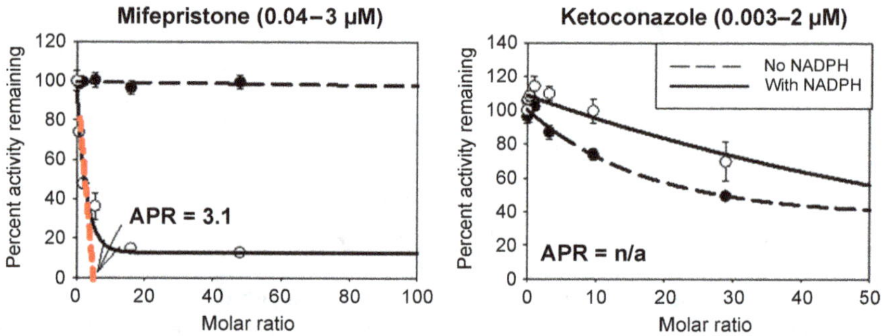

FIGURE 3-3 APR curves for mifepristone, a known CYP3A4 TDI, and ketoconazole, a potent CYP3A4 reversible inhibitor. Serially diluted concentrations of either compound were incubated with an NADPH (solid lines) or without an NADPH (dashed lines) regenerating system for 60min. Residual CYP3A4 activity was assayed by transferring an aliquot of each initial incubation to secondary incubations containing midazolam (CYP3A4 probe substrate) and NADPH. Residual activity was normalized to the activity in a vehicle control and was graphed as a function of inactivator to CYP3A4 molar ratio. The concentration- and NADPH-dependent loss of activity was consistent with a TDI. Ketoconazole, as a potent but reversible inhibitor, does not cause NADPH-dependent loss of activity and exhibits stronger loss of activity in the absence of NADPH. APR: apparent partition ratio; TDI: time-dependent inhibitor.

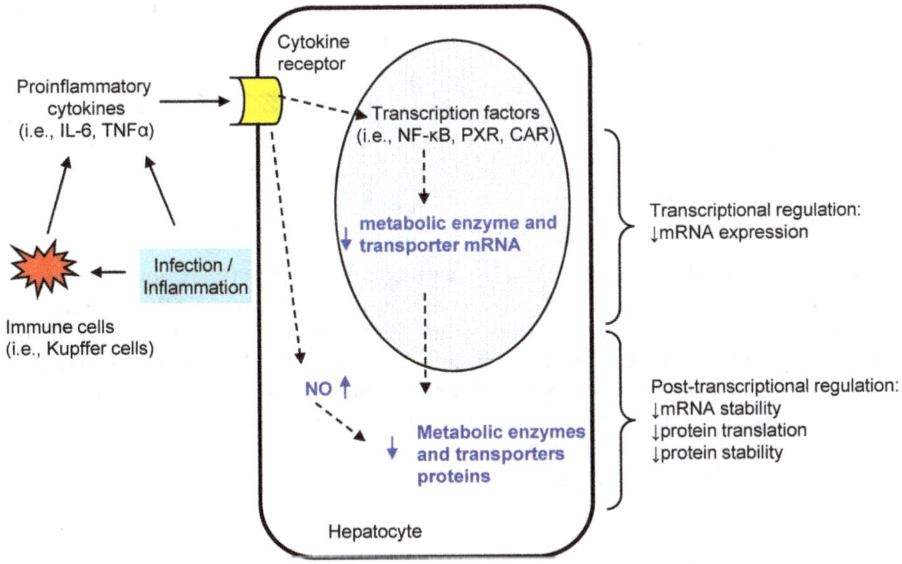

FIGURE 5-2 Proposed general mechanisms of cytokine regulation of metabolic enzymes and transporters. At the transcriptional level, cytokines can modulate gene expression of metabolic enzymes and transporters through interaction with various nuclear transcription factors—such as NF-kB, PXR, and CAR. At the post-transcriptional level, decreased enzyme and transporter proteins may be the result of reduction in mRNA stability owing to oxidative stress, suppression of protein translation, and/or increase in enzyme degradation due to elevated levels of NO associated with inflammation. Dotted lines: multiple intermediate steps may be involved; solid lines: a more direct effect. CAR: constitutive androstane receptor; IL: interleukin; NO: nitric oxide; PXR: pregnane X receptor; TNF: tumor necrosis factor.

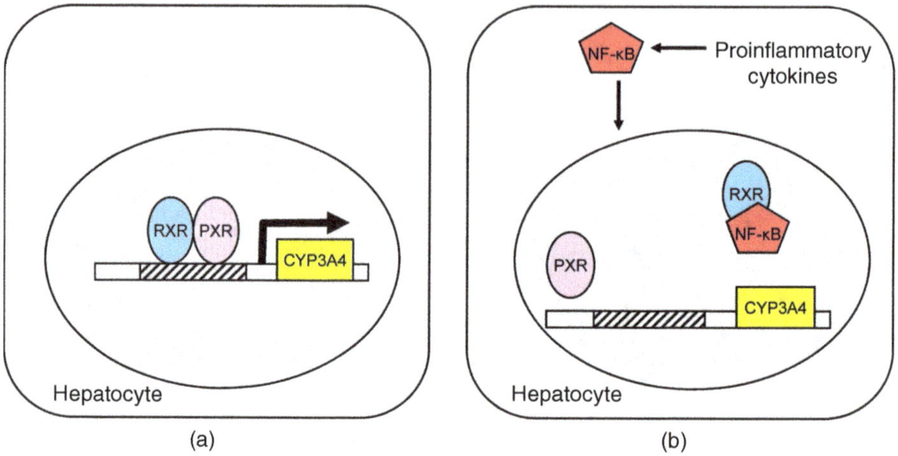

FIGURE 5-3 Regulation of CYP3A4 by pregnane X receptor (PXR) and its obligate partner retinoid X receptor (RXR). (a) Upon heterodimerization of PXR with RXR and co-activators, the complex binds to the consensus PXR response elements in target genes, leading to gene transcription. (b) The presence of proinflammatory cytokines will activate NF-κB, which in turn will bind to RXR and disrupt interaction of the RXR-PXR heterodimer to the PXR response element. The result is a suppression of gene transcription. It is hypothesized that the same mechanism could result in suppression of CAR-mediated transcription activity (CAR: constitutive androstane).

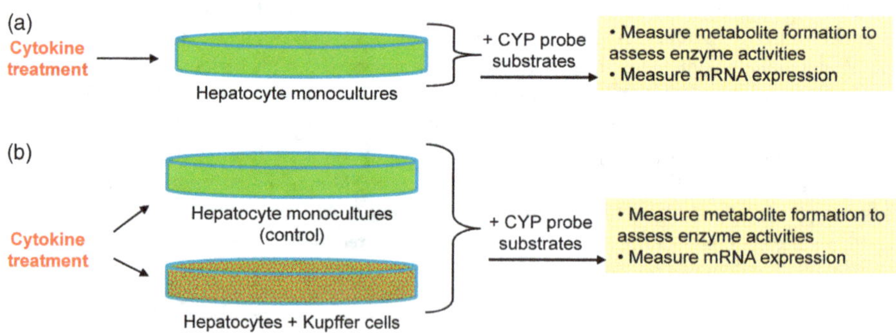

FIGURE 5-4 General experimental design of *in vitro assays*. Hepatocyte monocultures (a) and hepatocyte-Kupffer cell co-cultures (b) are exposed to cytokines of interest for 24 to 96 h. The cultures are then incubated with CYP probe substrates, followed by measurement of metabolite formation and mRNA levels to assess enzyme activities and gene expression.

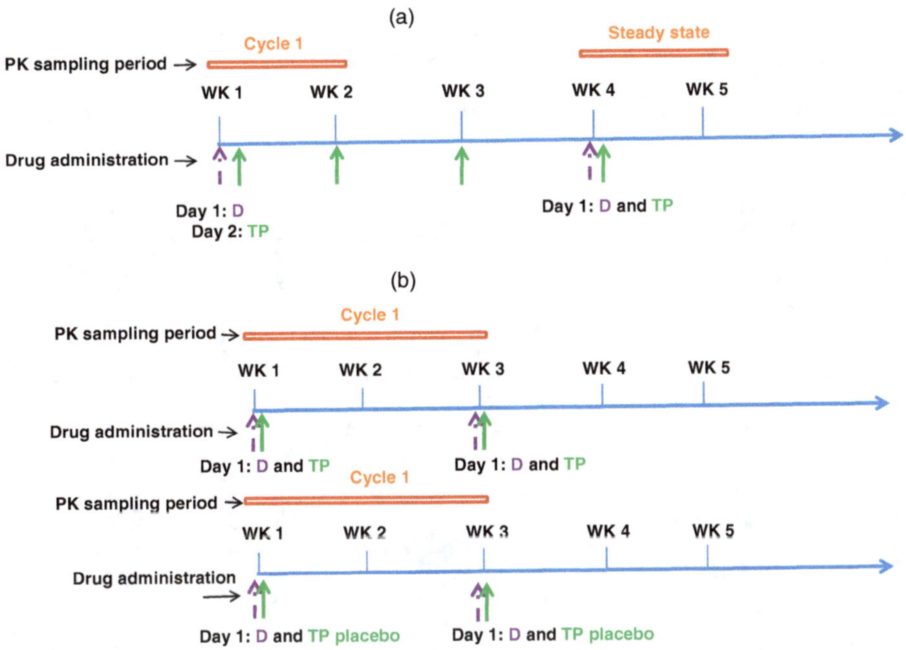

FIGURE 8-1 Examples of (a) one-sequence crossover and (b) parallel designs. D: small molecule drug; PK: pharmacokinetics; TP: therapeutic protein; WK: week.

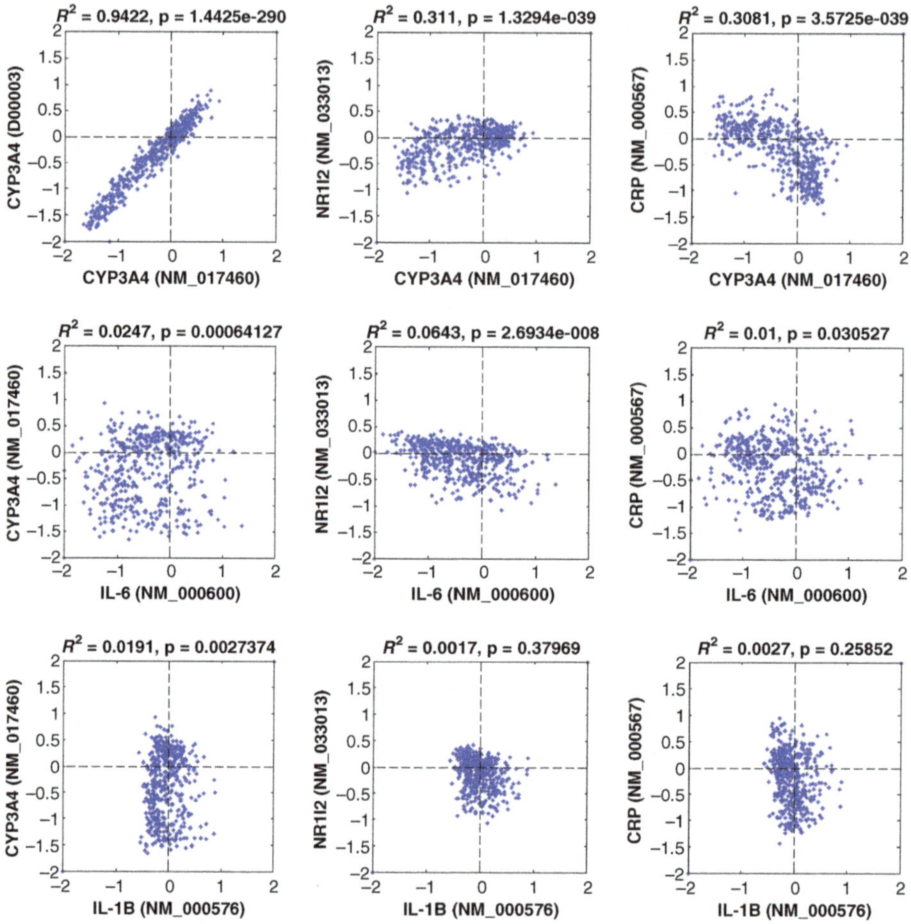

FIGURE 13-1 Gene to gene mRNA expression correlations. CYP3A4, IL-6, and IL1β are correlated with a second CYP3A4 probe, the nuclear pregnane X receptor (PXR; gene name NR1I2) and the acute phase response marker, C-reactive protein (CRP). *Top row, CYP3A4 correlations*: (left) Two unique CYP3A4 probes were present on the array and show the expected high correlation between different probes for the same enzyme across the donor population; (middle) expected positive correlation between CYP3A4 and PXR (NR1I2) expression; (right) negative correlation between CYP3A4 and CRP. *Middle row, IL-6 correlations*: (left) Lack of correlation with CYP3A4; (middle) negative correlation between IL-6 and PXR (NR1i2) expression; (right) lack of correlation between IL-6 and CRP. *Bottom row IL-1β correlations*: (left) Dynamic range of IL-1β expression is narrower than IL-6 ($< \pm$ 10-fold), no correlation with CYP3A4; (middle) no correlation between IL-1β and PXR (NR1i2) expression; (right) no correlation between IL-1β and CRP. Plots prepared by Dr. Yudong He (Amgen Corp), from open access data at sagebase.org and data from Ref. 55.

www.ingramcontent.com/pod-product-compliance
Lightning Source LLC
LaVergne TN
LVHW011835200926
848484LV00002B/360

9781118032169